7
Modern Mathematics for Schools

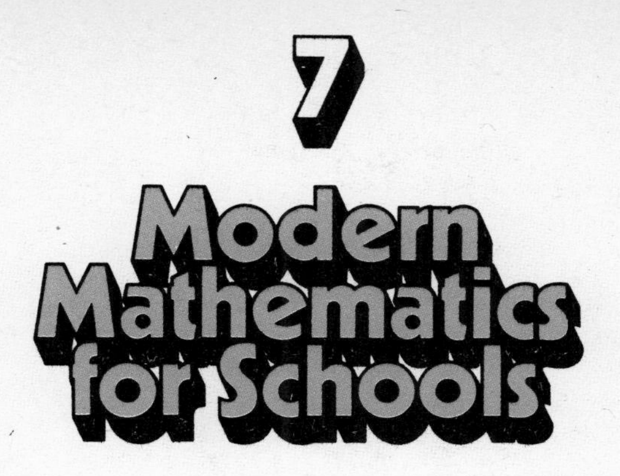

Second Edition

Scottish Mathematics Group

Blackie & Son Limited
Bishopbriggs · Glasgow G64 2NZ
14–18 High Holborn · London WC1V 6BX

W & R Chambers Limited
43–5 Annandale Street · Edinburgh EH7 4AZ

First Published 1974
Reprinted 1975. 1977, 1979, 1981,
1982, 1983, 1985

Designed by James W. Murray

International Standard Book Numbers
Pupils' Book
Blackie 0 216 89424 7
Chambers 0 550 75918 2
Teachers' Book
Blackie 0 216 89425 5
Chambers 0 550 75928 X

Printed in Great Britain by
Thomson Litho Ltd, East Kilbride, Scotland
Set in 10pt Monophoto Times Roman

Members associated with this book

W. T. Blackburn
Dundee College of Education
W. Brodie
Trinity Academy
C. Clark
Formerly of Lenzie Academy
D. Donald
Formerly of Robert Gordon's College
R. A. Finlayson
Jordanhill College School
Elizabeth K. Henderson
Westbourne School for Girls
J. L. Hodge
Madras College
J. Hunter
University of Glasgow
R. McKendrick
Langside College
W. More
Formerly of High School of Dundee
Helen C. Murdoch
Hutchesons' Girls' Grammar School
A. G. Robertson
John Neilson High School
A. G. Sillitto
Formerly of Jordanhill College of Education
A. A. Sturrock
Grove Academy
Rev. J. Taylor
St. Aloysius' College
E. B. C. Thornton
Bishop Otter College
J. A. Walker
Dollar Academy
P. Whyte
Hutchesons' Boys' Grammar School

Contributor of the Computer Studies
A. W. McMeeken
Dundee College of Education

Preface

Book 1 of the original series *Modern Mathematics for Schools* was first published in July 1965. This revised series has been produced in order to take advantage of the experience gained in the classroom with the original textbooks and to reflect the changing mathematical needs in recent years, particularly as a result of the general move towards some form of comprehensive education.

Throughout the whole series, the text and exercises have been cut or augmented wherever this was considered to be necessary, and nearly every chapter has been completely rewritten. In order to cater more adequately for the wider range of pupils now taking certificate-oriented courses, the pace has been slowed down in the earlier books in particular, and parallel sets of A and B exercises have been introduced where appropriate. The A sets are easier than the B sets, and provide straightforward but comprehensive practice; the B sets have been designed for the more able pupils, and may be taken in addition to, or instead of, the A sets. From Book 4 onwards a basic exercise, which should be taken by all pupils, is sometimes followed by a harder one on the same work in order to give abler pupils an extra challenge, or further practice; in such a case the numbering is, for example, Exercise 2 followed by Exercise 2B. It is hoped that this arrangement, along with the *Graph Workbook for Modern Mathematics*, will allow considerable flexibility of use, so that while all the pupils in a class may be studying the same topic, each pupil may be working examples which are appropriate to his or her aptitude and ability.

Each chapter is backed up by a summary, and by revision exercises; in addition, cumulative summaries and exercises have been introduced at the end of alternate books. A new feature is the series of Computer Topics from Book 4 onwards. These form an elementary introduction to computer studies, and are primarily intended to give pupils some appreciation of the applications and influence of computers in modern society.

Books 1 to 7 provide a suitable course for many modern Ordinary Level and Ordinary Grade syllabuses in mathematics, including the project examination SMG Mathematics administered by the Oxford and Cambridge Schools Examination Board. Books 8 and 9 complete the work for the Scottish Higher Grade Syllabus, and provide a good preparation for all Advanced Level and Sixth Year Syllabuses, both new and traditional.

Related to this revised series of textbooks are the *Teacher's Editions* of the textbooks, the *Graph Workbook for Modern Mathematics*, the *Three-Figure Tables for Modern Mathematics*, and the booklets of *Progress Papers for Modern Mathematics.* These new Progress Papers consist of short, quickly marked objective tests closely connected with the textbooks. There is one booklet for each textbook, containing A and B tests on each chapter, so that teachers can readily assess their pupils' attainments, and pupils can be encouraged in their progress through the course.

The separate headings of Algebra, Geometry, Arithmetic, and later Trigonometry and Calculus, have been retained in order to allow teachers to develop the course in the way they consider best. Throughout, however, ideas, material and methods are integrated *within* each branch of mathematics and *across* the branches; the opportunity to do this is indeed one of the more obvious reasons for teaching this kind of mathematics in the schools—for it is *mathematics* as a whole that is presented.

Pupils are encouraged to find out facts and discover results for themselves, to observe and study the themes and patterns that pervade mathematics today. As a course based on this series of books progresses, a certain amount of equipment will be helpful, particularly in the development of geometry. The use of calculating machines, slide rules, and computers is advocated where appropriate, but these instruments are not an essential feature of the work.

While fundamental principles are emphasised, and reasonable attention is paid to the matter of structure, the width of the course should be sufficient to provide a useful experience of mathematics for those pupils who do not pursue the study of the subject beyond school level. An effort has been made throughout to arouse the interest of all pupils and at the same time to keep in mind the needs of the future mathematician.

The introduction of mathematics in the Primary School and recent changes in courses at Colleges and Universities have been taken into account. In addition, the aims, methods, and writing of these books have been influenced by national and international discussions about the purpose and content of courses in mathematics, held under the auspices of the Organisation for Economic Co-operation and Development and other organisations.

The authors wish to express their gratitude to the many teachers who have offered suggestions and criticisms concerning the original series of textbooks; they are confident that as a result of these contacts the new series will be more useful than it would otherwise have been.

Algebra

Geometry

Arithmetic

Trigonometry

Computer Studies

Notation

Sets of numbers

Different countries and different authors
give different notations and definitions
for the various sets of numbers.
In this series the following are used:

E The universal set

ϕ The empty set

N The set of natural numbers $\{1, 2, 3, ...\}$

W The set of whole numbers $\{0, 1, 2, 3, ...\}$

Z The set of integers $\{..., -2, -1, 0, 1, 2, ...\}$

Q The set of rational numbers

R The set of real numbers

The set of prime numbers $\{2, 3, 5, 7, 11, ...\}$

Algebra

Number Systems and Surds

1 *Irrational numbers and surds*

We have studied the following number systems:

(i) The system of *whole numbers* W, consisting of the set $\{0, 1, 2, 3, \ldots\}$ together with the operations of addition and multiplication.

(ii) The system of *integers* Z, consisting of the set $\{\ldots, -3, -2, -1, 0, 1, 2, 3, \ldots\}$ together with the operations of addition and multiplication.

(iii) The system of *rational numbers* Q, consisting of the set of positive and negative fractions and zero, together with the operations of addition and multiplication.

(iv) The system of *real numbers* R, which has the same structure as the rational number system and includes the set of *irrational numbers*, i.e. those real numbers which are not rational.

We saw that in each system, the commutative, associative, and distributive laws held for the operations of addition and multiplication. Notice that each of these systems can be regarded as an *extension* of the previous ones, as illustrated by the Venn diagram in Figure 1.

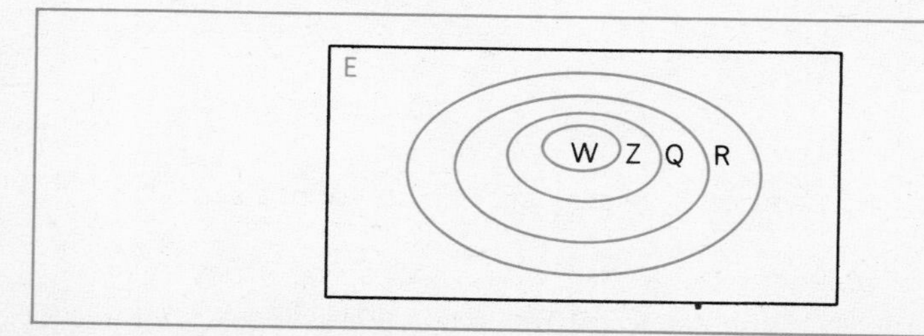

The following table describes the structure of the systems; a, b and c are members of each set of numbers in turn.

Addition	*Multiplication*	*Remarks*
$(a+b) \in$ the corresponding set	$(a \times b) \in$ the corresponding set	Each system is closed under addition and also under multiplication.
$a+b=b+a$	$a \times b = b \times a$	Addition and multiplication are both commutative.
$(a+b)+c=a+(b+c)$	$(a \times b) \times c = a \times (b \times c)$	Addition and multiplication are both associative.
$a+0=a=0+a$	$a \times 1 = a = 1 \times a$	Each operation of addition and multiplication has its own identity element.
$a+(-a)=0$, for Z, Q, R	$a \times \frac{1}{a} = 1$, for Q, R $(a \neq 0)$	Inverse elements.
$a(b+c)=ab+ac$		Multiplication is distributive over addition.

Throughout this chapter we shall be working with those irrational numbers which we call *surds*, i.e. those roots of rational numbers which cannot be expressed as rational numbers.

Examples of surds are $\sqrt{2}$, $\sqrt{0{\cdot}35}$ and $\sqrt[3]{21}$. Most irrational numbers are not surds; one you have already met is π.

You should note that the presence of a root sign does not necessarily mean that we are dealing with a surd. Thus, $\sqrt{49}$ and $\sqrt[3]{1{\cdot}728}$ are not surds because $7^2=49$ and $1{\cdot}2^3=1{\cdot}728$, and so $\sqrt{49}=7$ and $\sqrt[3]{1{\cdot}728}=1{\cdot}2$; i.e. $\sqrt{49}$ and $\sqrt[3]{1{\cdot}728}$ are rational numbers.

Remember that $\sqrt{a}$ denotes the positive square root of a, $a \geqslant 0$.

Exercise 1

Which of the following are surds and which are not surds?

1 $\sqrt{9}$ *2* $\sqrt{8}$ *3* $\sqrt[3]{1}$ *4* $\sqrt{12}$ *5* $\sqrt{0{\cdot}4}$

6 $\sqrt{0{\cdot}04}$ *7* $\sqrt{400}$ *8* $\sqrt[3]{10}$ *9* $\sqrt{121}$ *10* $\sqrt[3]{54}$

Solve the following equations, noticing that successive solutions require the successive extensions of the number systems which are listed at the beginning of the chapter:

11 $3x+5=14$ *12* $5x-7=-12$ *13* $3x-8=18$

14 $4x+10=3$ *15* $x^2=2$ *16* $2x^2+1=11$

Solve the following equations, stating which is the simplest number system required to give a solution set which is not the empty set:

17 $3(x+2)=7$ *18* $8x-9=23$ *19* $2x+10=4$

20 $x^2+3=19$ *21* $x^2-3=19$ *22* $3x-10=-1$

If a and b represent the number of units of length in the sides about the right angle in a triangle and c represents the number of units of length in the hypotenuse, in which of the following is the number represented by c a surd?

23 $a=2, b=3$ *24* $a=3, b=4$ *25* $a=1, b=1$

26 $a=5, b=10$ *27* $a=5, b=12$ *28* $a=24, b=7$

29 On 2-mm squared paper, draw the graph of the mapping $x \rightarrow x^3$, calculating values of x^3 at $-3, -2, -1, 0, 1, 2, 3$.
Scale: 2 cm to 1 unit for x and 2 cm to 10 units for x^3. Use the graph to give decimal approximations to three significant figures for:

a $\sqrt[3]{10}$ *b* $\sqrt[3]{20}$ *c* $\sqrt[3]{-16}$ *d* $\sqrt[3]{-22}$

2 *Simplifying surds; addition and subtraction*

We noted in Section 1 that the commutative, associative, and distributive laws held for the operations of addition and multiplication on the set of real numbers.

Let a and b be positive rational numbers.

Then $$[\sqrt{a}\sqrt{b}]^2 = \sqrt{a}\sqrt{b} \times \sqrt{a}\sqrt{b}$$
$$= \sqrt{a} \times \sqrt{a} \times \sqrt{b} \times \sqrt{b}$$
$$= ab$$

Taking the positive square root of each side, $\sqrt{a}\sqrt{b} = \sqrt{(ab)}$

Example 1. Simplify $\sqrt{20}$.

$$\sqrt{20} = \sqrt{(4 \times 5)} = \sqrt{4}\sqrt{5} = 2\sqrt{5}$$

This example shows that a surd may ultimately be expressed as the product of a rational number and a surd; when so reduced, the surd is said to be in its *simplest form*. Thus, the simplest form of $\sqrt{20}$ is $2\sqrt{5}$.

Example 2. Simplify $5\sqrt{3}-7\sqrt{3}+4\sqrt{3}$

$5\sqrt{3}-7\sqrt{3}+4\sqrt{3}$
$= (5-7+4)\sqrt{3}$, using the distributive law
$= 2\sqrt{3}$

Exercise 2

Express each of the following in its simplest form:

1 $\sqrt{8}$ *2* $\sqrt{12}$ *3* $\sqrt{27}$ *4* $\sqrt{50}$ *5* $\sqrt{20}$
6 $\sqrt{18}$ *7* $\sqrt{28}$ *8* $\sqrt{200}$ *9* $\sqrt{24}$ *10* $\sqrt{45}$
11 $\sqrt{75}$ *12* $\sqrt{300}$ *13* $\sqrt{72}$ *14* $\sqrt{54}$ *15* $\sqrt{147}$
16 $5\sqrt{8}$ *17* $3\sqrt{32}$ *18* $10\sqrt{40}$ *19* $2\sqrt{98}$ *20* $10\sqrt{1000}$

Simplify each of the following, using the distributive law:

21 $5\sqrt{2}+3\sqrt{2}$ *22* $9\sqrt{5}-7\sqrt{5}$ *23* $6\sqrt{10}-6\sqrt{10}$
24 $\sqrt{3}-6\sqrt{3}$ *25* $\sqrt{2}+\sqrt{2}-2\sqrt{2}$ *26* $2\sqrt{7}-3\sqrt{7}+4\sqrt{7}$

Triangle ABC is right-angled at A. Calculate the length of the third side in each of the following, expressing the answer as a surd in its simplest form:

27 $b = 3, c = 6$ *28* $b = 4, c = 4$ *29* $a = 2, b = 1$
30 $a = 12, b = 10$ *31* $b = 20, c = 20$ *32* $a = 18, b = 12$

33 Figure 2 shows two right-angled triangles; the unit of length is the centimetre. Calculate the lengths of AC and AD in surd form.

34 Figure 3 (i) shows a cube of side 2 cm. Calculate the lengths of:

a its face diagonals *b* its space diagonals, giving your answers as surds in their simplest form.

35 Repeat question ***34*** for the cuboid in Figure 3 (ii).

36 The length of a rectangle is three times its breadth, and its area is 9 m^2. Find its length, breadth, perimeter and diagonal length in surd form.

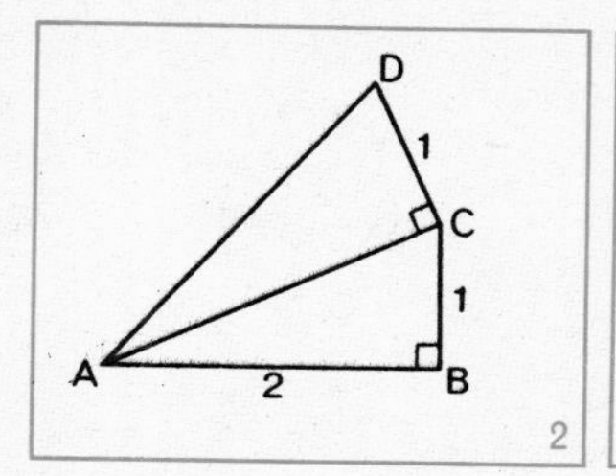

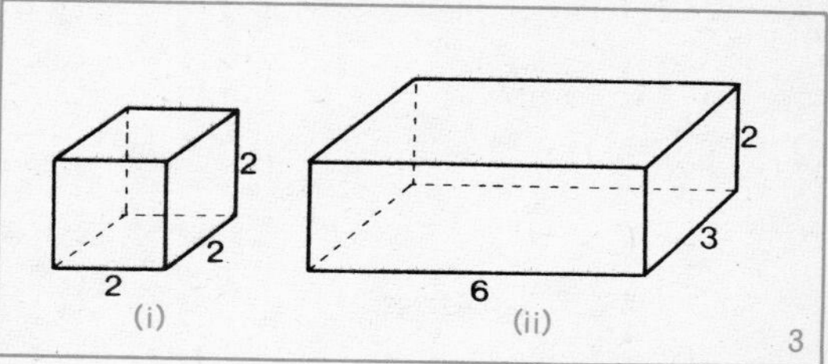

3 *Multiplication of surds*

Example 1. Simplify $\sqrt{12} \times \sqrt{8}$.

From Section 2, $\sqrt{a}\sqrt{b} = \sqrt{(ab)}$.

So $\sqrt{12} \times \sqrt{8} = \sqrt{96} = \sqrt{(6 \times 16)} = 4\sqrt{6}$

Alternatively, $\sqrt{12} \times \sqrt{8} = 2\sqrt{3} \times 2\sqrt{2} = 4\sqrt{6}$

Example 2. Simplify $(\sqrt{6}-\sqrt{2})(\sqrt{6}+3\sqrt{2})$

$(\sqrt{6}-\sqrt{2})(\sqrt{6}+3\sqrt{2})$

$= \sqrt{6}(\sqrt{6}+3\sqrt{2}) - \sqrt{2}(\sqrt{6}+3\sqrt{2})$, using the distributive law

$= 6 + 3\sqrt{12} - \sqrt{12} - 6$

$= 2\sqrt{12}$

$= 4\sqrt{3}$, since $\sqrt{12} = \sqrt{(4 \times 3)} = 2\sqrt{3}$

Exercise 3

Simplify:

1	$\sqrt{5} \times \sqrt{5}$	2	$\sqrt{2} \times \sqrt{2}$	3	$\sqrt{6} \times \sqrt{6}$	4	$\sqrt{1} \times \sqrt{1}$
5	$\sqrt{a} \times \sqrt{a}$	6	$\sqrt{3} \times \sqrt{2}$	7	$\sqrt{3} \times \sqrt{4}$	8	$\sqrt{9} \times \sqrt{a}$

9 $\sqrt{2}\times\sqrt{a}$ *10* $\sqrt{a}\times\sqrt{b}$ *11* $\sqrt{2}\times\sqrt{8}$ *12* $\sqrt{3}\times\sqrt{12}$

13 $\sqrt{6}\times\sqrt{3}$ *14* $\sqrt{10}\times\sqrt{20}$ *15* $2\sqrt{2}\times\sqrt{2}$ *16* $3\sqrt{2}\times 2\sqrt{3}$

17 $\sqrt{2}(1-\sqrt{2})$ *18* $\sqrt{3}(\sqrt{3}+1)$ *19* $\sqrt{5}(\sqrt{5}-1)$

20 $\sqrt{7}(1-\sqrt{7})$ *21* $\sqrt{2}(5+\sqrt{2})$ *22* $\sqrt{2}(3+4\sqrt{2})$

23 $(\sqrt{2}+1)(\sqrt{2}-1)$ *24* $(\sqrt{3}-1)(\sqrt{3}+1)$ *25* $(\sqrt{5}+2)(\sqrt{5}-2)$

26 $(\sqrt{3}+\sqrt{2})(\sqrt{3}-\sqrt{2})$ *27* $(\sqrt{5}-\sqrt{3})(\sqrt{5}+\sqrt{3})$

28 $(\sqrt{2}+\sqrt{5})(\sqrt{2}-\sqrt{5})$ *29* $(\sqrt{2}+1)^2$

30 $(\sqrt{3}+\sqrt{2})^2$ *31* $(\sqrt{5}-\sqrt{3})^2$

Given that $x=1+\sqrt{2}$ and $y=1-\sqrt{2}$, simplify:

32 $5x+5y$ *33* $2xy$ *34* x^2+y^2

Given that $p=\sqrt{5}+\sqrt{3}$ and $q=\sqrt{5}-\sqrt{3}$, simplify:

35 $2p-2q$ *36* $4pq$ *37* p^2-q^2

38 A rectangle has sides of length $(2+\sqrt{2})$ cm and $(2-\sqrt{2})$ cm. Calculate its area and also the length of a diagonal.

39 By squaring each side of the equation $2\sqrt{x}=x+2$, $x\in R$, deduce that the solution set of the given equation is ϕ.

40 P $(\sqrt{2}, k)$ is a point on the curve $y=2-\frac{1}{2}x^2$. Find k and the length of OP, where O is the origin.

---

4 *Rationalising denominators of fractions*

(i) *Expressions of the form* $\dfrac{a}{\sqrt{b}}$

Suppose we take 1·4142 as an approximation for $\sqrt{2}$ and wish to calculate $\dfrac{6}{\sqrt{2}}$, i.e. $\dfrac{6}{1{\cdot}4142}$ to four significant figures. This is easy with a calculating machine but time-consuming otherwise.

We can make use of the fact that $\sqrt{2}\times\sqrt{2}=2$, and write

$$\frac{6}{\sqrt{2}}=\frac{6}{\sqrt{2}}\times\frac{\sqrt{2}}{\sqrt{2}}=\frac{6\sqrt{2}}{2}=3\sqrt{2}=3\times 1{\cdot}4142=4{\cdot}2426$$

i.e. $\frac{6}{\sqrt{2}}=4{\cdot}243$ to four significant figures.

We are said to have *rationalised the denominator.*

As the above example shows, a fraction with a surd denominator may often be expressed most easily as a decimal approximation if it is first replaced by an equivalent fraction with a *rational denominator*. The following example gives a further illustration of the process of rationalising the denominator.

Example. $\sqrt{\frac{3}{2}}=\frac{\sqrt{3}}{\sqrt{2}}=\frac{\sqrt{3}}{\sqrt{2}}\times\frac{\sqrt{2}}{\sqrt{2}}=\frac{\sqrt{6}}{2}$

Exercise 4

Rationalise the denominator in each of the following, and using 1·414, 1·732, and 2·236 as approximations for $\sqrt{2}$, $\sqrt{3}$, and $\sqrt{5}$ respectively, calculate each answer to 3 significant figures:

1 $\frac{1}{\sqrt{2}}$ *2* $\frac{1}{\sqrt{3}}$ *3* $\frac{1}{\sqrt{5}}$ *4* $\frac{6}{\sqrt{3}}$ *5* $\frac{10}{\sqrt{5}}$

6 $\frac{2}{\sqrt{3}}$ *7* $\frac{3}{\sqrt{5}}$ *8* $\frac{20}{\sqrt{2}}$ *9* $\frac{3}{2\sqrt{5}}$ *10* $\frac{4}{5\sqrt{2}}$

Express each denominator in simplest form, and hence rationalise the denominator of each fraction:

11 $\frac{1}{\sqrt{20}}$ *12* $\frac{1}{\sqrt{50}}$ *13* $\frac{10}{\sqrt{12}}$ *14* $\frac{4}{\sqrt{8}}$ *15* $\frac{\sqrt{5}}{\sqrt{20}}$

Express each of the following in its simplest form with a rational denominator:

16 $\frac{\sqrt{4}}{\sqrt{3}}$ *17* $\frac{\sqrt{5}}{\sqrt{2}}$ *18* $\sqrt{\frac{9}{10}}$ *19* $\sqrt{\frac{1}{7}}$ *20* $\sqrt{\frac{3}{5}}$

(ii) *Expressions of the form* $\dfrac{1}{a+\sqrt{b}}$

$a+\sqrt{b}$ and $a-\sqrt{b}$, where a is rational and $\sqrt{b}$ is a surd, are *conjugate compound surds*, usually called *conjugate surds*. Their product is rational.

$$(a+\sqrt{b})(a-\sqrt{b})=a^2-b, \text{ which is rational.}$$

We use this property of conjugate surds to rationalise the denominator of expressions like $\dfrac{4}{\sqrt{3}-1}$ or $\dfrac{1-\sqrt{2}}{1+\sqrt{2}}$.

Example 1.

$$\frac{4}{\sqrt{3}-1}=\frac{4}{\sqrt{3}-1}\times\frac{\sqrt{3}+1}{\sqrt{3}+1}=\frac{4(\sqrt{3}+1)}{3-1}=\frac{4(\sqrt{3}+1)}{2}=2(\sqrt{3}+1)$$

Example 2.

$$\frac{1-\sqrt{2}}{1+\sqrt{2}}=\frac{1-\sqrt{2}}{1+\sqrt{2}}\times\frac{1-\sqrt{2}}{1-\sqrt{2}}=\frac{1-2\sqrt{2}+2}{1-2}=\frac{3-2\sqrt{2}}{-1}=2\sqrt{2}-3$$

Exercise 5B

1 Write down the conjugate surd for each of the following, and find the products of the conjugate pairs:

a $\sqrt{3}+1$ *b* $\sqrt{5}-2$ *c* $\sqrt{7}-3$ *d* $3-\sqrt{2}$

2 Use the quadratic formula to find the roots of each of the following equations, expressing the roots in simplest surd form:

a $x^2-4x+1=0$ *b* $x^2+2x-17=0$ *c* $2x^2-6x+1=0$

3 In question **2**, find the sum and product of the roots of each equation.

4 In each of the following multiply the numerator and denominator by the conjugate surd of the denominator, and hence simplify:

a $\dfrac{1}{\sqrt{2}-1}$ *b* $\dfrac{4}{\sqrt{5}+1}$ *c* $\dfrac{12}{2-\sqrt{3}}$ *d* $\dfrac{2-\sqrt{3}}{2+\sqrt{3}}$

5 Find the product of each of the following pairs of conjugate surds:

a $\sqrt{3}-\sqrt{2}, \sqrt{3}+\sqrt{2}$ *b* $\sqrt{5}+\sqrt{3}, \sqrt{5}-\sqrt{3}$

c $\sqrt{5}-\sqrt{2}, \sqrt{5}+\sqrt{2}$ *d* $\sqrt{7}+\sqrt{5}, \sqrt{7}-\sqrt{5}$

6 Simplify the following by rationalising the denominators.

a $\dfrac{1}{\sqrt{3}-\sqrt{2}}$ *b* $\dfrac{2}{\sqrt{5}+\sqrt{3}}$ *c* $\dfrac{9}{\sqrt{5}-\sqrt{2}}$ *d* $\dfrac{2}{\sqrt{7}+\sqrt{5}}$

7 Given that $\tan 22{\cdot}5° = \dfrac{1}{\sqrt{2}+1}$ and that $\sqrt{2} = 1{\cdot}4142$, calculate $\tan 22{\cdot}5°$ to four significant figures.

8 Given that $\tan 75° = \dfrac{\sqrt{3}+1}{\sqrt{3}-1}$, show that $\tan 75° = 2+\sqrt{3}$.

9 Express $\dfrac{6}{\sqrt{7}+1}$, and also its square, in the form $a+b\sqrt{c}$ where a, b and c are integers.

10 Show, by squaring, that if a and b are positive real numbers, $\sqrt{(a+b)} \neq \sqrt{a}+\sqrt{b}$.

Summary

1 A *surd* is a root of a rational number which cannot be expressed as a rational number.

Examples of surds are $\sqrt{2}, \sqrt{3}, \sqrt{5}, \sqrt{12}, \sqrt[3]{2}$.

In calculations, rational approximations to surds may be used; e.g. $\sqrt{7}$ is approximately 2·646 to 4 significant figures, or 2·645751 to 7 significant figures.

2 *Addition and subtraction of surds*

Example. $5\sqrt{3}+3\sqrt{3}-2\sqrt{3}=6\sqrt{3}$

3 *Multiplication of surds:* $\sqrt{a}\times\sqrt{b}=\sqrt{(ab)}$

Example. $\sqrt{12}\times\sqrt{8}=2\sqrt{3}\times 2\sqrt{2}=4\sqrt{6}$

4 *Rationalisation of denominator of* $\dfrac{a}{\sqrt{b}}$

Example. $\dfrac{10}{\sqrt{5}}=\dfrac{10}{\sqrt{5}}\times\dfrac{\sqrt{5}}{\sqrt{5}}=\dfrac{10\sqrt{5}}{5}=2\sqrt{5}$

5 *Rationalisation of denominator of* $\dfrac{1}{a+\sqrt{b}}$ and $\dfrac{1}{\sqrt{a}+\sqrt{b}}$

Example.

$$\frac{6}{\sqrt{5}+\sqrt{2}}=\frac{6}{\sqrt{5}+\sqrt{2}}\times\frac{\sqrt{5}-\sqrt{2}}{\sqrt{5}-\sqrt{2}}$$

$$=\frac{6(\sqrt{5}-\sqrt{2})}{5-2}=\frac{6(\sqrt{5}-\sqrt{2})}{3}=2(\sqrt{5}-\sqrt{2})$$

2 Indices

1 Positive integral indices

We have seen that 3^4 means $3 \times 3 \times 3 \times 3$, that the raised numeral 4 is called an *index*, and that 3^4 is read '3 to the power 4'. 3 is called the *base*.

In general, if a is a real number, and n is a positive integer, the nth power of a, denoted by a^n, is defined to be the product of n factors each which is a; i.e.

$$a^n = a \times a \times a \ldots \text{ to } n \text{ factors. We define } a^1 = a.$$

Exercise 1

1 By writing out the factors of each number as indicated above, show that:

a $3^4 \times 3^2 = 3^6$ *b* $2 \times 2^3 = 2^4$ *c* $10^3 \times 10^2 = 10^5$

d $(\frac{2}{3})^2 \times (\frac{2}{3})^4 = (\frac{2}{3})^6$ *e* $(-\frac{1}{2})^2 \times (-\frac{1}{2})^3 = (-\frac{1}{2})^5$

2 By writing out the factors of each of the following, show that:

a $a^4 \times a^2 = a^6$ *b* $x^3 \times x^3 = x^6$ *c* $z^4 \times z = z^5$

3 From questions ***1*** and ***2*** what appears to be the relation between the indices on the left side and the index on the right side?

4 Using the relation you gave in question ***3***, write down the simplest form of the products in the following:

a $2^3 \times 2^4$ *b* $3^5 \times 3^3$ *c* $7^2 \times 7^4$ *d* $2^7 \times 2^2$ *e* $10^{10} \times 10^{10}$

f $a^6 \times a^3$ *g* $x^4 \times x^4$ *h* $f^4 \times f^{15}$ *i* $z^6 \times z$ *j* $p^6 \times p^7$

The general identity is $a^p \times a^q = a^{p+q}$, where p and q are positive integers.

5 Use the above identity to simplify:

a $x^6 \times x^4$ *b* $a^2 \times a^9$ *c* $3^7 \times 3^7$ *d* $5^3 \times 5^5$ *e* $2^8 \times 2^8$

f $c^6 \times c^6$ *g* $z^9 \times z$ *h* $a^4 \times a^6$ *i* $x^{10} \times x^{20}$. *j* $y \times y^{11}$

6 Try to show that $a^p \times a^q = a^{p+q}$, using the definition $a^p = a \times a \times a \ldots$ to p factors, and the method you used in question **2**.

Exercise 2

1 By writing out the factors of each number, show that:

a $\frac{5^4}{5^2} = 5^2$ *b* $\frac{2^5}{2^3} = 2^2$ *c* $\frac{7^4}{7^3} = 7$

2 By writing out the factors of each of the following, show that:

a $\frac{a^3}{a^2} = a$ *b* $\frac{x^4}{x} = x^3$ *c* $\frac{b^4}{b^2} = b^2$

3 From questions ***1*** and **2** what appears to be the relation between the indices on the left side and the index on the right side?

4 Using the relation you gave in question **3**, write down the simplest form of the quotients in the following:

a $\frac{2^4}{2^2}$ *b* $\frac{3^7}{3^4}$ *c* $\frac{5^{12}}{5}$ *d* $\frac{x^{10}}{x^4}$ *e* $\frac{10^{15}}{10^9}$

f $\frac{c^3}{c^2}$ *g* $\frac{z^9}{z^3}$ *h* $\frac{k^{20}}{k^9}$ *i* $\frac{a^7}{a^7}$ *j* $\frac{x^{12}}{x^5}$

The general identity is $a^p \div a^q = a^{p-q}$, where p and q are positive integers and $p > q$.

5 Use the above identity to simplify:

a $x^6 \div x^3$ *b* $y^4 \div y^2$ *c* $z^9 \div z^4$ *d* $a^8 \div a$ *e* $c^8 \div c^4$

f $2^7 \div 2$ *g* $3^3 \div 3^3$ *h* $4^5 \div 4^4$ *i* $5^{20} \div 5^{10}$ *j* $a^{11} \div a^{10}$

6 Try to show that $a^p \div a^q = a^{p-q}$, using the definition $a^p = a \times a \times a \times \ldots$ to p factors, and the method you used in question **2**.

Exercise 3

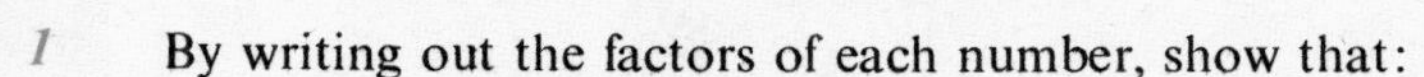

1 By writing out the factors of each number, show that:

a $(2^2)^3 = 2^6$ *b* $(3^4)^2 = 3^8$ *c* $(5^3)^3 = 5^9$

2 By writing out the factors of each of the following, show that:

a $(a^2)^2 = a^4$ *b* $(b^3)^2 = b^6$ *c* $(k^5)^2 = k^{10}$

3 From questions *1* and *2* complete the general relation $(a^p)^q = \ldots$, where p and q are positive integers.
Using this relation, express the following without brackets:

a $(5^2)^3$ *b* $(3^4)^5$ *c* $(4^6)^{10}$ *d* $(2^3)^2$ *e* $(10^{10})^{10}$

f $(a^6)^5$ *g* $(d^2)^2$ *h* $(x^7)^3$ *i* $(x^2)^3$ *j* $(p^3)^2$

The general identity is $(a^p)^q = a^{pq}$, where p and q are positive integers.

4 Use the above identity to simplify:

a $(a^2)^3$ *b* $(x^4)^4$ *c* $(y^2)^8$ *d* $(z^8)^2$ *e* $(c^5)^5$

f $(2^6)^2$ *g* $(2^7)^7$ *h* $(3^4)^3$ *i* $(4^2)^5$ *j* $(a^8)^{10}$

5 Try to show that $(a^p)^q = a^{pq}$, using the definition $a^p = a \times a \times a \ldots$ to p factors, and the method you used in question *2*.

Exercise 4

1 By writing out the prime factors of each number and using the associative and commutative laws, show that:

a $(2 \times 3)^2 = 2^2 \times 3^2$. $[(2 \times 3)^2 = (2 \times 3) \times (2 \times 3) = (2 \times 2) \times (\ldots) = \ldots]$

b $(3 \times 5)^3 = 3^3 \times 5^3$ *c* $2^3 \times 5^3 = (2 \times 5)^3 = 10^3$

d $2^{10} \times 3^{10} = (2 \times 3)^{10}$

2 By writing the factors of each of the following, show that:

a $a^3b^3 = (ab)^3$ *b* $(pq)^2 = p^2q^2$ *c* $x^5y^5 = (xy)^5$

The general identity is $(ab)^n = a^nb^n$, where n is a positive integer.

3 Use the above identity to simplify:

a $(ab)^4$ *b* $(ab^2)^3$ *c* $(x^2y^2)^5$ *d* $(x^3y^4)^2$ *e* $(m^5n^2)^3$

f $(3m)^2$ *g* $(2m^3)^2$ *h* $3(mn)^2$ *i* $(3mn)^2$ *j* $(2m^2n)^4$

Exercise 5

Use the following identities in this Exercise:

$$a^p \times a^q = a^{p+q}, \quad a^p \div a^q = a^{p-q}, \quad (a^p)^q = a^{pq}, \quad (ab)^n = a^n b^n$$

Simplify:

1 $a^2 \times a^3$ *2* $x^{10} \times x^6$ *3* $p^2 \times p^2$ *4* $y^{100} \times y^{100}$

5 $k^7 \div k^2$ *6* $a^{10} \div a^5$ *7* $b^4 \div b^3$ *8* $x^9 \div x^6$

9 $(a^2)^3$ *10* $(x^{10})^{10}$ *11* $(z^6)^2$ *12* $(xy^2)^3$

Write in a simpler form:

13 $2^3 \times 2^4$ *14* $4^3 \times 4^5$ *15* $7^4 \times 7$

16 $(\frac{2}{3})^2 \times (\frac{2}{3})^5$ *17* $(\frac{1}{2})^3 \times (\frac{1}{2})^4$ *18* $(0{\cdot}375)^4 \times (0{\cdot}375)^2$

19 $(3^4)^3$ *20* $(2^5)^4$ *21* $(13^3)^3$

22 $(15^2)^4$ *23* $2^6 \div 2^4$ *24* $3^4 \div 3$

25 $(5^2)^2 \div 5^3$ *26* $(10^{10})^{10} \div 10$ *27* $(3^2)^4 \times 3^5$

28 $(2^3)^4 \times (2^2)^5$

29 Express the following in the form 2^m:

a $(2^3 \times 2^4) \times 2^5$ *b* $2^3 \times (2^4 \times 2^5)$

30 Express in the form a^p:

a $3^2 \times 3^4 \times 3^7$ *b* $10 \times 10^3 \times 10^5$

31 Is it true that $2^3 \times 3^2 = 6^5$? If not, why not?

32 Which of the following are true and which are false?

a $2^3 = 3^2$ *b* $2^4 = 4^2$ *c* $3^3 \times 3^3 = 3^6$

d $(2^4)^2 = (2^2)^4$ *e* $2^3 \times 3^2 = 6^6$ *f* $2^2 \times 3^2 = 6^2$

g $10^6 \div 10^6 = 10^0$ *h* $2^3 \times 2^2 = 4^5$ *i* $5^3 \div 2^3 = 3^3$

33 Express the following in standard form, i.e. $a \times 10^n$, where $1 \leqslant a < 10$, and n is an integer:

a $6 \times 10^3 \times 10^5$ *b* $8 \times 7 \times 10^2$ *c* 120×8000

d $4000 \div 20$ *e* $5 \times (10^5 \div 10^2)$ *f* $93\,000\,000 \div 5000$

34 Solve the following equations, the variables being on the set of positive integers:

a $2^n = 8$ *b* $3^p = 81$ *c* $4^x = 64$ *d* $x^3 = 125$

e $5^n = 625$ *f* $x^4 = 1$ *g* $10^p = 1000$ *h* $x^6 = 64$

2 *Negative integral, and zero, indices*

We shall now extend the definition of powers to include the rest of the integers in such a way that the identities given at the beginning of Exercise 5 hold for all integers p and q. These identities are:

$$a^p \times a^q = a^{p+q}, \quad a^p \div a^q = a^{p-q}, \quad (a^p)^q = a^{pq}, \quad (ab)^n = a^n b^n$$

Exercise 6

1 In the sequence $10^4, 10^3, 10^2, 10^1, \ldots$, each term is obtained from the previous one by subtracting 1 from its index. Write down the next four terms in the sequence.

2 In the same sequence, $10^4, 10^3, 10^2, 10^1, \ldots$, each term is obtained from the previous one by dividing it by 10. Write down the next four terms in the sequence.

3 From your answers to questions ***1*** and ***2***, what appears to be the value of each of the following?

a 10^0 *b* 10^{-1} *c* 10^{-2} *d* 10^{-3}

4 Using the pattern of questions *1–3*, express with positive indices:

a 10^{-4} *b* 10^{-10} *c* 2^{-5} *d* 6^{-3}

5 If we suppose that $a^p \times a^q = a^{p+q}$ holds for all indices belonging to the set of integers, then we can easily find meanings for zero and negative indices.

a By considering $2^p \times 2^q = 2^{p+q}$, and putting $q = 0$, find a meaning for 2^0.

b By considering $a^p \times a^q = a^{p+q}$, and putting $q = 0$, find a meaning for a^0.

c By considering $2^3 \times 2^{-3} = 2^{3-3} = 2^0$, and using the result in ***a***, find a meaning for 2^{-3}.

d In the same way, find a meaning for 2^{-p} and a^{-p}, where p is a positive integer.

6 Express with positive indices:

a 2^{-5} *b* 3^{-2} *c* a^{-4} *d* b^{-10} *e* c^{-1}

f $2c^{-2}$ *g* ab^{-3} *h* $\dfrac{1}{x^{-2}}$ *i* $\dfrac{1}{2}x^{-2}$ *j* $\dfrac{1}{4x^{-3}}$

In question 5 you saw that if $a^p \times a^q = a^{p+q}$ for all integers, then $a^0 = 1$ and $a^{-p} = \dfrac{1}{a^p}$.

Conversely, it can be proved that if $a^0 = 1$ and $a^{-p} = \dfrac{1}{a^p}$, then $a^p \times a^q = a^{p+q}$ for all integers p and q. This is the reason that indices are so useful. The other identities given at the beginning of this section can also be proved to hold for all integers p and q. We shall *assume* that these identities are true for all integers p and q. If you wish to write out one of the proofs, try question 7.

7 Given that $a^m \times a^n = a^{m+n}$ for m, n positive integers, and that $a^{-m} = \dfrac{1}{a^m}$, prove that $a^p \times a^q = a^{p+q}$ for p, q negative integers. (Put $p = -m$, $q = -n$.)

Exercise 7

Use the following identities in this Exercise:

$$a^p \times a^q = a^{p+q}, \quad a^p \div a^q = a^{p-q}, \quad (a^p)^q = a^{pq}, \quad a^0 = 1, \quad a^{-p} = \frac{1}{a^p}$$

1 Write as powers of 2: 16, 8, 4, 2, 1

2 Write in the form 2^{-p}: $\frac{1}{2}$, $\frac{1}{4}$, $\frac{1}{8}$, $\frac{1}{16}$, $\frac{1}{256}$

3 Write as powers of 3: 81, 27, 9, 3, 1

4 Write in the form 3^{-p}: $\frac{1}{3}$, $\frac{1}{9}$, $\frac{1}{27}$, $\frac{1}{81}$, $\frac{1}{729}$

Express in a form not involving indices:

5 4^0, 4^{-1}, 4^{-2}, 4^{-3} *6* 6^0, 6^{-1}, 6^{-2}

7 2^0, 2^{-4}, $\frac{1}{2^{-2}}$, $\frac{1}{2^{-3}}$ *8* $(\frac{2}{3})^0$, $(\frac{2}{3})^{-1}$, $(\frac{2}{3})^{-2}$, $(\frac{2}{3})^{-3}$

Express with positive indices:

9 $3^2 \times 3^{-4}$ *10* $7^{-2} \times 7^{-5}$ *11* $2^4 \times 2^{-2}$ *12* $4^{-3} \times 5^2$

13 $3^4 \times 2^{-3}$ *14* $6^{-3} \times 5^{-2}$ *15* $3^3 \times x^{-4}$ *16* $2x^{-2}$

17 $5^2 \times y^{-1}$ *18* $25y^{-1}$ *19* $\frac{1}{2^{-2}}$ *20* $\frac{5}{3^{-3}}$

21 $(3^{-2})^3$ *22* $(2^{-3})^5$ *23* $(5^{-2})^{-3}$ *24* $(m^{-3})^{-2}$

25 Express in standard form:

a $3 \times 10^2 \times 10^{-4}$ *b* $28 \times 10^5 \times 10^3$ *c* $127 \times 10^{-8} \times 10^{-2}$

26a Draw the graph of the mapping $x \rightarrow 2^x$, calculating values at -3, -2, -1, 0, 1, 2, 3, and joining the points with a smooth curve. *Scales*: 2 cm to 1 unit for x, 1 cm to 1 unit for 2^x.

b On the same diagram draw the graph of the mapping $x \rightarrow 2^{-x}$.

c On the same diagram draw the graphs of the mappings $x \rightarrow 3^x$ and $x \rightarrow 3^{-x}$, calculating values at $x = -2, -1, 0, 1, 2$.
Write down any interesting features relating to the above graphs.

3 *Rational indices*

As in Section 2, if we assume the index relations to hold for all $p, q \in Q$, then we can find a meaning for a^p when p is rational.

Exercise 8

1 a Assuming that $a^p \times a^q = a^{p+q}$ for p, q rational, simplify $a^{1/2} \times a^{1/2}$.

b But $a^{1/2} \times a^{1/2}$ can also be written $(a^{1/2})^2$. Deduce a meaning for $a^{1/2}$.

2 Simplify $a^{1/3} \times a^{1/3} \times a^{1/3}$, and hence find a meaning for $a^{1/3}$.

3 Simplify $x^{1/4} \times x^{1/4} \times x^{1/4} \times x^{1/4}$, and hence find a meaning for $x^{1/4}$.

4 Simplify $a^{2/3} \times a^{2/3} \times a^{2/3}$, and hence find a meaning for $a^{2/3}$.

From the above examples you can see that for a real, and m and n positive integers, $a^{m/n} = (\sqrt[n]{a})^m = \sqrt[n]{a^m}$.

As in Exercise 6, question 7, it can be proved that if this meaning is given to $a^{m/n}$, then the index identities do hold.

5 Use the above identity to express the following with root signs of the form $\sqrt[n]{a^m}$:

a $a^{3/5}$ *b* $b^{5/2}$ *c* $c^{4/3}$ *d* $d^{3/2}$ *e* $e^{5/6}$ *f* $x^{1/2}$

6 Use the identity to express the following in index form:

a $\sqrt{c}$ *b* $\sqrt[3]{a}$ *c* $\sqrt[4]{x^3}$ *d* $\sqrt{x^3}$ *e* $\sqrt[3]{x^5}$ *f* $\sqrt[3]{x^6}$

By the method of Section 2 the use of indices is extended from indices taken from the set of positive rational numbers to indices taken from the set of all rational numbers.

Thus for every non-zero real number a, $a^{-(m/n)} = \dfrac{1}{a^{m/n}}$

7 Express the following with positive indices, and then with root signs:

a $a^{-1/2}$ *b* $b^{-2/3}$ *c* $c^{-3/4}$ *d* $d^{-1/3}$ *e* $e^{-4/3}$

8 Express in index form:

a $\dfrac{1}{\sqrt{p}}$ *b* $\dfrac{1}{\sqrt[3]{q}}$ *c* $\dfrac{1}{\sqrt{x^3}}$ *d* $\dfrac{1}{\sqrt[3]{y^6}}$ *e* $\dfrac{\sqrt{z}}{z}$

Note: (i) For all but the simplest examples, the index notation is usually much more convenient than the surd notation.
(ii) In a numerical example it is often helpful to express the number as a power of prime factors.

Example. Evaluate $16^{-3/4}$

$$16^{-3/4} = (2^4)^{-3/4} = 2^{-3} = \frac{1}{2^3} = \frac{1}{8}$$

Exercise 9

Evaluate:

1 $4^{1/2}$ 2 $100^{1/2}$ 3 $27^{1/3}$ 4 $64^{1/3}$ 5 $9^{3/2}$

6 $8^{2/3}$ 7 $16^{3/4}$ 8 $49^{3/2}$ 9 $36^{-1/2}$ *10* $125^{-1/3}$

11 $25^{-3/2}$ *12* $81^{3/4}$ *13* $81^{-3/4}$ *14* $(\frac{1}{4})^{-1/2}$ *15* $(\frac{1}{8})^{5/3}$

Write the following with root signs:

16 $a^{1/2}$ *17* $b^{1/3}$ *18* $c^{2/3}$ *19* $d^{4/5}$ *20* $e^{-3/4}$

Simplify:

21 $(a^4)^{1/2}$ *22* $(a^6)^{1/3}$ *23* $(a^6)^{2/3}$ *24* $(a^{-10})^{1/5}$

25 $(a^{-2})^{1/4}$ *26* $(a^{-2/3})^{-3}$ *27* $(a^{5/2})^{-2}$ *28* $(y^4)^{-1/2}$

29 $(y^{12})^{2/3}$ *30* $(y^{-3/2})^{-2}$ *31* $(x^{-2/3})^{-1/2}$ *32* $(x^{3/4})^{4/3}$

33 $(b^{2/3})^0$ *34* $(a^{4/5})^{1/2}$ *35* $(a^{3/2})^{4/3}$ *36* $(3^2+4^2)^{1/2}$

37 $(5^2+12^2)^{1/2}$ *38* $(0{\cdot}01)^{1/2}$ *39* $(1^3+2^3+3^3)^{1/2}$

4 *Miscellaneous examples*

We now use the following identities in some miscellaneous examples:

$$a^p \times a^q = a^{p+q}, \quad a^p \div a^q = a^{p-q}, \quad (a^p)^q = a^{pq},$$

$$(ab)^n = a^n b^n, \quad a^0 = 1, \quad a^{-p} = \frac{1}{a^p}, \quad a^{m/n} = \sqrt[n]{a^m}$$

Example 1. Simplify $4x^{1/2} \times 3x^{-1/6}$
$4x^{1/2} \times 3x^{-1/6} = 4 \times 3 \times x^{1/2-1/6} = 12x^{1/3}$

Example 2. Find the value of $2x^{-1/3}$ when $x=8$
$2x^{-1/3} = 2 \times 8^{-1/3} = 2 \times (2^3)^{-1/3} = 2 \times 2^{-1} = 1$

Exercise 10

Simplify:

1 $2x^{1/2} \times 2x^{-1/2}$ *2* $3a^{2/3} \times 4a^{1/3}$ *3* $6b^{1/2} \times b^{1/2}$

4 $2c^{3/2} \times c^{-1/2}$ *5* $4x^{3/2} \div 2x^{1/2}$ *6* $8z^{3/4} \div 2z^{-1/4}$

7 $12y^{1/2} \div 6y^{3/2}$ *8* $x^{-1/2} \div x^{-1/4}$ *9* $12x^{-1/3} \div 12x^{-1/3}$

Evaluate the following for $a = 16$ and $x = 27$:

10 $2a^{1/2}$ *11* $3a^{3/4}$ *12* $4x^{2/3}$

13 $2x^{-2/3}$ *14* $a^{-1/4} \times x^{1/3}$ *15* $3a^{-1/2} \times 4x^{-1/3}$

Express without root signs:

16 $\sqrt[3]{a}$ *17* $\sqrt{x^3}$ *18* $(\sqrt[3]{y})^5$ *19* $\sqrt[5]{x^3}$ *20* $\sqrt[4]{k}$

Express with root signs:

21 $x^{1/2}$ *22* $y^{-1/3}$ *23* $z^{-3/2}$ *24* $p^{4/5}$ *25* $q^{-7/4}$

Simplify the following, expressing your answers with positive indices:

26 $p^4 \times p^{-1/2}$ *27* $q^3 \div q^{-1/2}$ *28* $(r^{2/3})^{-3/2}$

29 $(x^{-3/4})^8$ *30* $2s^{3/4} \div 2s^{-3/4}$ *31* $2t^{1/2} \times 2t^{-3/2}$

32 Solve the following equations, the variables being on the set of rational numbers:

a $2^n = \frac{1}{16}$ *b* $9^x = 27$ *c* $5^{-x} = 125$ *d* $10^{3n} = 10\,000$

Simplify *33–35*:

33 $a^{1/2}(a^{1/2}+1)$ *34* $b^{-1/3}(b^{1/3}-b^{2/3})$ *35* $3a^{-1/2}(a^{1/2}+a^{-3/2})$

36 Light travels at 3×10^8 m/s. How many metres does it travel in 1 hour?

37 Which of the numbers 10^{15}, 10^{17} is nearer in value to $10^{15}+10^2$?

38 $A = 4p^2r^3$. Calculate A for $p = 10$ and $r = 1$.

39 $B = 2a^{1/2}b^{-1/3}$. Calculate B for $a = 100$ and $b = 64$.

40 Express in standard form:

a $3 \times 10^3 \times 10^{-6}$ *b* $46 \times 10^4 \times 10^{-1}$ *c* $789 \times 10^4 \times 10^{-4}$

41 The mass of 1 atom of oxygen is $2{\cdot}7 \times 10^{-23}$ g. What is the mass of 4×10^{28} atoms of oxygen? (Give the number of grammes in standard form.)

42 Find the positive rational number x in the following:

a $\sqrt{x}=2$ *b* $x^{1/2}=\frac{1}{2}$ *c* $\sqrt[3]{x}=3$ *d* $x^n=p$

43 If $Q = hd^3$, change the subject of this formula to d.

44 Given that $C = LD^2/25$, express D in terms of C and L.

Exercise 10B

Express the following with x in the numerator in index form.

For example, $\dfrac{2}{3\sqrt{x}} = \dfrac{2}{3x^{1/2}} = \dfrac{2x^{-1/2}}{3}$ or $\frac{2}{3}x^{-1/2}$

1 $\dfrac{1}{x}$ *2* $\dfrac{1}{x^2}$ *3* $\dfrac{5}{x^2}$ *4* $\dfrac{2}{x^3}$ *5* $\dfrac{1}{2x}$

6 $\dfrac{3}{4x}$ *7* $\dfrac{5}{2x^2}$ *8* $\dfrac{1}{\sqrt{x}}$ *9* $\dfrac{3}{2\sqrt{x}}$ *10* $\dfrac{2}{5x^{-3}}$

11 $\dfrac{1}{x^2}+\dfrac{1}{x^3}$ *12* $\dfrac{2}{x}-\dfrac{3}{x^4}$ *13* $x-\dfrac{1}{x}$ *14* $2x+\dfrac{1}{2x}$

15 $\sqrt{x}+\dfrac{1}{\sqrt{x}}$ *16* $x^3+\dfrac{1}{x^{-2}}$ *17* $\dfrac{3}{x^{-3}}+\dfrac{1}{2x^2}$ *18* $3\sqrt[3]{x}-\dfrac{1}{3\sqrt[3]{x}}$

Use the distributive law to express the following as a sum or difference of terms.

For example, $x^{1/2}(2-x^{-1/2}) = 2x^{1/2}-1$

19 $x^{1/2}(x^{1/2}+1)$ *20* $x^{-1/2}(x^{3/2}-1)$ *21* $x^{1/2}(x^{1/2}+x^{-1/2})$

22 $x^{-2}(x^3+x)$ *23* $x^{-1}(x+x^{-1})$ *24* $x^{2/3}(x^{4/3}-x^{1/3})$

25 $x^{-1}(x+1)$ *26* $x^{1/2}(x-2)$ *27* $u^{1/3}(u^{2/3}+u^{-2/3})$

28 $v^{1/3}(v^{-1/3}-v^{5/3})$ *29* $x^{-1}\left(x^{1/2}+\dfrac{1}{x^{1/2}}\right)\left(x^{1/2}-\dfrac{1}{x^{1/2}}\right)$

Expand the following.

For example, $(x^{1/2}+x^{-1/2})^2 = x+2+x^{-1}$

30 $(p+q)^2$ *31* $(r-1)^2$ *32* $(x^{1/2}+1)^2$ *33* $(x^{1/2}-1)^2$

34 $(x+x^{1/2})^2$ *35* $(x^{3/2}-x^{1/2})^2$ *36* $(x+x^{-1})^2$ *37* $(1-x^{-2})^2$

Express the following as a sum or difference of terms, with x in the numerator in index form.

For example, $\dfrac{1}{\sqrt{x}}(1-\sqrt{x})^2 = x^{-1/2}(1-2x^{1/2}+x)$

$= x^{-1/2}-2+x^{1/2}$

38 $\sqrt{x}(\sqrt{x}+1)^2$ *39* $\dfrac{1}{\sqrt{x}}(\sqrt{x}-1)^2$ *40* $\left(x+\dfrac{1}{x}\right)^2$

41 $\left(\sqrt{x}-\dfrac{1}{\sqrt{x}}\right)^2$ *42* $\left(2x+\dfrac{1}{2x}\right)^2$ *43* $\left(\dfrac{1}{3\sqrt{x}}-3\sqrt{x}\right)^2$

Express the following as a sum or difference of terms.

For example,
$$\frac{2x^2-3+4x^{-2}}{2x^2}=\tfrac{1}{2}x^{-2}(2x^2-3+4x^{-2})$$
$$=1-\tfrac{3}{2}x^{-2}+2x^{-4}$$

44 $\dfrac{x^2+x+1}{x}$ *45* $\dfrac{2x^{1/2}-x^{-1/2}}{x^{1/2}}$ *46* $\dfrac{x^4-2x^2+4}{2x^2}$

47 $\dfrac{(x+3)^2}{x}$ *48* $\dfrac{x^3+3x-6}{3x^3}$ *49* $\dfrac{x^{1/2}-x^{-1/2}}{x^{3/2}}$

Summary

Fundamental identities for p, q ∈ Q		*Illustrations*
1	$a^p \times a^q = a^{p+q}$	$a^5 \times a^{-2} = a^3$
2	$a^p \div a^q = a^{p-q}$	$a^{3/4} \div a^{1/4} = a^{1/2}$
3	$(a^p)^q = a^{pq}$	$(a^{1/2})^{-2} = a^{-1}$
4	$(ab)^n = a^n b^n$	$(ab)^3 = a^3 b^3$
5	$a^0 = 1$	$10^0 = 1$
6	$a^{-p} = \dfrac{1}{a^p}$	$x^{-2} = \dfrac{1}{x^2}$
7	$a^{m/n} = \sqrt[n]{a^m}$	$8^{2/3} = \sqrt[3]{8^2} = 4$; $\sqrt[3]{x} = x^{1/3}$

3 Linear Programming

1 A mathematical model

Problem

One kind of cake requires 200 g of flour and 25 g of fat, and another kind of cake requires 100 g of flour and 50 g of fat. Suppose we want to make as many cakes as possible but have only 4 kg of flour and 1·2 kg of fat available, although there is no shortage of the various other ingredients. How many cakes of each kind should we make?

The first step in solving a problem by mathematics is to translate the problem into the language of mathematics. This is called making a *mathematical model*.

Mathematical model

The data in the problem are set out concisely in the following table:

	Flour (g)	*Fat* (g)
First kind of cake	200	25
Second kind of cake	100	50

Suppose that x is the number of cakes of the first kind, and y is the number of the second kind.

Since we shall need $(200x+100y)$ g of flour, and we have 4000 g,

$$200x+100y \leqslant 4000$$
$$\Leftrightarrow \quad 2x+y \leqslant 40 \qquad . \quad . \quad . \quad . \qquad (1)$$

Since we shall need $(25x+50y)$ g of fat, and we have 1200 g,

$$25x+50y \leqslant 1200$$
$$\Leftrightarrow \quad x+2y \leqslant 48 \qquad . \quad . \quad . \quad . \qquad (2)$$

Since x and y cannot be negative integers,

$$x \geqslant 0 \quad . \quad . \quad . \quad . \quad . \quad (3)$$

$$y \geqslant 0 \quad . \quad . \quad . \quad . \quad . \quad (4)$$

The problem is to find replacements for x and y which will make $x+y$ as great as possible, i.e. we have to *maximise* $x+y$ subject to the conditions (1) to (4).

Notice that in the four inequations, the variables occur only in the first power and there are no products of the variables. Equations and inequations in which the variables appear in the first power only and in which there are no products of the variables, are called *linear*. When the problem is solved in Section 2, we shall have a programme for cake-making, hence the name *linear programming*.

In this chapter we are going to look at simple problems which can be solved readily by graphical methods which were first introduced in Book 4, Chapter 2. For complicated cases such as occur in industry, graphical methods are impracticable since more than a hundred variables may be required. Other techniques are therefore used and solutions are obtained by computer. Some of the ideas we shall use, however, also lie behind the more practical methods.

Example 1. Indicate on squared paper the solution set of the inequation $3x+5y \leqslant 30$, for $x, y \in R$.

Figure 1 shows the required solution set.

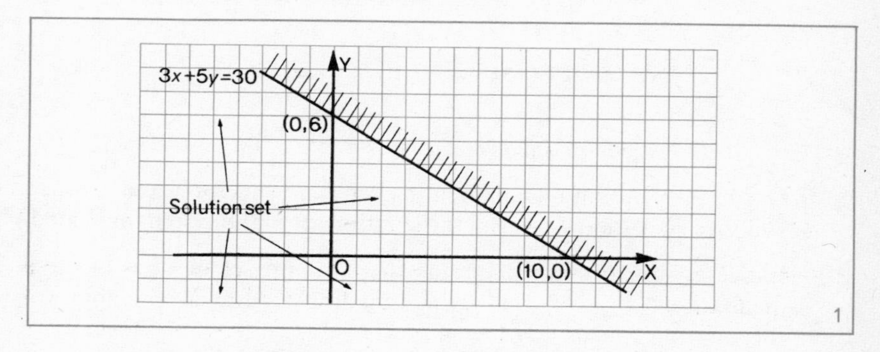

Method (i) Draw the line $3x+5y=30$ by joining the points $(10, 0)$ and $(0, 6)$ where it cuts the x- and y-axes. The line divides the coordinate plane into two half-planes.

(ii) For the inequation $3x+5y \leqslant 30$, points (x, y) must either lie on the line $3x+5y=30$ or in the half-plane on one side of the line. To decide which side, test with the point (0, 0). Since $0+0\leqslant 30$, (0, 0) belongs to the solution set, and so the half-plane required is on the origin side of the line.

(iii) *To show this, we shade the regions where points* ***cannot*** *lie, using 'fringe shading', so that the region indicating the solution set is left clear for further study.*

Example 2. Indicate on a Cartesian diagram the solution set of the system of inequations $x\geqslant 0$, $y\geqslant 0$, $x+2y\leqslant 8$ and $3x+2y<12$ for $x, y \in R$.

The solution set of the given system of inequations is indicated by the clear region in Figure 2. The line $3x+2y=12$ is shown as a broken line to indicate that points on the line do not give ordered pairs of the solution set.

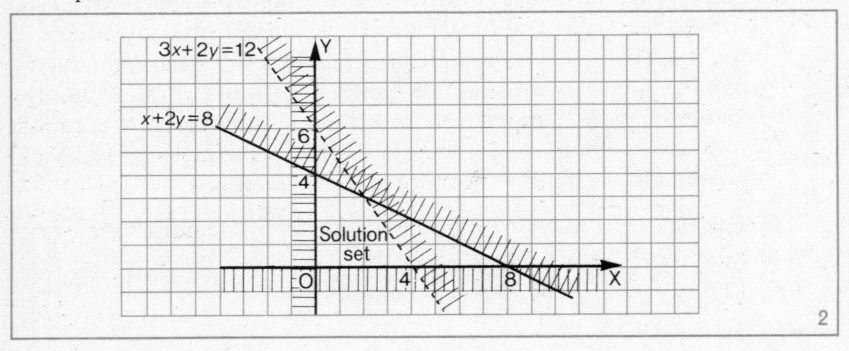

Exercise 1 (*mainly revision*)

(*5-mm squared paper will be suitable in this Exercise.*)

1 Use Figure 2 to find whole number replacements for x and y which will give *six* ordered pairs in the solution set of the system indicated.

2 Indicate on Cartesian diagrams the solution set of each of these inequations ($x, y \in R$). (Use 'fringe shading' to show where points of the solution *cannot* lie.)

a $x\geqslant 0$ *b* $y\geqslant 0$ *c* $0\leqslant x\leqslant 5$ *d* $0\leqslant y\leqslant 6$

3 Repeat question **2** for each of the following inequations:

a $x\leqslant 4$ *b* $x+y\leqslant 5$ *c* $x+y>7$ *d* $2x+y\leqslant 6$

4 Indicate on Cartesian diagrams the solution set of each of these *systems* of inequations for $x, y \in R$:

a $x \geqslant 0,\ y \geqslant 0,\ x+y \leqslant 5$

b $x \geqslant 0,\ y \geqslant 0,\ x+2y \leqslant 8$

c $x \geqslant 0,\ y \geqslant 0,\ 2x+3y \leqslant 12$

d $x \leqslant 8,\ y \geqslant 0,\ y \leqslant x,\ x+y \geqslant 8$

5 Indicate on Cartesian diagrams the solution set of each of the following systems of inequations for $x, y \in R$:

a $x \geqslant 0,\ y \geqslant 0,\ x+y \geqslant 4,\ x+y \leqslant 7$

b $x \geqslant 0,\ y \geqslant 0,\ x+y \leqslant 6,\ 3x+8y \leqslant 24$

c $x \geqslant 0,\ x \leqslant 6,\ y \geqslant 0,\ y \leqslant x+2$

d $x \leqslant 8,\ y \leqslant 6,\ x+4y \geqslant 8,\ 2x+y > 8$

6 Figure 3 (i) shows the set of '*lattice points*' which represents the solution set of the system of inequations $1 \leqslant x \leqslant 4$ *and* $1 \leqslant y \leqslant 4$, for $x, y \in W$. Copy the diagram.

a Mark the value of $x+y$ at each of the lattice points, e.g. at (2, 3) we have 5.

b What is the *maximum value* of $x+y$, and at which point does it occur?

c What is the *minimum value* of $x+y$, and at which point does it occur?

d What do you notice about the points for which $x+y=5$?

e Write down the solution set of $x+y<5$.

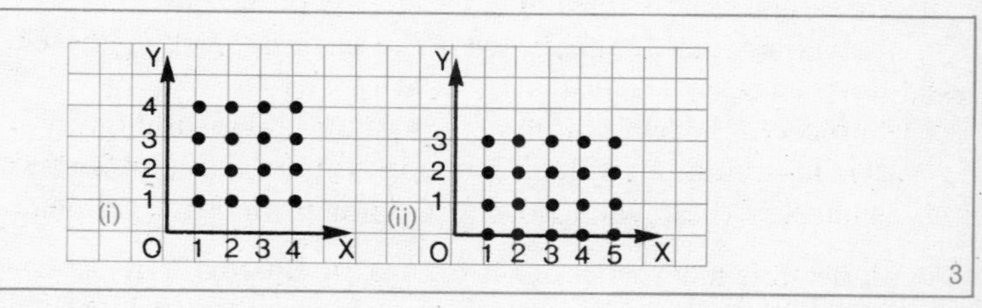

3

7 Figure 3 (ii) shows the solution set of the system of inequations $1 \leqslant x \leqslant 5$ and $0 \leqslant y \leqslant 3$, $x, y \in W$. Copy the diagram.

a Mark the value of $x+2y$ at each of the lattice points.

b Write down the pairs of replacements for x and y which make $x+2y=7$.

c Write down the solution set of $x+2y \leqslant 4$.

d Find the maximum value of $x+2y$, and state the corresponding replacements for x and y.

e What is the minimum value of $x+2y$, and for what x, y does it occur?

8 In Figure 4, the region OPQR represents the solution set of a system of inequations.

a Find the values of $2x+5y$ at O, P, Q and R.

b What are the values of $2x+5y$ at the points marked by black dots?

c Subject to the restrictions of the system on x and y, find (*1*) the minimum value (*2*) the maximum value of $2x+5y$, assuming $x, y \in W$.

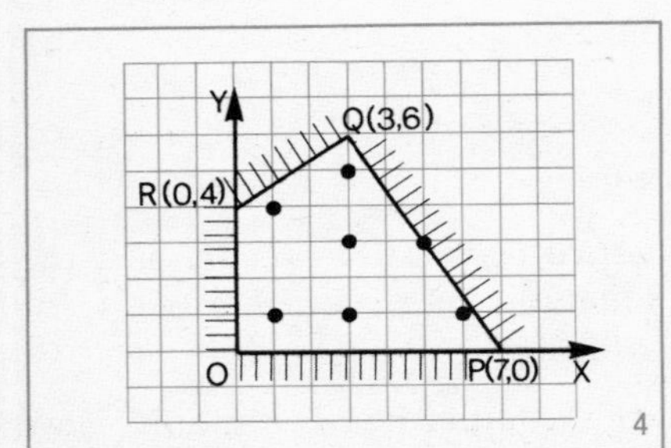

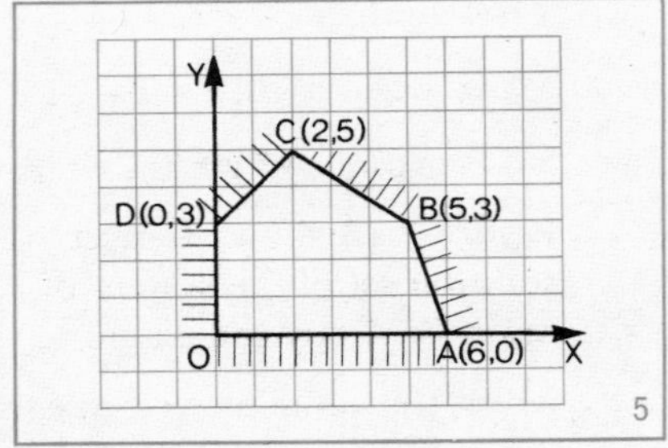

9 The solution set of a system of inequations is represented by the region bounded by the polygon OABCD shown in Figure 5:

a Find the values of $2x+3y$ at O, A, B, C and D.

b Investigate the values of $2x+3y$ at two, or more, points inside the polygon which are near each of the vertices, marking these with black dots.

c Subject to the restrictions of the system, and assuming $x, y \in W$, state: (*1*) the minimum value (*2*) the maximum value of $2x+3y$, and for what x, y these values occur. Comment on your results.

10a A triangle has vertices O(0, 0), A(6, 0), B(0, 6). Write down a system of inequations whose solution set is represented by the interior of triangle OAB.

b Repeat for a triangle whose vertices are O(0,0), P(6,0), Q(0,3).

2 *Using the mathematical model*

We shall now solve the problem in Section 1. First we require the solution set of the following system of inequations for $x, y \in R$:

$$2x+y \leqslant 40 \quad . \quad . \quad . \quad . \quad (1)$$

$$x+2y \leqslant 48 \quad . \quad . \quad . \quad . \quad (2)$$

$$x \geqslant 0 \quad . \quad . \quad . \quad . \quad (3)$$

$$y \geqslant 0 \quad . \quad . \quad . \quad . \quad (4)$$

The respective solution sets are shown in Figure 6. The solution set of the *system* of inequations (1) to (4) is the *intersection* of the four solution sets, and is represented by the region OABC. Notice that all points on the boundaries OA, AB, BC and CO belong to the intersection of the solution sets.

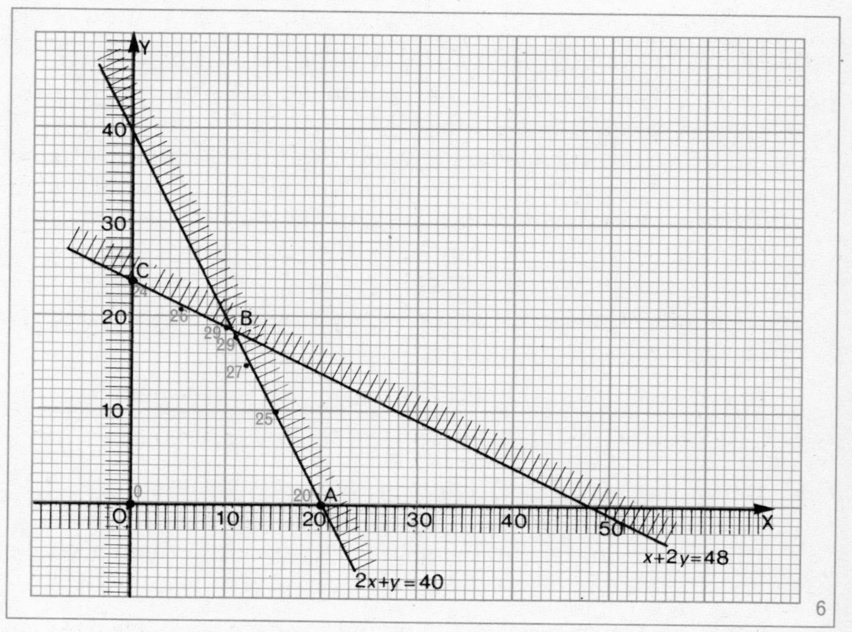

6

Next we have to look at the points belonging to the solution set of the system of inequations for which $x \in W$ and $y \in W$. Each member of this set is called a *feasible solution* of the problem. For example, the marked points, (0, 0), (0, 24), (5, 21), (10, 19), (11, 18), (12, 15), (15, 10), (20, 0) are feasible solutions.

To get a solution which is as good as possible, called an *optimal solution*, we have to find points whose coordinates satisfy the given conditions and which give a maximum value of $x+y$. Repeating the work of Exercise 1, we investigate values of $x+y$ at points on, or near, the boundary of polygon OABC, especially at a vertex. Some of these are marked in Figure 6. The coordinates and the corresponding values of $x+y$ are listed in the following table:

x	0	0	5	10	11	12	15	20
y	0	24	21	19	18	15	10	0
$x+y$	0	24	26	29	29	27	25	20

Optimal solutions are given by $x = 11$ and $y = 18$, and by $x = 10$ and $y = 19$. So the optimal value of $x+y$ is 29.

Hence baking 11 cakes of the first kind and 18 of the second kind, or 10 of the first and 19 of the second, will make the best use of the ingredients.

Note. The coordinates of vertex B do not give a feasible solution. Can you give a reason for this?

In linear programming, the linear form $ax+by$ which is to be maximised (or minimised) is called the *objective form*, and is an important part of the mathematical model.

Exercise 2

In questions *1–4*, use 5-mm squared paper and take a scale of 2 cm to 1 unit on each axis. Assume $x, y \in R$.

1 a Show the solution set of the following system of inequations:

$$x+y\leqslant 3,\ x+2y\leqslant 4,\ x\geqslant 0,\ y\geqslant 0$$

b Mark with dots the points (0, 0), (1, 0), (2, 0), (3, 0), (0, 1), (1, 1), (2, 1), (0, 2) and write in red the value of $x+y$ at each point.

c Which points give the maximum value of $x+y$ on the solution set, for $x, y \in W$?

2 For the system in question *1*, copy and complete the following table:

x	0	0	1	1	2	2	3
y	1	2	0	1	0	1	0
$2x+3y$	3			5			

What is the maximum value of $2x+3y$ on the solution set?

3 a Show the solution set of the system of inequations:

$$x+y \leqslant 4,\ 2x+y \leqslant 6,\ x \geqslant 0,\ y \geqslant 0.$$

b On the diagram, mark with dots the points (3, 0), (2, 1), (1, 2), (2, 2), (0, 3), (1, 3), (0, 4), and write in red the value of $2x+y$ at each point.
c Which points, for which $x, y \in W$, make $2x+y$ a maximum?
d Find the maximum value of $2x+y$ on the solution set for $x, y \in W$.

4 a Show the solution set of the system of inequations:

$$x+y \geqslant 3,\ x+2y \geqslant 4,\ x \geqslant 0,\ y \geqslant 0.$$

b Mark the points (4, 0), (4, 1), (3, 1), (2, 1), (2, 2), (1, 2), (1, 3), (0, 3), (0, 4), and write in red the value of $2x+3y$ at each point.
c Deduce the *minimum* value of $2x+3y$ on the solution set for $x, y \in W$, and the corresponding replacements for x and y.

In questions ***5–8***, use 2-mm squared paper, and take a scale of 2 cm to 10 units on each axis.

5 a Graph the set determined by:

$$x+3y \geqslant 30,\ 5x+y \geqslant 50,\ 5x+3y \geqslant 90.$$

b Examine values of: (*1*) $3x+y$ (*2*) $2x+3y$ at and near the points (0, 50), (5, 20), (15, 5), (30, 0) and hence find the *minimum* value on the solution set of each of these linear forms.

6 One cake requires 150 g flour and 50 g fat, and another requires 75 g flour and 75 g fat. We want to make as many cakes as possible of the two kinds when 2·25 kg flour and 1·5 kg fat are available.

a Let x be the number of cakes of the first kind, and y the number of cakes of the second kind. Write down two obvious inequations for x and y, and show that two further inequations reduce to $2x+y \leqslant 30$ and $2x+3y \leqslant 60$.
b Show on a Cartesian diagram the solution set of the above system of inequations.
c Find replacements for x and y on the solution set which will maximise $x+y$, and hence write down the number of each kind of cake which gives as good a solution as possible.

7 A cycle dealer wishes to buy up to 25 cycles for stock. He intends to purchase touring models at £30 each and racing models at £40 each, and has planned an outlay of not more than £840.

a Suppose he buys x touring models and y racing models. Write down four inequations that must be satisfied by x and y.

b Show graphically the solution set of the system of inequations, taking replacements for x and y from 0 to 30.

c If he expects to make a profit of £10 on each touring model and £12 on each racing model, write down an expression in x and y for his total profit. Hence find the number of each model he should buy to give maximum profit. What is the maximum profit?

8 An aircraft has seats for not more than 48 passengers. Those willing to pay first-class fares can take 60 kg of baggage each, but tourist-class passengers are restricted to 20 kg each. Only 1440 kg of baggage can be carried altogether.

a If x is the number of first-class passengers and y is the number of tourist passengers, write down in simplest form four inequations satisfied by x and y.

b Illustrate on squared paper the region in which feasible solutions of these inequations can be found.

c If first-class fares are £100 and tourist fares £50 each, use the diagram to find the number of passengers of each kind that should be carried for maximum profit, and the gross sum that will be obtained.

3 *Using a search line* $ax+by=k$

The shaded region in Figure 7 shows the solution set of the system of inequations:

$$x+3y\leqslant 9,\ 2x+y\leqslant 8,\ x\geqslant 0,\ y\geqslant 0 \text{ for } x, y\in R.$$

To find the maximum value of a linear form such as $x+2y$, subject to the above conditions, consider the set of parallel lines corresponding to the equation $x+2y=k$ given by different replacements for k, where $k\in R$.

In Figure 7, the coloured lines have equations given by $k=0, 2, 4, 7$. Notice that the greater the value of k, the further the line is from the origin. Examining the properties of some of these lines:

(1) $k=0$ gives the line through the origin, $x+2y=0$, which gives *minimum* value zero of $x+2y$ in the region.

(2) $k=2$ gives the line $x+2y=2$ which cuts off a triangular portion of the feasible region in which $x+2y$ is less than or equal to 2.

(3) To find an optimal value of $x+2y$, we must draw a line parallel to the line $x+2y=0$, through the point P at the extremity of the feasible region. The required line is $x+2y=7$, and so the *maximum* value of $x+2y$ in the region is 7.

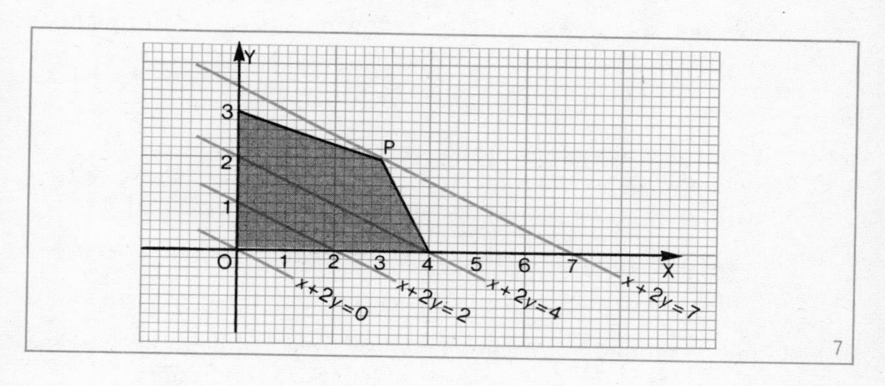

7

Note. A set of parallels $x+2y=k$, $k \in R$, can easily be constructed by first drawing the member through the origin, $x+2y=0$ *or* any other suitable member, e.g. $x+2y=2$ through the points (2, 0) and (0, 1). Parallel lines can then be drawn using a ruler and setsquare. For $x, y \in R$ the optimal point, or points, always come on the edge of the feasible region, and usually at a vertex. When $x, y \in W$ this is not necessarily so, as we have seen in the worked example in Section 2.

Exercise 3

In questions *1–5*, use a scale of 2 cm to 1 unit on each axis.

1 Show the set of lattice points which represent the solution set of the system $1 \leqslant x \leqslant 3$, $1 \leqslant y \leqslant 3$ for $x, y \in W$.

a Using a coloured pencil, mark in the value of $x+y$ at each of the lattice points, e.g. at (2, 3), $x+y=5$.

b Draw the set of lines $x+y=k$ for $k=2, 3, 4, 5, 6$. What do you notice?

c Write down the maximum and also the minimum value of $x+y$ on the solution set.

2 Show the solution set of the following system of inequations ($x, y \in R$):

$$2x+y \leqslant 6,\ x+2y \leqslant 6,\ x \geqslant 0,\ y \geqslant 0.$$

a Mark with dots the points (0, 0), (1, 0), (2, 0), (3, 0), (0, 1), (1, 1), (2, 1), (0, 2), (1, 2), (2, 2), (0, 3) and write in red the value of $x+y$ at each point.
b Draw the set of lines $x+y=k$ for $k=0, 1, 2, 3, 4$.
c State the maximum and minimum values of $x+y$ on the solution set.

3 Show the solution set of the following system of inequations ($x, y \in R$):

$$x+y \leqslant 4,\ x+3y \leqslant 6,\ x \geqslant 0,\ y \geqslant 0.$$

a On your diagram, draw the set of lines $x+2y=k$, $k=0, 1, 3, 5$.
b Deduce the maximum value of $x+2y$ subject to the above restrictions, and state the corresponding replacements for x and y.

4 Show the solution set of the system of inequations ($x, y \in R$):

$$x+y \leqslant 6, 2x+y \geqslant 3, x \geqslant 1, x \leqslant 4, y \geqslant 0.$$

Using ruler and setsquare, draw suitable parallels to the line $4x+y=0$ to show that, subject to the above conditions:

(1) the minimum value of $4x+y$ is 5
(2) the maximum value of $4x+y$ is 18.

5 *a* Re-draw the feasible region in question **4**.
b Draw the line $2x+4y=0$, i.e. $y=-\frac{1}{2}x$.
c Subject to the restraints on x, y in question **4**, draw parallels to the line $2x+4y=0$ to find the maximum and minimum values of $2x+4y$.

6 The parking area in a small car park is 360 m^2. The average area for a car is 6 m^2 and for a bus 24 m^2. Not more than 30 vehicles can be accommodated.

a Supposing that x is the number of cars and y is the number of buses to be accommodated, write down a system of inequations in x and y which models the given facts.
b Show on a graph the solution set of the system of inequations.
c If the parking charge for a car is 25p and for a bus 75p, the income I pence is given by $I=25x+75y$. By maximising I, find how many of each should be parked for maximum income, and state this income.

7 Solve the problem of question ***6*** for a car park with area 500 m^2, average areas of 5 m^2 for cars and 20 m^2 for buses, and costs of 50p and £1 respectively, not more than 60 vehicles being accommodated. (Decide which vertex you want, and solve a system of two linear equations to find the coordinates of this vertex. Now find the best x, y for $x, y \in W$.)

8 A doctor advises a patient to take daily at least 10 units of vitamin B_1 and at least 15 units of vitamin B_2. The patient finds he can buy tablets containing 2 units of B_1 and 1 unit of B_2, or capsules containing 1 unit of B_1 and 3 units of B_2.

a Suppose that the patient uses x tablets and y capsules daily. Write down a system of four inequations in x and y which must be satisfied.

b Taking a scale of 1 cm to 1 unit on each axis, show the solution set of this system.

c Assuming that a tablet costs $\frac{1}{2}$p and a capsule 1p, write down the daily cost in terms of x, y, and hence find how many of each the patient should take to make the cost as small as possible. State this daily cost.

9 A development corporation plans to build in 80 weeks a number of semi-detached houses and bungalows on 12 000 m^2 of ground. Semi-detached houses each take up 500 m^2 of ground and bungalows 400 m^2. A semi-detached house requires 360 man-weeks, (by 360 man-weeks we mean 1 man working for 360 weeks, 6 men for 60 weeks and so on), and a bungalow 160 man-weeks. The work force available is 90 men.

Find, by a graphical method, the number of each kind of house that should be built to give the greatest profit, assuming that the profit on a semi-detached house is £1500 and on a bungalow is £1000. Write down the total profit.

10 £100 worth of product A needs 30 kg of raw material and 18 hours of machine time. Product B needs 20 kg of the same raw material and 24 hours of machine time for £100 worth. Find, by a graphical method, the maximum value of products that can be made with 720 hours of machine time and 750 kg of raw material. (Let x be the number of £100 worth of A and y the number of £100 worth of B that are required.)

11 The composition per tonne (1000 kg) of two metal alloys X and Y is shown in the following table:

	Copper (kg)	*Metal A (kg)*	*Metal B (kg)*
Alloy X	500	300	200
Alloy Y	200	300	500

Another metal alloy is to be produced *as cheaply as possibie* from a mixture of X and Y so as to contain at least 6 tonnes of copper, 7·2 tonnes of A and 6 tonnes of B.

a If x tonnes of X are mixed with y tonnes of Y, show that $5x+2y \geqslant 60$, $x \geqslant 0$, $y \geqslant 0$, and write down in simplest form two other inequations that must be satisfied.

b Show on a Cartesian diagram, with scale 2 cm to 5 units, the region of feasible solutions for the problem.

c Given that X costs £400 a tonne and Y costs £200 a tonne, write down an expression for the total cost. Hence find the number of tonnes of X and Y which must be mixed and the total cost.

d If the cost of Y is doubled, what would the minimum cost have been?

Summary

1 *Linear programming (L.P.)* is a method of solving certain problems in which the mathematical model consists of a system of linear inequations which admits many solutions. Of all the feasible solutions, one (or more) may give the best solution (*optimal solution*).

2 The *solution set of a system of inequations* is the *intersection* of the solution sets of the separate inequations, and is best given graphically in the case of two variables.

3 *Method for solving an L.P. problem in two variables*:

(i) Translate the problem into mathematical language and set up a mathematical model consisting of a system of inequations, together with an *objective form* $ax + by$ which is to be maximised (or minimised).

(ii) Show the solution set of the system of inequations on a Cartesian graph. Points inside, or on the boundary of the polygon give feasible solutions.

(iii) Choose the point (or points) which will give the best solution by exploring the feasible region for points which maximise (or minimise) the objective form, *or* by using a search line.

4 *The optimal point, or points*, for $x, y \in R$ always come at a *vertex* or on the *edge* of the feasible region. When $x, y \in W$ this is not necessarily so (see cake-making problem in Section 2).

Graphs, Gradients and Areas

1 *The gradient of a straight line*

In the chapter on Similar Shapes in the geometry of Book 5 we saw that the gradient of a line segment AB could be defined as

$$\frac{y\text{-component of AB}}{x\text{-component of AB}}$$

If A is the point (x_1, y_1) and B is (x_2, y_2), the gradient of AB, which is often denoted by m_{AB}, is given by:

$$m_{AB} = \frac{y_2 - y_1}{x_2 - x_1} \quad \text{(see Figure 1 (i))}$$

In Figure 1 (ii) and (iii),

$$m_{PQ} = \frac{5-3}{6-2} = \frac{2}{4} = \frac{1}{2}; \quad m_{RS} = \frac{1-5}{1-(-1)} = \frac{-4}{2} = -2$$

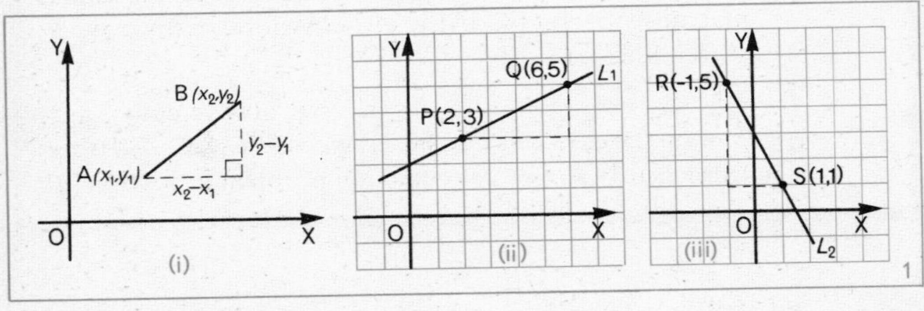

(i) (ii) (iii) 1

For a line such as L_1 (Figure 1 (ii)) which rises from left to right the gradient is a positive number; for a line such as L_2 (Figure 1 (iii)) which falls from left to right the gradient is a negative number.

If $x_2 = x_1$, then $x_2 - x_1 = 0$, and the gradient is undefined; the line is parallel to the y-axis.

If $y_2 = y_1$, then $y_2 - y_1 = 0$, and the gradient is zero; the line is parallel to the x-axis.

Exercise 1

1 Write down the gradient of each line in Figure 2, if the gradient exists.

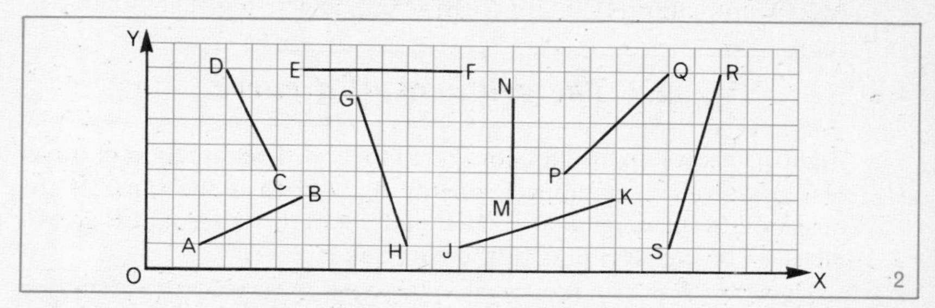

2 Plot the points G(−6, −4), K(9, −4), Q(9, 5), R(6, 8), S(−2, 6) and P(4, 2) on squared paper. Find the gradient of the following, where possible:

a PQ *b* PR *c* PS *d* PG *e* PK *f* KQ *g* GK

3 Find the gradients of OA, AB, BC and CD in Figure 3.

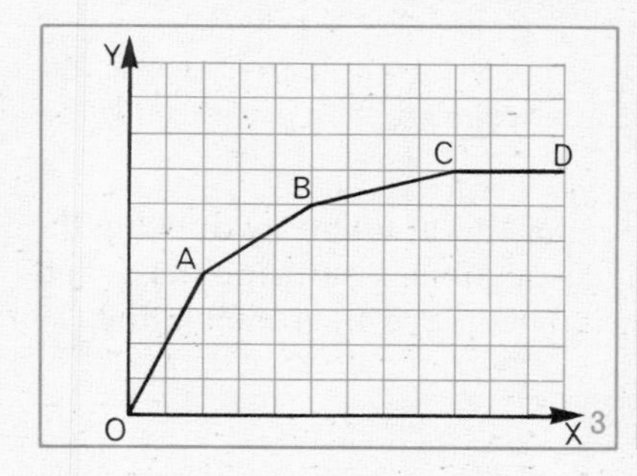

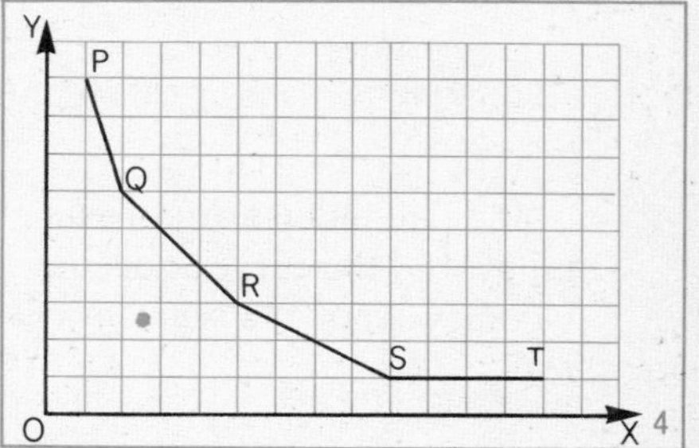

4 Find the gradients of PQ, QR, RS and ST in Figure 4.

5 Find the gradients of the lines through these pairs of points:

a A(−1, −4), B(5, −2) *b* P(1, 3), Q(4, 2)

c K(6, 5), L(−2, 3) *d* R(−3, 6), S(1, 4).

6 On squared paper draw lines with gradients:

a 1 *b* $\frac{1}{2}$ *c* $\frac{1}{4}$ *d* 3 *e* $\frac{2}{3}$ *f* $-\frac{3}{4}$

Which of these lines has the greatest slope, and which has the least slope?

7 Sketch the lines through the points:

a O(0,0), and rising '1 in 3' (i.e. 1 unit vertically for 3 horizontally)
b A(5,0), and falling '1 in 2' *c* B(−5,3), and rising '1 in 10'.

2 *The gradient of a curve*

Figure 5 shows an aircraft coming in to land. The direction of travel of the aircraft at a point P on the flight path is the same as the direction of the tangent at P. The gradient of the tangent at P gives a measure of the steepness of the curve at P.

5

The gradient of a curve changes from point to point, and the gradient of the curve at P is defined to be the *gradient of the tangent to the curve* at P.

We can obtain an approximation for this gradient by drawing the tangent to the curve as accurately as possible, measuring the x- and y-components of a segment of the tangent, and then calculating the ratio

$$\frac{y\text{-component}}{x\text{-component}}$$

as indicated in Figure 6.

Gradient of $AB=\frac{3}{4}$; gradient of $CD=-\frac{4}{2}=-2$; gradient of $EF=\frac{3}{3}=1$.

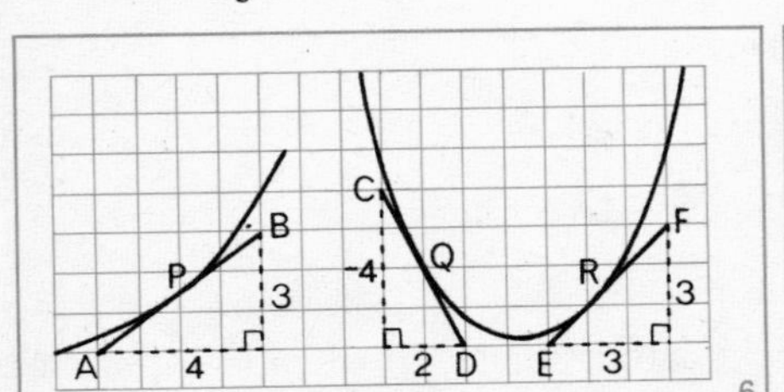

6

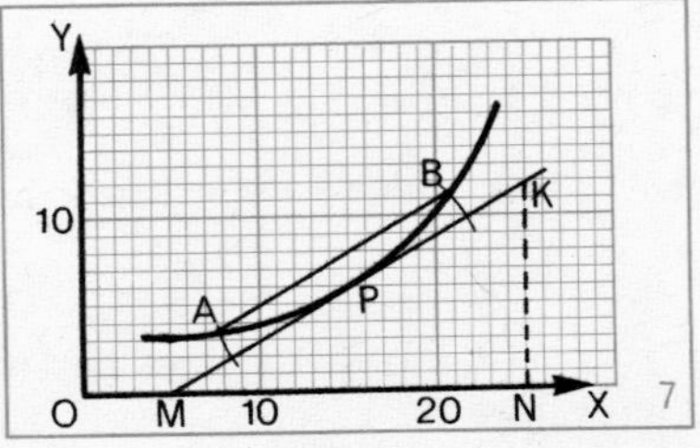

7

Figure 7 shows how a tangent can be drawn at a point P on a curve fairly accurately. With centre P and radius about 1 cm two arcs are drawn to cut the curve at A and B. The line through P parallel to AB is a good approximation for the tangent at P. Alternatively, given points on the curve on either side of P may be joined, and the parallel line through P drawn. (For example, if in question *1* of Exercise 2 the points (1, 0·1) and (3, 1) are joined, the parallel line through (2, 0·4) can be taken as the tangent at this point.)

In Figure 7, suitable x- and y-components MN and NK are 20 and 12 units long. So the gradient of the curve at P, i.e. the gradient of the tangent at P, is approximately $\frac{12}{20}$, i.e. 0·6.

Exercise 2

1 Taking a scale of 2 cm to 1 unit on each axis, plot the points given by the following table on 2-mm squared paper, and draw a smooth curve through them.

x	0	1	2	3	4	4·6	4·9	5
y	0	0·1	0·4	1	2	3	4	5

a Draw the tangents to the curve at the points where $x = 2$ and $x = 4$.
b Find the gradient of the curve at each of these points (to 1 decimal place.).

2 Copy and complete this table for the curve $y = \frac{1}{10}x^2$.

x	0	1	2	3	4	5	6	7
y	0	0·1			1·6			4·9

a Taking a scale of 2 cm to 1 unit on each axis, plot the set of eight points and draw a smooth curve through them.
b Draw the tangents at the points for which $x = 2$, 4, 6 and find the gradient of the curve at each of these points (to 1 decimal place).

3 Copy and complete this table for the curve $y = x - \frac{1}{10}x^2$.

x	0	1	2	3	4	5	6	7	8
y	0	0·9		2·1		2·5			

a Draw the graph of $y = x - \frac{1}{10}x^2$, to a scale of 2 cm to 1 unit on each axis.
b Find the gradient of the curve at the points where $x = 1$, 3, 5 and 7.

4 Draw the graph of $y = 12/x$ from $x = 1$ to 12, taking 1 cm to 1 unit on each axis, and plotting points at $x = 1$, 1·5, 2, 3, 4, 6, 8, 10, 12.
Find the gradient of the curve at the points where $x = 2$, 4, 8.

5 Figure 8 shows a graph in which *the two scales are not equal.* Use the definition of the gradient given in Section 1 to calculate the gradients of AB, CD and EF.

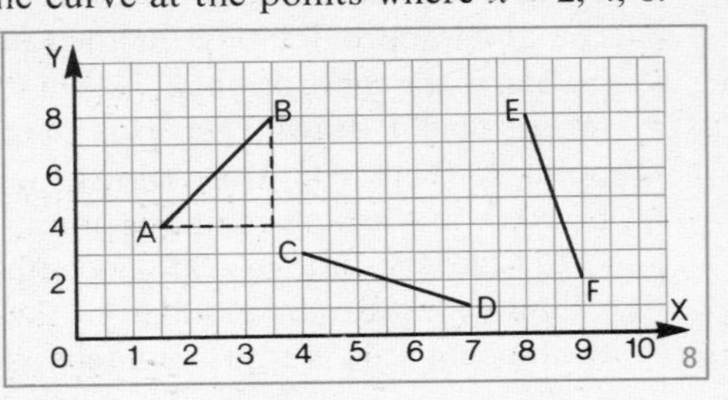

6 Draw the graph of $y = x(6-x)$ from $x = 0$ to $x = 6$, taking 2 cm to 1 unit on the x-axis, and 1 cm to 1 unit on the y-axis, and plotting points for which $x = 0$, 1, 2, 2·5, 3, 3·5, 4, 5, 6.

a Find the gradient of the curve at the points (1, 5) and (4, 8).
b State the coordinates of the point at which the gradient of the curve is zero.

7 Draw the graph of $y = 8x - x^2$ from $x = 0$ to $x = 8$, taking 2 cm to 1 unit on the x-axis, and 1 cm to 5 units on the y-axis.

a Find the gradient of the curve at the points where $x = 1$, 2, 3 and 4.
b Use the answers to ***a*** to give the gradients where $x = 5$, 6 and 7.

8 P is a point on the parabola $y = px^2 + qx + r$ for which $x = a$. A and B are neighbouring points on the parabola with x-coordinates $a-h$ and $a+h$, h being small.

a Find expressions for y_A and y_B, and show that $y_B - y_A = 4aph + 2qh$.
b Deduce that the gradient of AB is given by $2ap + q$. Use this formula to check your answers to questions *6* and *7*.

3 Gradient as a rate measurer

(i) *Average speed*

The table below gives the distances from the starting point of a car at various times:

Time (seconds)	0	2	4	6	8
Distance (metres)	0	30	60	90	120

The ordered pairs (0, 0), (2, 30), (4, 60),... are plotted to give the *distance–time graph* shown in Figure 9.

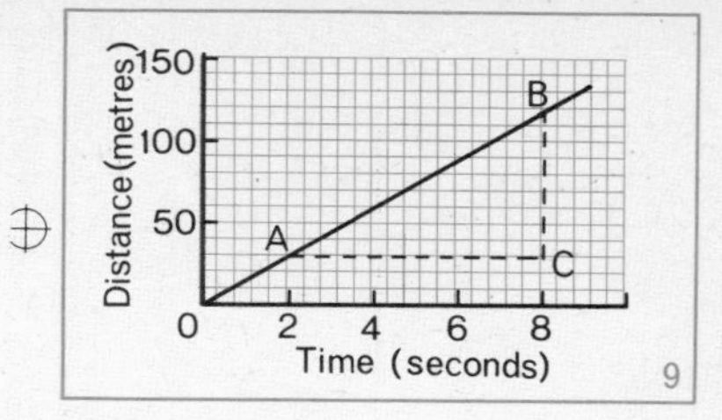

9

The *average speed*, which is a *rate*, is given by

$$\frac{\text{distance travelled}}{\text{time taken}}$$

the rate being measured in metres per second.

For example, in the time interval from 2 seconds to 8 seconds, corresponding to A and B on the graph, the distance travelled is (120–30) metres, so that

$$\text{the average speed} = \frac{(120\text{–}30)\,\text{m}}{(8\text{–}2)\,\text{s}} = \frac{90\,\text{m}}{6\,\text{s}} = \frac{15\,\text{m}}{1\,\text{s}} = 15\,\text{m/s}.$$

Notice that the distance 90 m corresponds to CB and the time 6 s corresponds to AC. So the average speed 90 m/6 s corresponds to CB/AC, i.e. to the gradient of AB.

The average speed of the car corresponds to the gradient of AB.

By taking different time intervals we see that the average speed in each case is the same over the 8-second period. The car is said to move with *constant* speed. If the distance–time graph is a straight line, the object is moving with constant speed.

* * *

Here is another set of observations for a car starting from rest.

Time (s)	0	0·5	1	1·5	2	2·5	3	3·5	4
Distance (m)	0	0·5	2	4·5	8	12·5	18	24·5	32

The distance–time curve is shown in Figure 10. The gradient of the curve shows a continuous increase which corresponds to the increase in speed of the car.

We first calculate the average speed of the car in the interval represented by $\overrightarrow{MN}$.

KQ corresponds to a distance of 24 m, and PK corresponds to a time of 2 s.

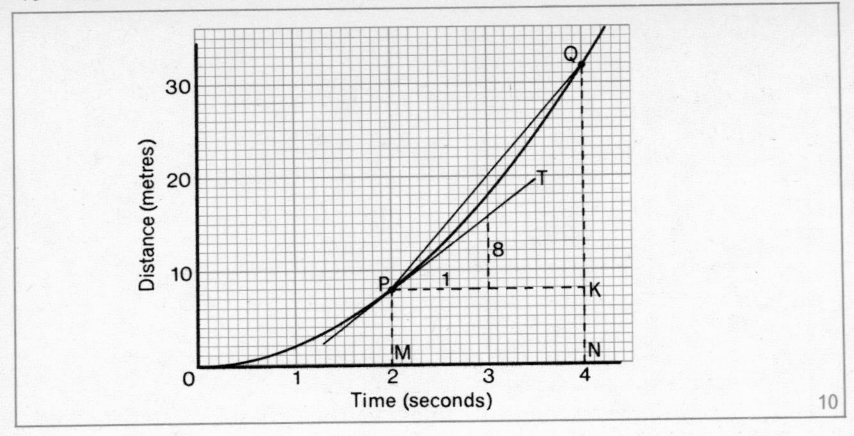

So the average speed 24 m/2 s corresponds to KQ/PK, i.e. to the gradient of PQ.

Average speed in a time interval corresponds to the gradient of the corresponding chord in the distance–time curve.

(ii) *Speed at an instant*

Can we now calculate the speedometer reading 2 seconds from the start? Assuming that the smaller that the time interval is from 2 seconds, the nearer the average speed will be to the speedometer reading, we proceed as follows, using the graph as a guide.

Time interval (*s*)	*Distance travelled* (*m*)	*Average speed* (*m/s*)
2 to 3	$18-8=10$	$\frac{10}{1}=10$
2 to 2·5	$12{\cdot}5-8=4{\cdot}5$	$\frac{4{\cdot}5}{0{\cdot}5}=9$
2 to 2·2	$9{\cdot}7-8=1{\cdot}7$	$\frac{1{\cdot}7}{0{\cdot}2}=8{\cdot}5$

From Figure 10, the gradient of the tangent at P gives the rate of 8 m/s, to which the average speed approaches; so it is reasonable to take 8 m/s as the speedometer reading after 2 seconds. Since we cannot talk about the average speed at an instant, we say that *the speed of the car 2 seconds from the start is 8 m/s.*

Summing up

If P and Q are points on a distance–time curve corresponding to times t_1 and t_2 seconds, then *the average speed* in the interval $(t_2 - t_1)$ seconds is given by *the gradient of the chord* PQ, and the *speed at the instant* t_1 is given by *the gradient of the tangent* at P.

As question **2** of Exercise 3 shows, rate need not be associated with time.

Exercise 3

1 Figure 11 shows a graph of the distance travelled by a motor boat and the time taken.

a How many kilometres had it gone after 1 hour? 2 hours? 3 hours?
b What was its average speed?
c Write down the gradient of OA as the ratio MA/OM.

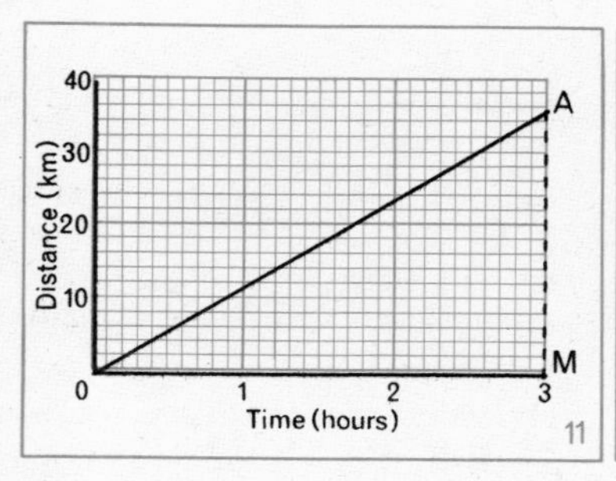

11

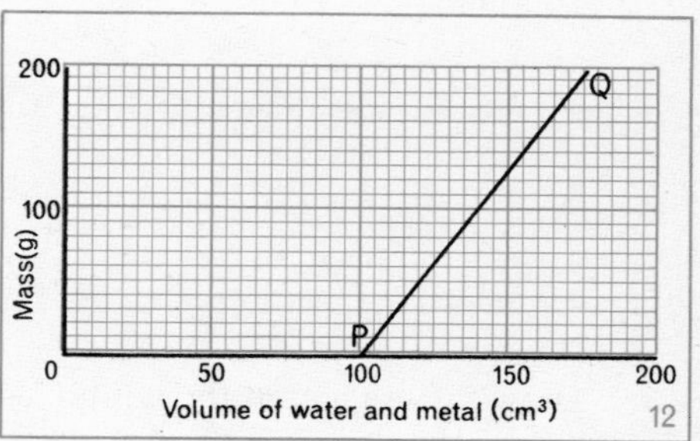

12

2 Blocks of aluminium of known mass are added to water in a container, and the volume (water + metal) is read. The graph in Figure 12 shows the result of plotting the masses against the volumes and drawing a best-fitting straight line.

a Write down the gradient of PQ.
b Write down the mass in grammes per cm^3 of the metal (to 1 decimal place).

3 Figure 13 shows the population of the United Kingdom during the period 1921–1965.

a Write down the population in 1921 and 1951, and hence calculate the *rate of growth* of the population in *people per year.*

b Find the rate of growth of the population between 1955 and 1965.

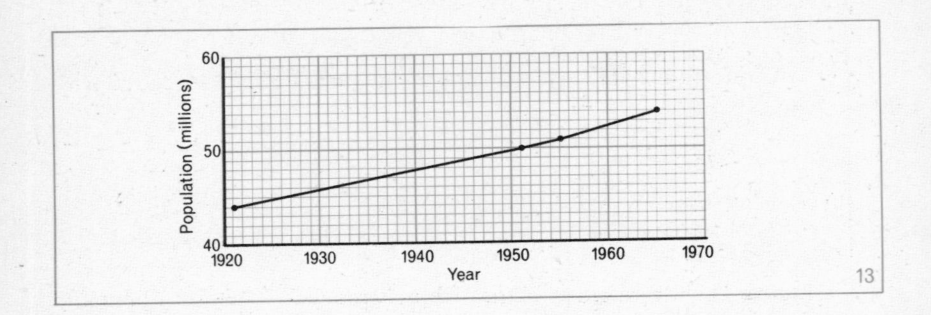

4 Use Figure 10 to estimate the average speed of the car during the first, second, third and fourth seconds.

5 The table shows the distance fallen by a stone dropped from a high building.

Time (t seconds)	0	1	2	3	4	5
Distance (h metres)	0	5	20	45	80	125

Draw a graph with scale 2 cm to 1 second for t, and 1 cm to 25 metres for h, joining the points with a smooth curve.

a Calculate the average speed of the stone in the interval from $t = 2$ to $t = 4$.

b Draw the tangent at the point (3, 45). Hence estimate the speed of the stone when $t = 3$. Comment on the results.

6 The distances of a train from its starting point at 5-minute intervals are given in the table.

Time (t minutes)	0	5	10	15	20	25	30	35	40
Distance (s kilometres)	0	2	5	11	19	24	26	27	27

Draw a distance–time graph, taking scales of 2 cm to 5 minutes on the t-axis and 2 cm to 5 km on the s-axis.

Use the gradient of the tangent to estimate the speed of the train in km/h after:

a 10 minutes *b* 20 minutes *c* 30 minutes.

4 *Calculating the area under a curve*

Engineers, surveyors and scientists often have to calculate the area of shapes which have curved boundaries. Two methods for obtaining approximations for such areas are given below.

(i) *Area by counting squares*

In this basic method, the figure whose area is required is drawn on squared paper, to some suitable scale. The number of unit squares enclosed by the boundary is then counted, one half of a square or more counting as one, less than half of a square being ignored. The smaller the unit squares in the grid, the more accurate the result will be. An example of the method is shown in Figure 14 for a semicircle of radius 3 cm, the unit square being a square of side $\frac{1}{2}$ cm.

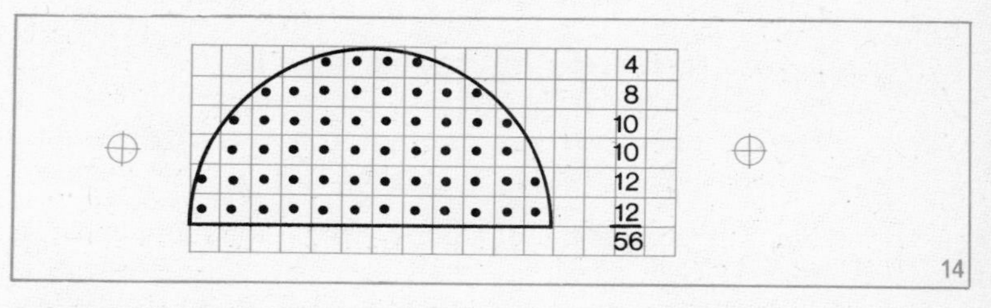

14

The area of a unit square is $(\frac{1}{2}\times\frac{1}{2})$ cm^2, i.e. $\frac{1}{4}$ cm^2, so the area of the semicircle is approximately $(56\times\frac{1}{4})$ cm^2, i.e. 14 cm^2. Checking by the formula $A=\pi r^2$, the area of the semicircle is $\frac{1}{2}\times 3{\cdot}14\times 3^2$ cm^2 $=14{\cdot}1$ cm^2, to 3 significant figures. The answers agree to within 1 per cent.

(ii) *Area by trapezium rule*

To find the area of the shape shown in Figure 15, we divide the base into a convenient number of equal parts (say 5) of width h units, and draw vertical lines of lengths $y_1, y_2, \ldots, y_6$ (called *ordinates*), so dividing the area into five strips.

Each strip is approximately equal in area to the *trapezium* formed by drawing the chord of the arc, so the total area under the curve is approximately equal to the sum of the areas of the trapezia.

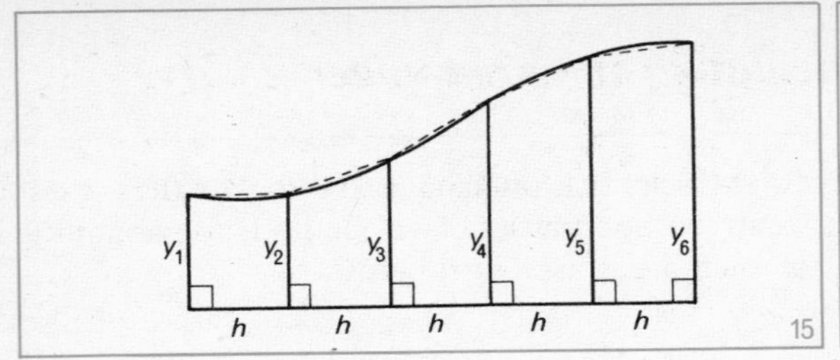

15

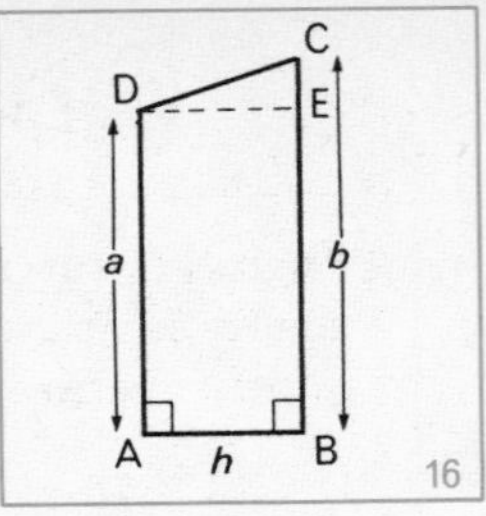

16

From Figure 16, the area of trapezium ABCD

$$= \text{area of rectangle ABED} + \text{area of triangle DEC}$$
$$= ah + \tfrac{1}{2}(b-a)h = ah + \tfrac{1}{2}bh - \tfrac{1}{2}ah = \tfrac{1}{2}(a+b)h$$

Hence the total area under the curve in Figure 15

$$\doteqdot \tfrac{1}{2}(y_1+y_2)h + \tfrac{1}{2}(y_2+y_3)h + \tfrac{1}{2}(y_3+y_4)h + \tfrac{1}{2}(y_4+y_5)h + \tfrac{1}{2}(y_5+y_6)h$$
$$= \tfrac{1}{2}h(y_1+y_2+y_2+y_3+y_3+y_4+y_4+y_5+y_5+y_6)$$
$$= \tfrac{1}{2}h[(y_1+y_6)+2(y_2+y_3+y_4+y_5)]$$
$$= h[\tfrac{1}{2}(y_1+y_6)+(y_2+y_3+y_4+y_5)]$$

This is a special case of the Trapezium Rule for finding an area bounded by a curve. The rule may be stated as follows:

The area under a curve $\doteqdot$ width of strip $\times$ (average of first and last ordinates + sum of remaining ordinates).

Note. The greater the number of strips, the more accurate is the result.

Example. Find an approximation for the area of the shape shown in Figure 17 by taking 8 strips of equal width and using the trapezium rule.

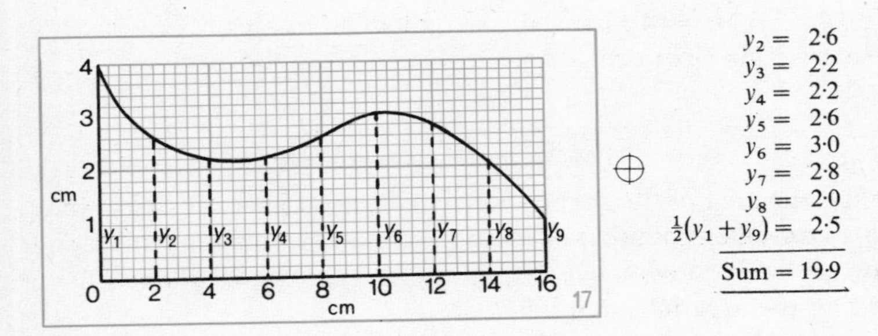

17

$$\begin{aligned} y_2 &= 2{\cdot}6 \\ y_3 &= 2{\cdot}2 \\ y_4 &= 2{\cdot}2 \\ y_5 &= 2{\cdot}6 \\ y_6 &= 3{\cdot}0 \\ y_7 &= 2{\cdot}8 \\ y_8 &= 2{\cdot}0 \\ \tfrac{1}{2}(y_1+y_9) &= 2{\cdot}5 \\ \hline \text{Sum} &= 19{\cdot}9 \end{aligned}$$

The width of the strip, h, $=2$.
Hence the required area $\doteqdot 2 \times 19{\cdot}9\ \text{cm}^2 = 39{\cdot}8\ \text{cm}^2$

Exercise 4

1 Construct △ ABC on $\frac{1}{2}$-cm squared paper, with base BC=8 cm, AB=6 cm and AC=7 cm.

a Find the area of ΔABC in square centimetres by counting squares of side $\frac{1}{2}$ cm.
b Calculate the area by measuring the length of a suitable altitude.

2 Draw a semicircle of radius 5 cm on $\frac{1}{2}$-cm squared paper.

a Find the area of the semicircle in square centimetres by counting $\frac{1}{2}$ cm squares.
b Calculate the area from the formula for the area of a circle.

3 A curve passes through the points given by the following table:

x	0	2	4	6	8	10
y	4	4	3·5	4	5	6·5

Construct the curve on $\frac{1}{2}$-cm squared paper, taking a scale of 1 cm to 1 unit on each axis. Shade the area enclosed by the curve, the x-axis and the two end ordinates.
a Find the shaded area by counting the number of $\frac{1}{2}$-cm squares.
b Estimate the area by the trapezium rule, taking 5 strips.

4 Draw a curve on 2-mm squared paper from the following data, scale 1 cm to 1 unit on each axis:

x	1	3	4	5	7	9	10
y	6·2	5·4	5·2	5·4	6	6·8	6·8

Draw the ordinates at $x=1, 2, 3, \ldots, 10$, so dividing the area under the curve into 9 strips. Use the trapezium rule to estimate the area under the curve.

5 a Draw the graph of $y=4x^2$ from $x=0$ to $x=5$, using a scale of 2 cm to 1 unit on the x-axis and 1 cm to 10 units on the y-axis. Draw ordinates at $x=1, 2, 3, 4, 5$.

b Use the trapezium rule to estimate the area under the curve.
c Is this approximation for the area too large or too small?

6 a Draw the graph of the hyperbola $y=12/x$ from $x=2$ to $x=10$, taking a scale of 1 cm to 1 unit on each axis.

b Taking ordinates at $x = 2, 3, 4, \dots, 10$, use the trapezium rule to find an approximation for the area under the curve.

c Is the true area greater or less than your approximation?

7 The depth of a stream 20 m wide measured from one bank to the other at intervals of 2 m is given in this table:

Distance (m)	0	2	4	6	8	10	12	14	16	18	20
Depth (cm)	10	22	28	30	30	35	40	34	20	9	4

a Without using a graph, estimate the area of the cross-section of the stream by the trapezium rule, giving your answer in square metres.

b If the stream is flowing at 3 km/h, calculate the flow in cubic metres of water per minute.

8 Use the trapezium rule to estimate the area under each of the curves given by the following experimental data.

a

x	1	2	3	4	5	6	7
y	5·4	9·6	14·5	16·7	12·6	11·8	10·6

b

x	2	4	6	8	10	12
y	1·6	2·2	3·4	5·5	8·2	12·8

9 Copy and complete the table below for the parabola $y = 4x - x^2$

x	0	0·5	1	1·5	2	2·5	3	3·5	4
$4x$	0	2·0	4						
x^2	0	0·25	1						
y	0	1·8	3·0						

Use the trapezium rule to estimate the area cut off between the parabola and the x-axis.

10 A sketch plan of a field APBC is shown in Figure 18. The offsets on AB are 26 m apart. Use the trapezium rule to estimate the area bounded by AB and APB, and hence calculate the total area of the field.

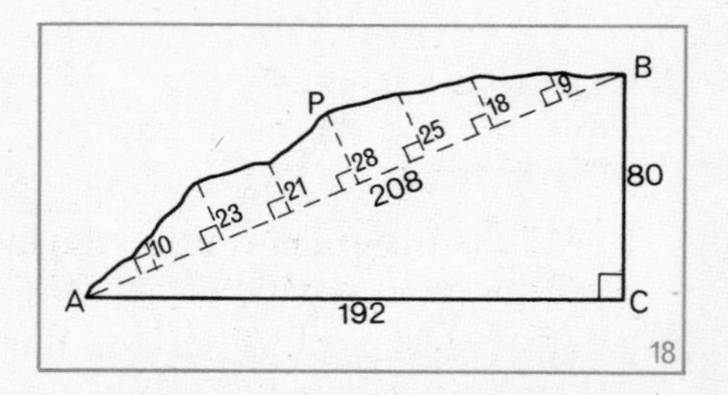

5 *Using the area under a curve*

The speed of a train at the end of 50-second intervals is shown in the following table:

Time (s)	0	50	100	150	200
Speed (m/s)	2	12	18	22	24

From this information, we can use the definition of average speed in Sections 3 to estimate the distance travelled in each 50-second interval, and hence obtain an idea of the total distance travelled in the 200 seconds. For example,

$$\begin{aligned}\text{distance travelled in first } 50\,\text{s} &= \text{average speed} \times \text{time}\\ &= \tfrac{1}{2}(2+12)\,\text{m/s} \times 50\,\text{s}\\ &= 350\,\text{m}\end{aligned}$$

Figure 19 shows the *speed–time graph* for this train over the 200 seconds. Notice that the estimated distance travelled in the first 50 seconds corresponds to the area of the shaded trapezium. The distance travelled in the remaining intervals can be estimated in the same way.

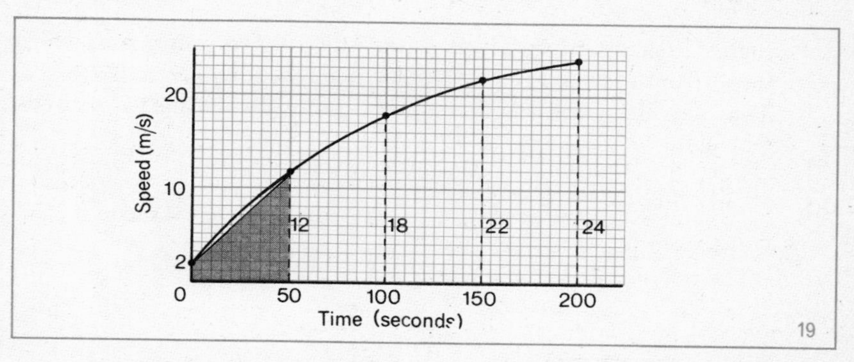

19

We can now say that the total distance travelled by the train in the 200 seconds corresponds to the area under the speed–time graph, which can be estimated using the trapezium rule. Hence,

the area under a speed–time curve gives the distance travelled in the given time, and may be estimated by the trapezium rule.

Exercise 5

1 For the illustration in Section 5, use the trapezium rule to estimate the distance travelled by the train in 200 seconds.

2 Draw a curve to fit the following data:

Time (seconds)	0	5	10	15	20	25
Speed (m/s)	0	6·2	11	15	17·8	20

Estimate the distance travelled in 25 seconds by means of the trapezium rule.

3 The speed of a car at 10-second intervals is as follows:

Time (s)	0	10	20	30	40	50	60	70	80	90	100	110	120
Speed (m/s)	16	20	25	28	31	32	30	25	20	17	14	8	0

Without drawing a speed–time graph, use the trapezium rule to estimate the total distance in kilometres travelled in the 2 minutes.

4 The speed of an electric train travelling between two stations is as follows:

Time from start (s)	0	20	40	60	80	100	120
Speed (m/s)	0	10·3	15·5	15·8	15·5	10·4	0

a Estimate the distance in kilometres between the stations.
b Calculate the average speed in kilometres per hour for the journey.

5 Figure 20 shows the speed–time graph for a car travelling west which starts from rest and whose speed increases steadily to 20 m/s in 4 seconds. Its speed then remains constant for the next 8 seconds. Calculate:

a the distance travelled in the first 12 seconds
b the *rate* at which the speed is changing during the first 4 seconds. This rate gives the *acceleration* of the car.

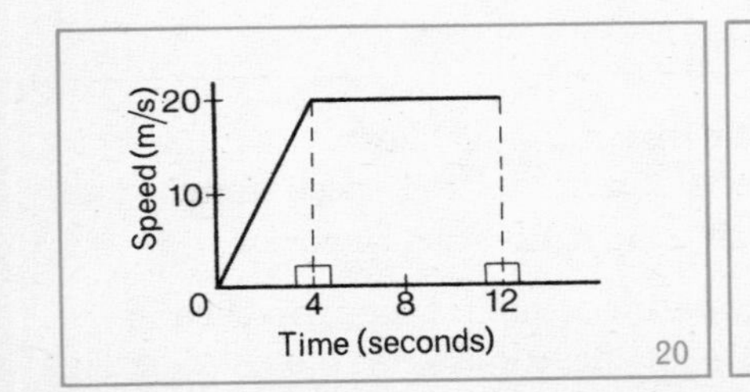

20

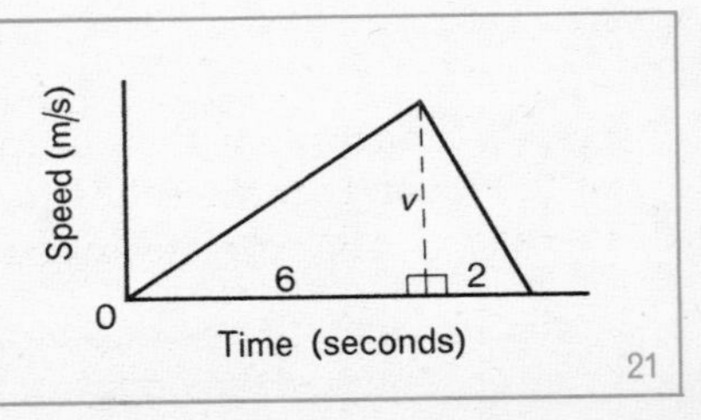

21

Note. When the speed of a body is increasing, we say that the body is *accelerating*. When the speed is decreasing, we say that the body is *decelerating*.

6 Figure 21 shows the speed–time graph for an object moving in a straight line. The object accelerates uniformly for 6 seconds, then decelerates uniformly for 2 seconds. If the total distance travelled is 40 m, calculate the maximum speed v m/s attained, and hence work out the acceleration and deceleration (using gradients, or otherwise).

7 The water discharged in litres per minute from a pipe was as follows:

Time (minutes)	0	1	2	3	4	5
Discharge (litres/minute)	0	60	80	88	86	44

Estimate the total volume of water discharged in this period.

8 The consumption of electricity in one large region on Christmas Day 1973 was:

Time	00 00	03 00	06 00	09 00	12 00	15 00	18 00	21 00	24.00
Millions of units/hour	3·8	2·3	2·2	2·8	4·6	4·2	3·4	3·6	3·8

a Draw a graph of the consumption, and suggest reasons for its shape.
b Calculate the total consumption of electricity in the 24-hour period.

Summary

1 *The gradient of a straight line* AB is $\dfrac{y\text{-component of AB}}{x\text{-component of AB}}$

If A is the point (x_1, y_1) and B is (x_2, y_2), $m_{AB} = \dfrac{y_2 - y_1}{x_2 - x_1}$.

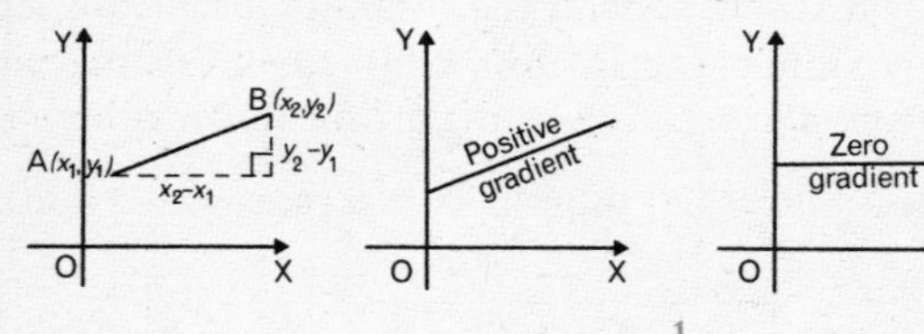

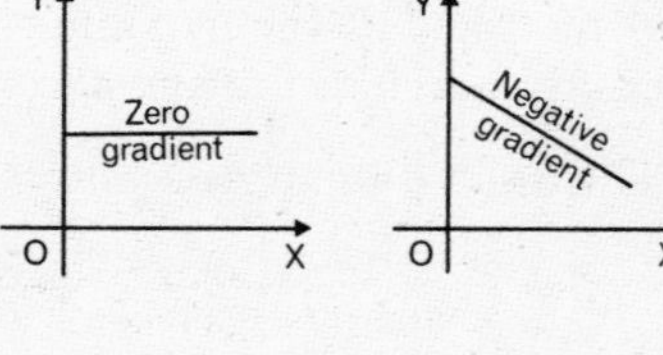

1

2 *The gradient of a curve* at a point on the curve is the *gradient of the tangent* at the point (Figure 2).

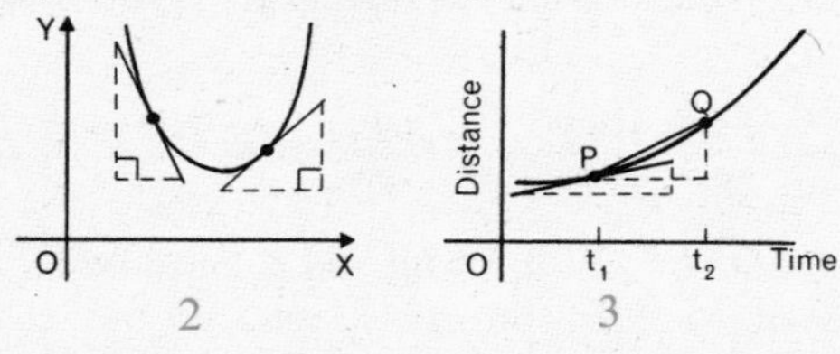

2 3

3 *On a distance–time curve* the *average speed* in the time interval $(t_2 - t_1)$ is given by the *gradient of the chord* PQ. The *speed at the instant* t_1 is given by the *gradient of the tangent* at P (Figure 3).

4 *The area under a curve* can be found approximately by:
(i) counting squares (ii) the trapezium rule.
Area $\doteqdot$ width of strip $\times$ (average of first and last ordinates + sum of other ordinates)

$$= h[\tfrac{1}{2}(y_1 + y_6) + (y_2 + y_3 + y_4 + y_5)] \text{ in Figure 4.}$$

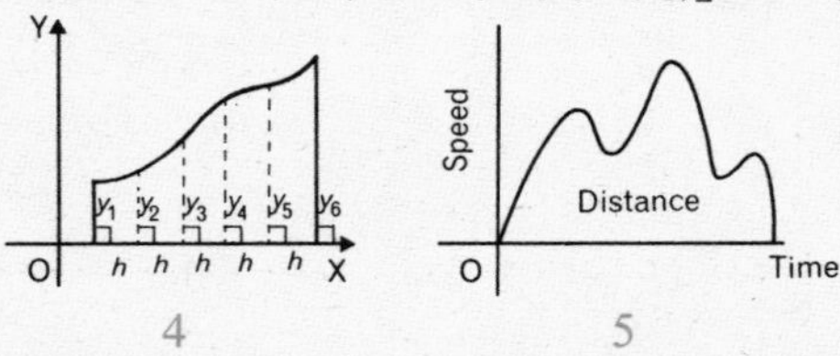

4 5

5 *The area under a speed–time curve* represents the *distance* travelled.

Revision Exercises

Revision Exercise on Chapter 1
Number Systems and Surds

Revision Exercise 1

1. $E = \{\sqrt{0}, \sqrt{1}, \sqrt{2}, \sqrt{3}, \sqrt{4}, \sqrt{5}, \sqrt{6}, \sqrt{7}, \sqrt{8}, \sqrt{9}, \sqrt{10}\}$;
$P = \{x : x \in E, \text{and } x \text{ is rational}\}$, $Q = \{x : x \in E, \text{and } x \text{ is a surd}\}$.
List:

   a P *b* Q *c* $P \cap Q$ *d* $P \cup Q$

2. Find the solution sets of the following, $x \in R$:

   a $3x^2 + 7 = 10$ *b* $\frac{1}{2}x^2 = 72$ *c* $5x^2 = 10$ *d* $x^2 = 0{\cdot}01$

3. Which of the following are true and which are false?

   a If $x^2 = 25$ and $x \in Z$; then $x = 5$.

   b If $x^2 = 25$ and $x \in Z$, then $x = 5$ or -5.

   c $\sqrt{3} = 1{\cdot}73$ exactly *d* $\sqrt{(-1)} = 1$ or -1 *e* $3\sqrt{5} = \sqrt{45}$

   f $(\sqrt{5})^2 = 25$ *g* $(2\sqrt{3})^2 = 12$ *h* $\sqrt{(-4)} = -2$

4. Simplify:

   a $\sqrt{100}$ *b* $\sqrt{500}$ *c* $\sqrt{48}$ *d* $\sqrt{150}$ *e* $\sqrt{540}$ *f* $\sqrt{1000}$

5. In $\triangle$ABC, AB = AC = 12 cm and BC = 8 cm. Express the length of the altitude from A to BC as a surd in its simplest form.

6. An equilateral triangle has each side $2a$ metres long. Find the length of an altitude of the triangle, and hence find the area of the triangle in terms of a.

7 A cuboid has sides of length 5 cm, 10 cm and 15 cm. Express the lengths of its face diagonals and its space diagonals as surds in their simplest form.

8 Simplify as far as possible:

a $\sqrt{5}+2\sqrt{5}-3\sqrt{5}$ *b* $\sqrt{8}+\sqrt{2}$ *c* $\sqrt{27}-\sqrt{3}$

d $2\sqrt{20}-\sqrt{80}$ *e* $3\sqrt{32}-2\sqrt{50}$ *f* $(3\sqrt{2}+2\sqrt{2})^2$

9 Simplify as far as possible:

a $\sqrt{6}\times\sqrt{12}$ *b* $3\sqrt{3}\times\sqrt{75}$ *c* $5\sqrt{2}\times 4\sqrt{2}$ *d* $\sqrt{5}(\sqrt{5}+\sqrt{20})$

e $(5+\sqrt{2})(5-\sqrt{2})$ *f* $(\sqrt{7}+\sqrt{2})(\sqrt{7}-\sqrt{2})$ *g* $(\sqrt{6}-\sqrt{3})^2$

10 Given $q=3+\sqrt{2}$ and $\bar{q}=3-\sqrt{2}$.

a Find $q+\bar{q}$, and $q\bar{q}$. *b* Are they both rational numbers?

11 Evaluate p^2-2p if: *a* $p=2\sqrt{2}$ *b* $p=\sqrt{5}+1$.

12 Show that $3-\sqrt{2}$ is a solution of the equation $x^2-6x+7=0$.

13 Rationalize the denominator of each of the following:

a $\dfrac{9}{\sqrt{3}}$ *b* $\dfrac{12}{\sqrt{32}}$ *c* $\dfrac{\sqrt{3}}{\sqrt{5}}$ *d* $\sqrt{\dfrac{1}{11}}$ *e* $\dfrac{1}{\sqrt{3}-1}$ *f* $\dfrac{12}{\sqrt{7}-\sqrt{3}}$

14 An equilateral triangle is drawn on one side of a square with side x mm long. Show that the area of the whole figure is $\frac{1}{4}x^2(4+\sqrt{3})\,\text{mm}^2$.

15 Four holes of radius r cm are cut from a circular metal plate of radius R cm. If the remaining area is one third of the original area, show that $R:r=\sqrt{6}:1$.

16 A right pyramid with altitude 3 cm has a rectangular base with sides 2 cm and $2\sqrt{2}$ cm long. Calculate in surd form the length of: *a* a diagonal of the base *b* a sloping edge of the pyramid *c* the altitude drawn to the base on each sloping face.

Revision Exercises on Chapter 2
Indices

Revision Exercise 2

1. Simplify: *a* $a^3 \times a^4$ *b* $a^5 \div a^3$ *c* $(a^2)^3$ *d* $2^9 \times 2^8 \div 2^7$

2. Which of the following are true and which are false?

   a $2^{m+n} = 2^m \times 2^n$ *b* $(2 \times 3)^n = 2^n + 3^n$ *c* $(2^m)^n = (2^n)^m$

   d $\dfrac{3^m}{3^n} = 3^{n-m}$ *e* $3000 \div 20 = 1{\cdot}5 \times 10^2$ *f* $(10^3 \div 10^3) \times 10^3 = 1000$

3. Simplify: *a* $a^{-2} \times a^{-2}$ *b* $a^{-2} \div a^{-3}$ *c* $(a^{-2})^4$ *d* $(a^{-4})^{-2}$

4. Express with root signs: *a* $3^{1/2}$ *b* $a^{3/4}$ *c* $b^{2/3}$ *d* $(a+b)^{1/2}$

5. Express in index form: *a* $\sqrt[3]{p}$ *b* $\sqrt[3]{p^2}$ *c* $\sqrt{a}$ *d* $\sqrt[4]{(ab)^2}$

6. Simplify:

   a $4a^{1/2} \times 2a^{-1/2}$ *b* $2a^{1/2} \div 4a^{-1/2}$

   c $(a^6)^{-3}$ *d* $(a^{2/3})^{3/2}$

7. Given that $x = 9$ and $y = 16$, simplify

   a $10\,(xy)^0$ *b* $x^{1/2} + y^{3/4}$ *c* $6x^{-1/2}\,y^{1/2}$ *d* $(x+y)^{3/2}$

8. Express in standard form $a \times 10^n$:

   a $12 \times 8 \times 1000$ *b* $(8{\cdot}5 \times 4) \div 1000$ *c* $1{\cdot}5 \times 10^{-5} \times 8 \times 10^3$

9. Evaluate: *a* $(-27)^{1/3}$ *b* $(-32)^{4/5}$ *c* $(-125)^{-1/3}$

10. Find x in each of the following, if $x \in R$:

    a $2^x = 128$ *b* $3^x = 1$ *c* $2^x = \frac{1}{2}$ *d* $3^x = \frac{1}{243}$

11. Simplify: *a* $x^{-2}(x^4 + x^2)$ *b* $x^3(x^3 + x^{-6})$ *c* $x^{1/2}(x^{1/2} + x^{-1/2})$

12. If $T = \frac{1}{8}\pi f d^3$, find a formula for d in terms of T, π and f.

13. A function f is defined by $f: x \to 8x^2 - 1$. Find the images of $2^{1/2}$ and 2^{-1} under f.

14. From the formula $H = \dfrac{v^3 d^{2/3}}{120}$, find H when $d = 1728$ and $v = 10$.

Revision Exercise 2B

1 Simplify: *a* $(a^{1/2} \times a^{-1/2}) \div a^{1/2}$ *b* $(a^{2/3}b^{3/2})^{-6}$ *c* $(a^2b^{-3}c^4)^0$

2 Express the square root of each of the following in the form $a \times 10^n$:

a 9×10^4 *b* $0{\cdot}36 \times 10^{-8}$ *c* $1{\cdot}6 \times 10^{-3}$ *d* $0{\cdot}0009$

3 Find a positive integer n if: *a* $1-(\frac{1}{2})^n = \frac{15}{16}$ *b* $2^n < 10^3 < 2^{n+1}$

4 Change the subject of the formula $T = 2\pi\sqrt{\dfrac{l}{g}}$ to l, and of $V = \frac{4}{3}\pi r^3$ to r.

5 In nuclear physics the formula $E = kT$ occurs. If $E = 4{\cdot}14 \times 10^{-14}$ and $T = 300$, find k.

6 Find x, when $x \in R$, if: *a* $4^x = 16$ *b* $16^x = 4$ *c* $4^x = \frac{1}{8}$

7 If x is small, then $(1+x)^n \doteqdot 1+nx$, and $(1-x)^n \doteqdot 1-nx$. Use these results to calculate the following, and check from three-figure tables:

a $\sqrt{1{\cdot}08} = (1+0{\cdot}08)^{1/2} = \ldots$ *b* $\sqrt{1{\cdot}12}$ *c* $\sqrt{0{\cdot}98}$

8 Express each of the following with x in the numerator in index form:

a $\dfrac{1}{x^3}$ *b* $\dfrac{2}{x}$ *c* $\dfrac{3}{x^{-3}}$ *d* $\dfrac{1}{2\sqrt[3]{x}}$ *e* $2\sqrt{x} - \dfrac{1}{2\sqrt{x}}$

9 Use the distributive law to expand the following:

a $a^{1/3}(a^{2/3} - a^{-1/3})$ *b* $(a^{1/2} + a^{-1/2})(a^{1/2} - a^{-1/2})$ *c* $(a^{1/4} + a^{-1/4})^2$

10 Express each of the following with x in the numerator in index form:

a $\left(\sqrt{x} + \dfrac{1}{\sqrt{x}}\right)^2$ *b* $\dfrac{1}{x}(x^4 - x^{-4})$ *c* $\dfrac{1}{x}\left(\dfrac{1}{x} + \dfrac{1}{x^2}\right)$ *d* $\dfrac{x}{2}\left(\dfrac{x}{2} + \dfrac{2}{x}\right)^2$

e $\dfrac{1-x-x^2}{x}$ *f* $\dfrac{x^2-2x-1}{2x}$ *g* $\dfrac{x^{1/2} - x^{-1/2}}{x^{3/2}}$ *h* $\dfrac{(2x^{1/2}-1)^2}{x^{1/2}}$

11 Evaluate $\dfrac{a^{1/2} + a^{-1/2}}{a^{1/3} - a^{-1/3}}$ when $a = 64$.

12 Show that $\left(\dfrac{1}{\sqrt{x}} - \dfrac{\sqrt{x}}{2}\right)^2$ can be expressed in the form $\dfrac{(x-2)^2}{4x}$

13 Given that (i) $h/e = 4{\cdot}14 \times 10^{-15}$ and (ii) $e/m = 1{\cdot}76 \times 10^{11}$, where $e = 1{\cdot}60 \times 10^{-19}$, calculate h and m to 3 significant figures, giving your result in standard form.

14a If $pv^{3/2} = k$, express v in terms of p and k.

b Given that $S_n = \dfrac{a(1-r^n)}{1-r}$, show that $S_n = 3 - \dfrac{1}{3^{n-1}}$ when $a = 2$ and $r = \frac{1}{3}$.

15 If $x = \frac{1}{2}(e^c + e^{-c})$ and $y = \frac{1}{2}(e^c - e^{-c})$, express $x^2 + xy + y^2$ in terms of e and c in its simplest form.

Revision Exercise on Chapter 3
Linear Programming

Revision Exercise 3

1 Show on squared paper the solution sets of these systems of inequations ($x, y \in R$):

a $x \geqslant 0,\ y \geqslant 0,\ x \leqslant 10,\ x + y \leqslant 15.$

b $x \geqslant 0,\ y \geqslant 0,\ 2x + y \geqslant 10,\ x + 2y \geqslant 10.$

2 a Show in a Cartesian diagram the solution set ($x, y \in R$) of:

$$x \geqslant 0,\ y \geqslant 0,\ x + 2y \leqslant 12 \text{ and } 2x + y \leqslant 12.$$

b Given that $P = x + y$ and $Q = 5x + y$, find the maximum values of P and Q, subject to the restrictions in ***a***.

3 a Show on squared paper the solution set ($x, y \in R$) of:

$$2 \leqslant x \leqslant 8,\ 0 \leqslant y \leqslant 6,\ 3x + 4y \leqslant 36.$$

b Subject to the above restrictions find the coordinates of the points which maximise:

(1) $x + y$ (2) $y - x$ (3) $x + 3y$

c Find the coordinates of the points which will minimise these expressions.

4 a Show the intersection of the system of inequations:

$$x + y \leqslant 50,\ 2y \leqslant x + 40,\ x \leqslant 30,\ x \geqslant 0,\ y \geqslant 0.$$

b List in a table the coordinates of the points (0, 20), (20, 30), (30, 20) and (30, 0) and of two other points in the neighbourhood of each of these.

c Deduce the maximum value in the region of:

$$(1)\ 3x+5y \quad (2)\ 3x+2y \quad \text{for } x, y \in W.$$

5 The diet for some animals is to consist of two foods A and B. The animals must have at least 20 units per day of a certain ingredient, of which 3 units occur in each kilogramme of A but only 1 unit in each kilogramme of B. They also need 30 units per day of another ingredient, of which 1 unit occurs in each kilogramme of A and 2 units in each kilogramme of B.

a Suppose that x kg of A and y kg of B are required. Write down three inequations, in addition to $x \geqslant 0$, $y \geqslant 0$, that must be satisfied by x and y.

b Illustrate the solution sets of the inequations by a graph.

c What is the least mass of A and of B to provide the necessary diet?

6 A factory turns out two articles A and B, each of which is processed by two machines M and N. A requires 2 hours of M and 4 hours of N; B requires 4 hours of M and 2 hours of N.

a If x is the number of A and y is the number of B produced daily, write down two inequations in x and y, noting that neither M nor N can work more than 24 hours a day.

b Assuming that all articles produced are sold, if each article A yields a profit of £3 and each article B yields a profit of £5, find a relation giving the daily profit £P.

c From graphical considerations, find how many of each article should be produced daily for maximum profit. Calculate this profit.

7 A building corporation decides to develop a new housing area which is planned to accommodate up to 540 persons. It proposes to build at most 120 houses, details of which are as follows:

Type	*Number of persons per house*	*Proposed annual rent*
3-apartment	4	£90
4-apartment	6	£107

a If the corporation builds x 3-apartment houses and y 4-apartment houses, write down two inequations in x and y based on: (1) the maximum number of houses (2) the maximum number of persons to be accommodated.

b Write down a linear form $(ax+by)$ which expresses the gross annual rental.

c Subject to the above conditions, what number of each type of house should be built so as to ensure maximum income? Find this income.

8 Two types of paper strips are available for making paper chains. One is white at 2 pence per dozen and makes a link 6 cm long, and the other is coloured at 3 pence per dozen and makes a link 8 cm long. For a good display, the number of coloured links must be at least double the number of white ones. We wish to find the cheapest way to make a chain 15·84 m long.

a Make a mathematical model by considering x dozen white and y dozen coloured, assuming the strips are sold by the dozen.

b Hence or otherwise, find how many dozen of each will give the necessary length at minimum cost.

c What is this minimum cost?

d What would the minimum cost be if strips could be sold individually?

Revision Exercise on Chapter 4
Graphs, Gradients and Areas

Revision Exercise 4

1 *a* Plot the points A(4,0), B(10,0), C(10,6), D(2,2), E(−2,8).

b Find the gradients of AB, BC, CD, DE, EA.

2 *a* Taking a scale of 2 cm to 5 units on each axis, plot the points (0,0), (5,4), (10,7·5), (15,10·5), (20,12·5), (25,14), (30,15) on 2-mm squared paper, and draw a smooth curve through them.

b Draw the tangents to the curve at the points where $x = 5$, $x = 15$ and $x = 25$.

c Find the gradient of the curve at each of these points.

3 *a* Draw the graph of the parabola $y = \frac{1}{4}x^2$ for $x = 0, 1, 2, 3, 4, 5$ with a scale of 2 cm to 1 unit on each axis.

b Draw the tangents at the points where $x=1, 3, 5$ and find the gradients of these tangents.

4 *a* Draw the graph of $y = x(x-4)$ for $x = -1, 0, 1, 2, 3, 4, 5$ with scales of 2 cm to 1 unit on the x-axis and 1 cm to 1 unit on the y-axis.

b Find the gradient of the curve at $x = 0$, 1, 2. Hence write down the gradients at $x = 3$ and 4.

5 The distance of an object from a given point after certain times is:

Time (t seconds)	0	1	2	3	4	5
Distance (metres)	0	10	30	60	100	150

a Draw a distance–time graph with scales of 2 cm to 1 second and 1 cm to 25 metres.

b Calculate the average speed of the object from $t = 0$ to $t = 1$, $t = 0$ to $t = 3$ and $t = 0$ to $t = 5$.

c Find the speed of the object at $t = 2$ and $t = 4$.

6 *a* Draw a distance–time curve for the following data:

Time (seconds)	0	5	10	15	20	25	30
Distance (metres)	0	3	8	15	24	30	32

b Find the speed after 5, 15 and 25 seconds.

7 Use the trapezium rule to estimate the areas under the curves given by:

a

x	0	4	8	12	16	20
y	0	16	25	20	24	12

b

x	0	5	10	15	20	25	30
y	6	12	18	15	30	25	10

8 Use the trapezium rule to estimate the area of the quarter-circle bounded by the x-axis, the y-axis and the curve $x^2 + y^2 = 25$, using ordinates at $x = 0$, 1, 2, 3, 4, 5.

9 Use the trapezium rule to estimate the area cut off between the parabola $y = 6x - x^2$ and the x-axis.

10*a* Draw a speed–time curve for the following data concerning an underground train travelling between three stations.

Time (s)	0	30	60	90	120	150	180	210	240	270	300
Speed (m/s)	0	10	20	25	15	0	12	15	20	8	0

b Estimate the distances between pairs of adjacent stations.

c Calculate the average speed for each journey.

Cumulative Revision Section

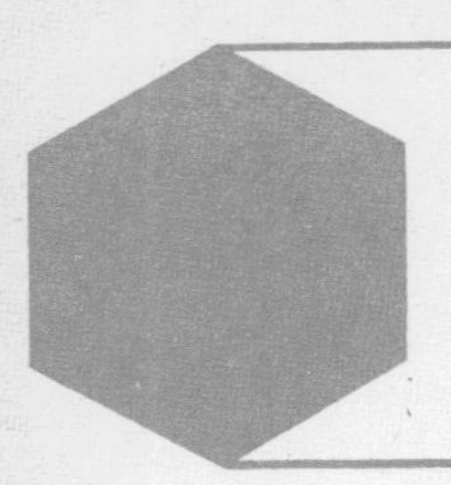

Cumulative Revision Section (Books 1–7)

Revision Topic 1 Sets

Reminders

1

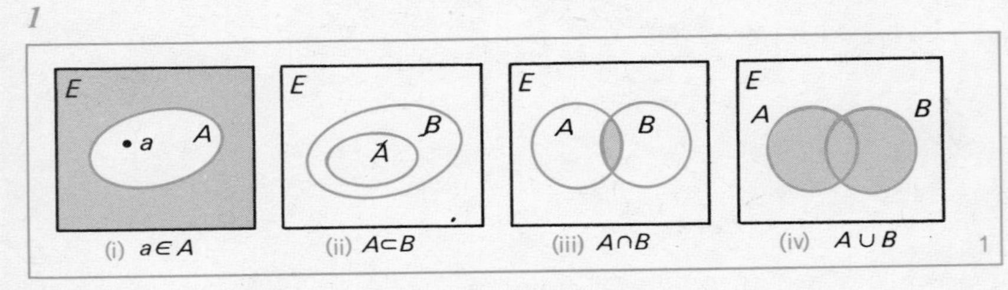

(i) $a \in A$ (ii) $A \subset B$ (iii) $A \cap B$ (iv) $A \cup B$ 1

In (i), A', the complement of A, is shaded; $A \cap A' = \phi$.

2 $\{(x, y): x + y = 4, x \in N, y \in N\} = \{(1, 3), (2, 2), (3, 1)\}$.

Exercise 1

1 *a* List the set of letters of the word *parallelogram.*
b List all the subsets of $\{0, 1, 2\}$.

2 $A = \{$odd numbers between 10 and 20$\}$, $B = \{$prime numbers between 6 and 16$\}$. List the numbers of the sets:
a A *b* B *c* $A \cap B$ *d* $A \cup B$.

3 *a* Use set-builder notation to describe the set $\{-3, -2, -1, 0, 1, 2, 3\}$.
b Given $A = \{x : x \text{ is an integer, and } -1 \leqslant x \leqslant 4\}$ and $B = \{-1, 0, 1, 2, 3, 4\}$, is $A = B$?

4 $P = \{4, 5, 6\}$, $Q = \{1, 2, 3, 4\}$, $R = \{4, 5, 8\}$.
List the members of: *a* $P \cap Q$ *b* $P \cap Q \cap R$ *c* $Q \cup R$.

5 $E = \{1, 2, 3, \ldots, 10\}$; $A = \{1, 2, 3, 4, 5\}$, $B = \{1, 2, 6, 7\}$.

a Show the relationship between the sets in a Venn diagram.
b List the following subsets of E: (1) $A \cap B$ (2) $A \cup B$ (3) A' (4) B' (5) $A' \cup B'$ (6) $(A \cup B)'$ (7) $(A \cap B)'$ (8) $A' \cap B'$.

6 a Copy the Venn diagrams in Figure 2. Shade horizontally the region representing $A \cap B$, vertically the region representing $A \cup B$, and in dots the region representing $(A \cup B)'$.

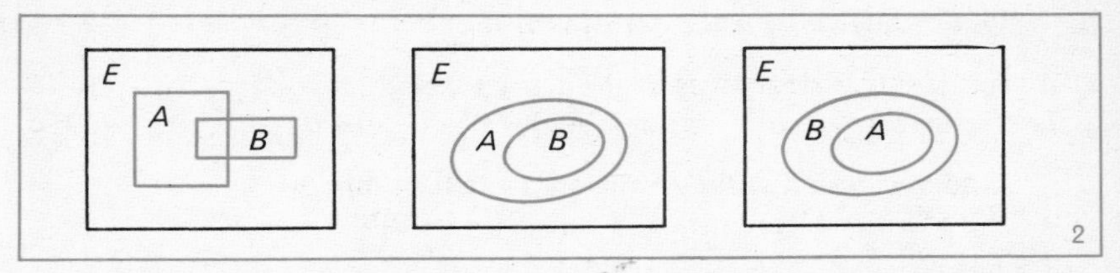

b $X = \{p, q, r, s\}$ and $Y = \{p, q\}$. Show $X \cap Y$ in a Venn diagram.

7 a Show the relationship between the following sets in a Venn diagram. $E = \{1, 2, 3, \ldots, 10\}$; $A = \{1, 2, 3, 4, 5\}$, $B = \{4, 5, 6, 7, 8\}$.

b Show by shading: (*1*) $A \cap B'$ (2) $A' \cap B$.

8 In a class of 35 boys, A is the set of boys who take athletics and C is the set who play cricket $n(A) = 15$, $n(C) = 16$, $n(A \cap C) = 5$. Using the whole class as the universal set, draw a Venn diagram and mark appropriate numbers in the various regions. How many boys take neither athletics nor cricket? Express this last set of boys in set notation.

9 A school has a rugby fifteen, a cricket eleven and a swimming eight. 3 boys are in all three teams, 9 are in the rugby team only and 5 are in both the rugby and cricket teams; 2 boys are in the swimming team only. Show these facts in a Venn diagram, and deduce the number of boys who represent the school in the cricket team only.

10 In a form of 100 boys, 65 take games, 25 take crafts and 20 take other activities only. How many take games *and* crafts? If one of the boys is chosen at random, what is the probability that he takes games and crafts?

11 A, B and C are sets. $n(A) = 17$, $n(B) = 29$ and $n(C) = 14$. Also, $n(A \cup B \cup C) = 45$, $n(A \cap B \cap C') = 6$, $n(B \cap C \cap A') = 1$ and $n(A \cap C \cap B') = 4$. With the aid of a Venn diagram find $n(A \cap B \cap C)$.

Revision Topic 2 Laws of Operation

Reminders

1 Under *addition* on the set of real numbers:

a the identity element is 0; $0+\frac{2}{3}=\frac{2}{3}+0=\frac{2}{3}$
b the inverse of a is $-a$; $a+(-a)=0=(-a)+a$.

2 Under *multiplication* on the set of real numbers:
a the identity element is 1; $1\times(-9)=(-9)\times 1=-9$
b the inverse of a $(\neq 0)$ is $\frac{1}{a}$; $a\times\frac{1}{a}=1=\frac{1}{a}\times a$.

3 The *product* of a positive real number and a negative real number is a negative real number: $3\times(-1{\cdot}2)=-3{\cdot}6$

4 The *product* of two negative real numbers is a positive real number: $-0{\cdot}6\times(-0{\cdot}8)=0{\cdot}48$

5 The *commutative laws*: $a+b=b+a$; $ab=ba$.

6 The *associative laws*: $(a+b)+c=a+(b+c)$; $(ab)c=a(bc)$.

7 The *distributive law*: $a(b+c)=ab+ac$.
$(a+b)^2=a^2+2ab+b^2$; $(a-b)^2=a^2-2ab+b^2$;
$(a+b)(a-b)=a^2-b^2$; $(2x-5)(4x+3)=8x^2-14x-15$

Exercise 2

1 Simplify: *a* $7+(-2)-6$ *b* $-13-8-(-24)$
c $4\times(-8)$ *d* $-8\times(-13)$ *e* $(-2)(-3)(-4)$

2 a Write down the additive inverse of: (*1*) -2 (*2*) $\frac{2}{3}$.
b Write down the multiplicative inverse of: (*1*) -2 (*2*) $\frac{2}{3}$.
c Using the symbol <, arrange $-\frac{5}{8}$, $-\frac{7}{12}$, $-\frac{2}{3}$ in order.

3 If $x=-1$, $y=-2$, $z=3$, $t=0$, evaluate:
a $x^2+y^2+z^2$ *b* $x^2-2xy+5xyz$ *c* zy^2x *d* $(xyt)/z$

4 Simplify as far as possible:
a $2+4(n-3)$ *b* $7x-(x-y)$ *c* $(a^2-a)-3(a^2+a)$

5 $A=3x+4y$ and $B=2x-5y$. Express the following in terms of x and y, in their simplest forms:

a $A+B$ *b* $2A+3B$ *c* $A-B$ *d* $3A-2B$

6 Which of these are true and which false, a, b, c being real numbers?

a $a \times b = b \times a$ *b* $a \div b = b \div a$ *c* $a+(b \times c) = (a+b) \times c$

d $(a \times b) \times c = (c \times b) \times a$ *e* $(a \times b) \div c = a \times (b \div c)$

7 Express the following products as sums or differences:

a $(2a+1)(5a-2)$ *b* $(3m-7)(2m-3)$ *c* $(4p-9)(2p+5)$

d $(3c+4)^2$ *e* $(4x-1)^2$ *f* $(5x-4y)(3x-10y)$

g $(e^2+2)(e^2-2)$ *h* $(3n^2+1)(4n^2-5)$ *i* $(3ab-7)(ab+1)$

8 A cuboid is $3x$ cm long, x cm broad and $2x$ cm high. Find:

a the total length of its edges *b* its area *c* its volume.

9 A set $G = \{a, b, c\}$ has an operation $\circ$ defined by the table shown.

a Name the identity element in the system.

b Is the operation $\circ$ commutative?

c Verify that $(a \circ b) \circ c = a \circ (b \circ c)$. Do you think that the operation $\circ$ is associative? Try some other cases.

d Show that each element of the system has an inverse.

$\circ$	a	b	c
a	a	b	c
b	b	c	a
c	c	a	b

10 Use the distributive law to express the following as sums or differences: *a* $(2x-1)(2x^2-3x+5)$ *b* $(a+b)(a^2-ab+b^2)$

11 A rectangular plate $3x$ mm long and $2x$ mm wide has a square piece of side y mm cut out at one corner.

a Find expressions for the perimeter and area of the remaining shape.

b Given that the remaining area is $23y^2$ mm², show that $x = 2y$.

12 Simplify: *a* $\dfrac{5 \pm 2{\cdot}6}{2}$ *b* $\dfrac{2 \pm 5{\cdot}8}{2}$ *c* $\dfrac{-1 \pm 3{\cdot}48}{4}$ *d* $\dfrac{-8 \pm 1{\cdot}53}{10}$

13 If $a = p+2q$ and $b = 2p-q$, express a^2+b^2 and a^2-b^2 in terms of p and q in simplified form.

14 Simplify: *a* $\left(x-\dfrac{1}{x}\right)\left(x+\dfrac{1}{x}\right)$ *b* $\left(a+1+\dfrac{1}{a}\right)\left(a-1-\dfrac{1}{a}\right)$

15 Given that $\dfrac{x+y}{x-y} = \dfrac{2}{3}$, find the value of the ratio $\dfrac{x}{y}$.

Revision Topic 3
Equations and Inequations in One Variable

Reminders

1

$$3x+8=2, x \in Z$$
$$\Leftrightarrow 3x+8+(-8)=2+(-8)$$
$$\Leftrightarrow 3x=-6$$
$$\Leftrightarrow \tfrac{1}{3}\times 3x=\tfrac{1}{3}\times(-6)$$
$$\Leftrightarrow x=-2$$

Solution set is $\{-2\}$

2

$$\tfrac{1}{3}(x-2)<\tfrac{1}{2}x+1, x \in R$$
$$\Leftrightarrow 2(x-2)<3x+6$$
$$\Leftrightarrow 2x-4<3x+6$$
$$\Leftrightarrow -x<10$$
$$\Leftrightarrow x>-10$$

Solution set is $\{x: x>-10, x \in R\}$

Exercise 3

1 In this question the variables are on the set $S=\{1,2,3,\ldots,10\}$. Find the solution sets of the following:

a $x+4<7$ *b* $x-2\geqslant 5$ *c* $2x-1=6$

d $a^2+15=8a$ *e* $7(b-7)=0$ *f* $y+6$ is not in S

g $x>3$ and $x<6$ *h* $x>3$ or $x<6$ *i* $a^2 \in S$

2 a Make an addition table and a multiplication table for 'clock arithmetic' with a clock showing the numbers 0, 1, 2, 3, 4; e.g. $2+3=0$, $4\times 3=2$.

b If x is a variable on $\{0,1,2,3,4\}$, with the tables in ***a***, solve:

(*1*) $x+3=0$ (*2*) $x+2=1$ (*3*) $3x+2=4$ (*4*) $4x+3=1$

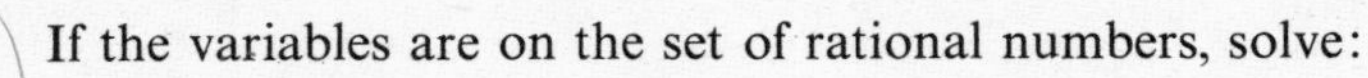

3 If the variables are on the set of rational numbers, solve:

a $3(x-5)-(2-x)=7$ *b* $4+2(x-3)=18$

c $5{\cdot}5a+5{\cdot}1=2{\cdot}8a+15{\cdot}9$ *d* $7(2x+3)-3(8-2x)=27$

e $(2t+3)(t+3)=(2t+1)(t+3)$ *f* $(3y+1)(2y-4)-y(6y-7)=0$

g $(c+2)^2-(c-1)^2=15$ *h* $2(x-4)^2-7=(x-1)(2x-3)$

4 Solve the following equations, where $x \in Q$:

a $\frac{3}{5}(2y-1)=\frac{1}{3}(y+6)$ *b* $\frac{1}{2}(a+4)-\frac{1}{3}(a-3)=1$

c $\frac{1}{2}(x+3)-2(x-1)=-14$ *d* $\frac{3}{4}(x+3)-\frac{1}{3}(2x-3)=\frac{5}{6}$

5 Find the solution sets of these inequations:

a $4r+1\geqslant 97, r\in W$ *b* $3(n-2)-2(n-4)\geqslant 8, n\in W$

c $4(s-1)-6(s-3)<17, s\in Q$ *d* $\frac{1}{3}(2x-1)-\frac{1}{4}(2x+1)<\frac{5}{6}, x\in Q$

e $4-2(s-3)>s, s\in Q$ *f* $\frac{1}{5}(x-1)+\frac{1}{4}(1-2x)\geqslant 0, x\in Q$

6 a Represent the solution set of $2x+9\geqslant 5$, $x\in R$, on the number line.
b Show $P=\{x:x\leqslant 3, x\in R\}$ and $Q=\{x:x\geqslant -1, x\in R\}$ on the number line. What is $P\cap Q$?

7 Draw Venn diagrams for the following, in which A and B are subsets of E.

a $n(E)=60$, $n(A)=28$, $n(B)=37$, $n(A\cap B)=10$. Find $n(A\cup B)$ and $n(A'\cap B')$
b $n(E)=80$, $n(A)=32$, $n(B)=14$, $n(A'\cap B')=46$. Find $n(A\cap B)$. *Hint*: Let $n(A\cap B)=x$.
c $n(E)=48$, $n(B)=19$, $n(A\cap B)=5$, $n(A'\cap B')=13$. Find $n(A)$.

8 The perimeter of a rectangle is 116 cm. If the length is 22 cm more than the breadth, form an equation and use it to find the length and breadth of the rectangle.

9 The least of three consecutive integers is x, and their sum is 453. Form an equation, and use it to find the integers.

10 A man is x years old and his son is 30 years younger. In 2 years the father's age will be twice the son's age. Form an equation, and hence find the present ages of father and son.

11 A motorist drove 180 km in two stages. The first part took 2 hours, and the second part, at twice the average speed of the first part, took $1\frac{1}{2}$ hours. Find his speeds.

12 Solve the following, where $x\in R$:

a $(x-1)^2-(x+1)^2=1-x$ *b* $3(x+3)^2-3(x-3)^2<x$

13 A motorist travelled for 6 hours at a certain average speed, then for 2 hours at an average speed which was 8 km/h greater than the previous one. If he covered 368 km altogether, find his two average speeds.

Revision Topic 4
Relations, Functions and Graphs

Reminders

1 A *relation* from a set A to a set B is a pairing of elements of A with elements of B.

2 A *function*, or *mapping*, f from a set A to a set B is a special kind of relation in which every element of A is paired with exactly one element of B. The set A is called the *domain* of the function.

3 If $f: a \to f(a)$, $f(a)$ is called the *image* of a under f, or the *value* of f at a. The set of values $f(a)$ is called the *range* of the function f.

4 Two sets A and B are in *one-to-one correspondence* if each element of A corresponds to one element of B and vice versa.

5 Relations and functions can be *described* by means of:

a Sets of ordered pairs.
b The solution sets of open sentences, e.g. x *is the square of* y.
c Arrow diagrams and Cartesian graphs.

Exercise 4

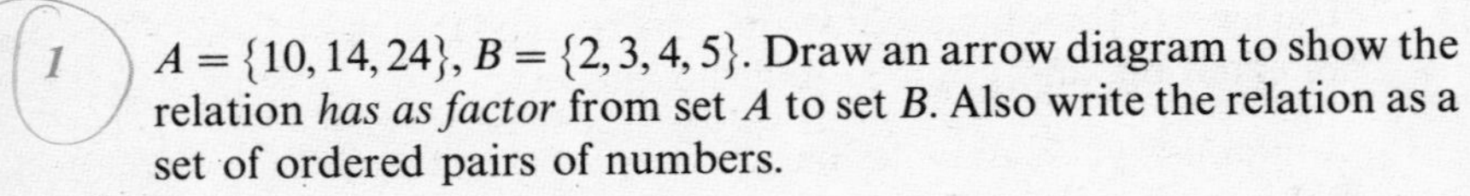

1 $A = \{10, 14, 24\}$, $B = \{2, 3, 4, 5\}$. Draw an arrow diagram to show the relation *has as factor* from set A to set B. Also write the relation as a set of ordered pairs of numbers.

2 Which relations in Figure 3 are:

a mappings *b* $1-1$ correspondences?

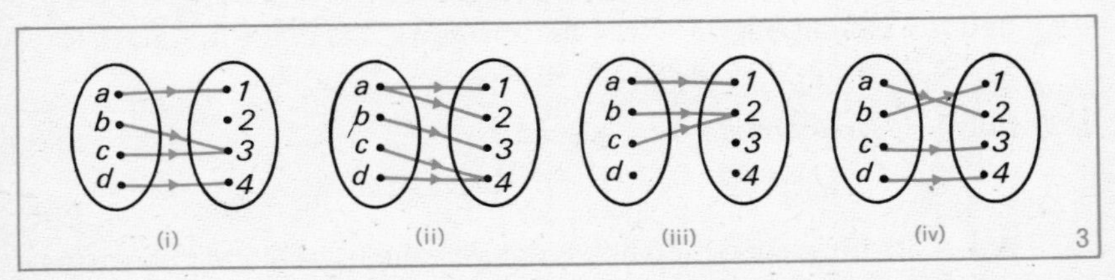

(i) (ii) (iii) (iv) 3

3 a List the set of ordered pairs defining the relation *is the square of* on the set $A = \{-10, -9, -8, \ldots, 8, 9, 10\}$.

b Show the relation in a Cartesian graph.
c Is the relation a function? Explain your answer.

4 A function f is defined by the formula $f(x)=3x-1$, $x \in R$.

a What is the image of -3? *b* What is the value of $f(\frac{1}{3})$?
c Find the element of the domain whose image is 20.
d If $f(a)=-19$, find a.

5 A function g is defined by $g:x \rightarrow x^2+1$, $x \in R$.

a Calculate the value of: (*1*) $g(-1)$ (*2*) g at 2.
b Find the image of zero under g.
c If $g(r)=101$, find r.

6 A function t is defined by $t(x)=a(x+2)$, $x \in R$, a being a real number. If the image of 1 under t is -30, calculate the value of $t(4)$.

7 A function h is defined by the formula $h(x)=ax+b$, where $x \in R$, and a and b are real numbers. If $h(0)=-8$ and $h(6)=22$, find a and b, and hence calculate $h(-1)$.

8 Which of the following are true and which are false?

a Every relation is a function.
b $(-3, -5)$ and $(2, 5)$ are ordered pairs belonging to the mapping $x \rightarrow 2x+1$.
c The relation *is the square root of* from the set of positive real numbers to the set of real numbers is a function.

9 Try to show in an arrow diagram the relation *has as reciprocal* (*or multiplicative inverse*) on the set $A=\{2, 1\frac{1}{3}, \frac{3}{4}, \frac{1}{2}, 0\}$. Why is this relation not fully defined on the set? List the largest subset B of A for which the relation is on the whole of B. Would the relation then be a function on B?

10 A function f is defined by $f:x \rightarrow 3^x-3$, $x \in R$.

a Evaluate $f(0)$, $f(-1)$ and $f(\frac{1}{2})$.
b Find a such that $f(a)=6$.
c What is the zero of the function?
d Draw the graph with equation $y=3^x-3$ on 2-mm squared paper for $x=-2, -1, 0, \frac{1}{2}, 1, 2$, taking scales of 2 cm to 1 unit on the x-axis and 1 cm to 1 unit on the y-axis.

Revision Topic 5 Factors and Fractions

Reminders

1 *Common factor*: $ax+ay=a(x+y)$;

$$x(a+b)+y(a+b)=(x+y)(a+b)=(a+b)(x+y)$$

2 *Difference of two squares*: $a^2-b^2=(a-b)(a+b)$;

$$(a+x)^2-y^2=(a+x-y)(a+x+y)$$

3 *Quadratic expressions*: $2x^2-x-3=(2x-3)(x+1)$

4 To add *fractions*: (i) factorise the denominators, if possible

(ii) find the LCM of the denominators.

$$\frac{1}{x-2}-\frac{2}{x^2-4}=\frac{1}{x-2}-\frac{2}{(x-2)(x+2)}=\frac{(x+2)-2}{(x-2)(x+2)}=\frac{x}{(x-2)(x+2)}$$

Exercise 5

1 Factorise the following:

a $4x-4y$ *b* $20x-16$ *c* $3x^2-6xy$

d x^3-2x^2 *e* $9mn^2-16m^2n$ *f* $4(a-b)^2-(a-b)$

2 Factorise:

a $16x^2-81$ *b* $1-9y^2$ *c* $(a+b)^2-25$

d $2x^2-2$ *e* ab^2-a *f* $\frac{1}{2}mv^2-\frac{1}{2}mu^2$

3 Factorise:

a $px^2+py^2-pz^2$ *b* $16x^2-9y^2$ *c* $16pq-24q^2$

d $a^2b^3+a^3b^2$ *e* $(x-1)^2-4$ *f* $2\pi rh+\pi r^2$

4 Factorise:

a x^2-x-2 *b* $a^2+17a+72$ *c* $y^2-12y+36$

d $4x^2+23x+15$ *e* $2b^2-5b-3$ *f* $6x^2+19x+10$

Simplify the expressions in questions ***5–10***:

5 $\dfrac{ax+a}{a}$ *6* $\dfrac{x^2-4}{4x-8}$ *7* $\dfrac{x}{x^2+2x}$

8 $\dfrac{mn-n^2}{m^2-n^2}$ *9* $\dfrac{a^2-ax}{ax-a^2}$ *10* $\dfrac{3+3a}{6+6b}$

Simplify the expressions in questions ***11–16***:

11 $4+\dfrac{1}{x}-\dfrac{1}{x^2}$ *12* $2+\dfrac{1}{2x}+\dfrac{4}{5x}$ *13* $\dfrac{2y-1}{2}-\dfrac{3y-2}{3}$

14 $\dfrac{1}{x-1}-\dfrac{1}{x+1}$ *15* $\dfrac{1}{x-1}-\dfrac{1}{x^2-1}$ *16* $\dfrac{3}{x-5}-\dfrac{3}{x-2}$

17 Factorise the right-hand side of the formula $A=\pi(x^2-y^2)$. Then calculate A when 3·14 is used as an approximation for π, and:

a $x=5{\cdot}2$, $y=4{\cdot}8$ *b* $x=65$, $y=35$.

18 Given $x^2-y^2=77$ and $x+y=11$, find the values of $x-y$, x and y.

19 Factorise:

a $4(p+q)^2-9r^2$ *b* $1-(r-s)^2$ *c* $16-9(x+y)^2$

d $1-16x^4$ *e* x^2-x^6 *f* $x^2(a-b)^3-(a-b)$

20 A car travels the first 20 km of a journey at an average speed of r km/h, and the next 30 km at an average speed of s km/h. Calculate the total time for the journey.

If $rs=600$, prove that this time is given by $\dfrac{r^2+400}{20r}$ hours.

21 Simplify:

a $\dfrac{a^2-b^2}{x-y}\times\dfrac{2x-2y}{a-b}$ *b* $\dfrac{2p^2-5p-3}{4p^2-1}\times\dfrac{2p-1}{p^2-9}$

22a Use the distributive law to multiply out $(x-1)(x^2+x+1)$

b Hence write down the factors of x^3-1.

c Find the factors of a^2-1, a^3-1, a^4-1 and a^5-1.

23 Simplify: *a* $\dfrac{3a^2-5a-12}{3a^2-9a}$ *b* $\dfrac{x+2}{2}-\dfrac{2}{x+2}$

24 Which of the following are true and which are false?

a $x-y$ is a factor of: (1) x^2-y^2 (2) $x^3y^2-x^2y^3$ (3) $x^2+xy-2y^2$

b $\dfrac{a-b}{b-a}=1$ *c* $\dfrac{x+2}{x}=2,\ x\neq 0$ *d* $\dfrac{x+2}{x}=1+\dfrac{2}{x},\ x\neq 0$

Revision Topic 6

Formulae and Literal Equations

Reminders

1 A *formula* is an equation which expresses one symbol (called the subject of the formula) in terms of other symbols:

e.g. $P = 2d + 2\pi r$

2 The *subject of a formula* may be changed in order to arrange the formula in a more suitable form for the calculation required:

e.g. $d = \frac{1}{2}(P - 2\pi r)$

3 A *literal equation* is an equation in which letters are used to represent numerals and variables:

e.g. $y = mx + c$; x and y are variables, m and c are numerals.

Exercise 6

1 The perimeter of a regular hexagon is $P = 6x$. Change the subject of the formula to x, and hence find x when $P = 96$.

2 The road resistance to a car is $R = kv^2$. Change the subject to v, and find v when $k = 4$ and $R = 324$.

3 The exposure time for a photograph is $e = f^2/16s$. Change the subject to s, and find s when $f = 8$ and $e = \frac{1}{50}$.

4 The rate at which water flows from a valve is $v = 14\sqrt{h}$. Change the subject to h, and find h when $v = 28$.

5 Check that for the sequence 5, 8, 13, 20, 29,... the first term can be written $1^2 + 4$, the second term $2^2 + 4$, and so on. Write down a formula for T, the nth term; hence find the 100th term.

6 The volume of a cube of edge p cm is V cm^3. Write down a formula for V, and use it to find V when $p = 2{\cdot}2$.

7 Change the subject in each of the following formulae to the variable stated:

a $C = 2\pi r; r$ *b* $y = 2x + q; x$ *c* $s = \dfrac{2\pi}{r^2}; r$

d $M = 3(2d + 1); d$ *e* $v = u + at; t$ *f* $v^2 = u^2 + 2as; s$

8 Show that the area A square metres of a square field of side $2x$ metres with a semicircular piece of grass on each of two sides as diameter is given by $A=x^2(4+\pi)$. Change the subject to x, and calculate x if $A=6000$ and $\pi \doteqdot 3{\cdot}14$.

9 A rectangular lawn a metres long and b metres wide is surrounded by a path of uniform width t metres. Show that the area, A square metres, of the path is given by $A=2t(a+b+2t)$. Change the subject of this formula to a.

10 Given that $Q=\dfrac{3ab}{a-b}$, $a \neq b$, express a in terms of b and Q. What restrictions must be placed on b and Q?

11 Change the subject in each of the following to the variable stated:

a $y=\dfrac{x}{ax+b}$; x *b* $\dfrac{1}{y}=\dfrac{1}{x}+\dfrac{2}{u}$; u

c $p=2\sqrt{r+1}$; r *d* $2\pi fL=\dfrac{T}{fC}$; f

12 Find the gradients of the straight lines with equations:

a $2x+3y=12$ *b* $x-y-1=0$ *c* $5x-6y+2=0$

13 Find a formula for the area A cm² of a cylinder closed at both ends with radius r cm and height h cm, assuming that the curved surface area is $2\pi rh$ cm². Change the subject of the formula to h. Also express the formula as a quadratic equation in r, in standard form.

14 Make x the subject in each of the following:

a $ax=bx+c$ *b* $y=1+\dfrac{1}{x}$ *c* $1+\sqrt{x}=k$

15 Express each of the following in the form $ax^2+bx+c=0$:

a $px-q=rx^2$ *b* $\dfrac{x}{a}+\dfrac{a}{x}=1$ *c* $\dfrac{a}{x-b}+1=x$

16 Simplify the right-hand side of $S=\dfrac{a(1-r^2)}{1-r}$.

Then change the subject to: *a* a *b* r

Revision Topic 7

Systems of Equations and Inequations

Reminders

1 An equation of the form $ax+by+c=0$ is a *linear equation.* Its *graph is a straight line* which can be drawn by finding the points where it cuts the x- and y-axes.

2 The *solution set of a system of equations* can be found:

(i) approximately, by finding the point of intersection of the graphs of the equations
(ii) by eliminating one of the variables
(iii) by substitution.

3 The *solution set of a system of inequations* is best represented graphically.

Exercise 7

(Assume that the variables are on R unless stated otherwise.)

1 Sketch the graphs of the following linear equations:

a $y=4x-8$ *b* $y-7x=14$ *c* $2x+3y=12$

d $4x+5y+20=0$ *e* $y=-2$ *f* $y=-2x$

2 On separate diagrams show by shading the solution sets of:

a $y\leqslant 8$ *b* $y>x$ *c* $y\geqslant 4+2x$ *d* $3x-4y>24$

3 $A=\{(x,y):x\geqslant 3\}$, $B=\{(x,y):y\leqslant x\}$. On the same diagram, show:

a A *b* B *c* $A\cap B$.

4 Find graphically the solution sets of the following systems:

a $y=2x-1$ and $y=5-x$ *b* $x+y=4$ and $2x-y+1=0$

5 Solve the following systems by elimination or substitution:

a $\left.\begin{array}{r}x+y=1\\3x+2y=8\end{array}\right\}$ *b* $\left.\begin{array}{r}x+y=6\\3x-2y+7=0\end{array}\right\}$ *c* $\left.\begin{array}{r}3x-4y=26\\5x+6y+20=0\end{array}\right\}$

d $\left.\begin{array}{l}5x-2y=17\\2x+3y=3\end{array}\right\}$ *e* $\left.\begin{array}{r}3x-2y=12\\4x+y=5\end{array}\right\}$ *f* $\left.\begin{array}{r}\frac{1}{2}x+\frac{1}{3}y=2\\4x-y=5\end{array}\right\}$

g $\frac{1}{6}(x-1)+y=6$, $\frac{1}{4}(y-1)+x=8$ *h* $y=\frac{1}{4}(x-2)$, $\frac{1}{3}x+y=10$

6 Figure 4 shows the line with equation $y=3-2x$.

a Give the coordinates of the points A and B.
b State the regions through which the line $y=-2x$ passes.
c In which region is the point $(4, -4)$?
d Which regions are defined by $\{(x, y): y>0, y<3-2x, \text{ and } x, y \in R\}$?
e List $\{(x, y): y=3-2x, x>0, y>0 \text{ and } x, y \in Z\}$

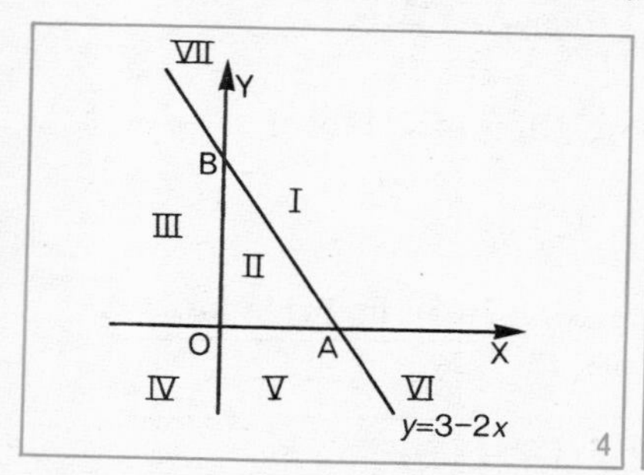

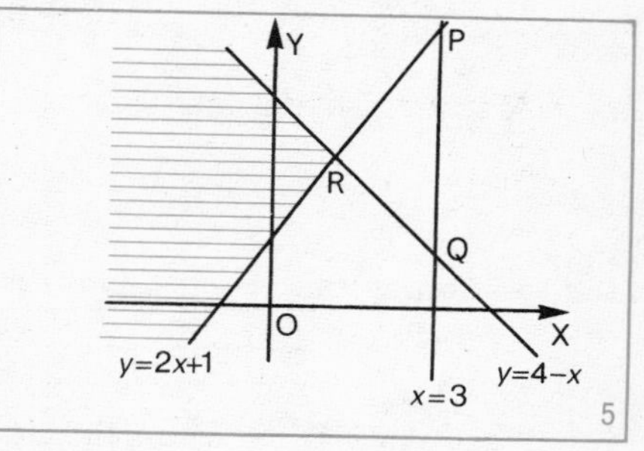

7 Figure 5 shows the lines $x=3$, $y=4-x$ and $y=2x+1$.

a Find the coordinates of the points P, Q and R.
b Give a set definition of the region which is:
(*1*) shaded (*2*) the interior of $\triangle$PQR.

8 $ax+by=25$ when $x=1$ and $y=-3$, and $ax+by=20$ when $x=-2$ and $y=-4$. Find a and b.

9 The length of a rectangle is 2 cm more than the breadth, and the perimeter is 84 cm. Form two equations; find length and breadth.

10 A pile of 40 coins consists of 5p coins and 50p coins. If the total sum of money is £12·80, find the number of each kind of coin in the pile.

11 Show that the straight lines with equations $3x+4y=2$, $5x-2y=12$ and $2x+3y=1$ all intersect at the same point. Illustrate in a sketch.

12 The points $(-4, 2)$ and $(1, -3)$ lie on the line with equation $y=mx+c$. Find m and c, and hence the equation of the line.

13 Show the solution set of the system $y \geqslant 0$, $y \leqslant x+3$, $-6 \leqslant x \leqslant 6$ on squared paper. Find the maximum value of $x+y$ subject to these restrictions.

Revision Topic 8 Variation

Reminders

1 If y *varies directly* as x, i.e. $y \propto x$, then $y = kx$, or $\frac{y_1}{x_1} = \frac{y_2}{x_2} = \ldots$

2 If y *varies inversely* as x, i.e. $y \propto \frac{1}{x}$, then $y = \frac{k}{x}$, or $xy = k$.

3 If y *varies as* x and *inversely as* z, then $y = \frac{kx}{z}$.

4 A straight-line graph indicates a relation of the form $y = ax + b$.

Exercise 8

1 a Assuming that $V \propto h$, find the missing entry in Table I.

Table I

h	125	275
V	25	

Table II

S	1	2	5
W	2	8	50

b Verify that $W \propto S^2$ for the entries in Table II, and find an equation containing W and S^2.

2 Express each of the following as an equation:

a s varies as the cube of t *b* p varies inversely as q^4

c m varies as the square of n *d* u varies as the square root of v.

3 v varies directly as w, and $w = 5$ when $v = 40$. Find:

a v when $w = 8$ *b* w when $v = 4$.

4 y varies inversely as x, and $y = 6$ when $x = 6$. Find:

a y when $x = 24$ *b* x when $y = 32$.

5 Q varies as x and inversely as z^2, and $Q = 10$ when $x = 30$ and $z = 14$. Find:

a the formula connecting Q, x and z *b* Q when $x = 36$ and $z = 42$.

6 The time of swing of a pendulum varies as the square root of its length. If the time for a pendulum 25 cm long is 1 second, calculate:

a the time of swing of a pendulum 4 metres long
b the length of a pendulum that swings in 1·5 seconds.

7 The diameter d cm of a sphere varies as the cube root of its mass m kg, and the mass of a sphere of diameter 15 cm is 8 kg. Find the formula for d in terms of m, and calculate the mass of a sphere of diameter 30 cm.

8 The volume V of a given mass of gas varies directly as the absolute temperature T and inversely as the pressure P. When $V = 490$, $T = 350$ and $P = 750$. Calculate V when $T = 400$ and $P = 560$.

9 The table gives the pull on a wire and the corresponding stretch.

Pull in newtons (P)	0·5	1·0	1·5	2·0	2·5	3·0
Stretch in mm (S)	2	3	4	5	6	7

Show graphically that $S = aP + b$; hence find a and b.

10 The time taken by a box to slide down a sloping delivery chute varies directly as the length of the chute and inversely as the square root of the height of one end above the other. For a chute 12 m long with one end 2·25 m above the other the time taken is 10 seconds. Calculate the time taken for a chute 18 m long with one end 4 m above the other.

11 As a model train runs round a circular track, the thrust, T newtons, of the rails on the wheels varies directly as the mass m kg of the train and as the square of its velocity v m/s, and inversely as the radius r cm of the circle. $T = 1{\cdot}5$ when $m = 2$, $v = 1{\cdot}5$ and $r = 3$. Calculate T when $m = 5$, $v = 2{\cdot}4$ and $r = 6$.

12 The distance d metres travelled by a falling object in t seconds is given by the formula $d = kt^2$. When $t = 4$, $d = 80$.

a Find d when $t = 8$, and t when $d = 1125$.
b Change the subject of the given formula to t, and copy and complete: 'The time in seconds an object takes to fall varies ... ...'

13 The volume of a cone varies as the height and as the square of the radius of the base of the cone. Find the ratio of the volumes of two cones with heights 8 cm and 12 cm, and radii 6 cm and 9 cm respectively.

Revision Topic 9

Quadratic Functions, Equations and Inequations

Reminders

1 A function f for which $f(x) = ax^2 + bx + c$, where $a, b, c \in R$ and $a \neq 0$, is called *a quadratic function.*

2 The Cartesian graph of every quadratic function is a *parabola.*

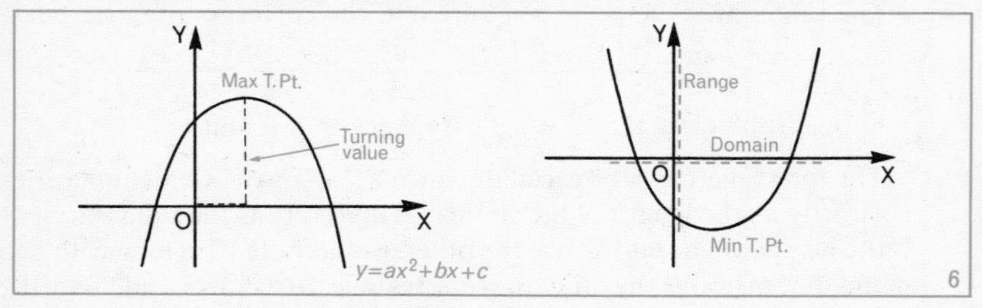

3 To *solve the quadratic equation* $ax^2 + bx + c = 0$,

a attempt to factorise the quadratic expression on the left-hand side
b if ***a*** is not possible, use the formula

$$x = \frac{-b \pm \sqrt{(b^2 - 4ac)}}{2a}.$$

Exercise 9

1 Write down the solution sets of the following:

a $(x+2)(x+5) = 0, x \in Z$ *b* $(x+2)(x-2) = 0, x \in W$

c $(x-10)(2x+1) = 0, x \in Q$ *d* $5x(x+10) = 0, x \in Q$

2 Use factors to solve these equations, $x \in R$:

a $x^2 - 6x + 5 = 0$ *b* $x^2 + 9x + 20 = 0$ *c* $4x^2 - 9 = 0$

d $4x^2 + 4x + 1 = 0$ *e* $40x - 10x^2 = 0$ *f* $x(x-8) = 33$

3 The sides of a rectangle with area 135 cm^2 are $3x$ cm and $(x+4)$ cm long. Form an equation in x, and solve it. Write down the length and breadth of the rectangle.

4 A parabola has equation $y = 15 + 14x - x^2$.

a Find the coordinates of the points where it cuts the axes.
b The point $(h, 15)$ is on the parabola. Find two replacements for h.
c The turning point of the parabola is $(7, k)$. Find k.
d Sketch the parabola on squared paper, using your answers to ***a–c***.

5 Solve the following equations, where $x \in R$:

a $(x+5)^2 = 81$ *b* $(x+\frac{1}{4})^2 = 9$ *c* $(2x-1)^2 = \frac{1}{4}$

6 Find the solution sets of the following equations, in which $x \in R$, rounding off the roots to 1 decimal place:

a $x^2+2x-6=0$ *b* $2x^2-5x+1=0$ *c* $x(5x+1)=3$

d $3+3x-2x^2=0$ *e* $5x^2=12-10x$ *f* $x^2=2(x+1)$

7 Draw the graphs of the following functions on 2-mm squared paper, and estimate:

(*1*) the coordinates of the turning point of the graph
(*2*) the maximum or minimum value of the function
(*3*) the zeros of f
(*4*) the range of f for the given domain
(*5*) the equation of the axis of symmetry of the parabola.

	a	*b*	
Function	$f(x)=x^2-x-12$	$f(x)=x^2+x+1$	$f(x)=2-3x-2x^2$
Scale for x	1 cm per unit	1 cm per unit	1 cm per unit
Scale for f(x)	2 cm per 5 units	1 cm per unit	1 cm per 2 units
Domain	$\{x: -4 \leqslant x \leqslant 5\}$	$\{x: -3 \leqslant x \leqslant 2\}$	$\{x: -3 \leqslant x \leqslant 2\}$

8 The hypotenuse of a right-angled triangle is $(2x+3)$ cm long, and the other sides have lengths x cm and $(x+7)$ cm. Find x, and calculate the area of the triangle.

9 Find the solution set of each of the following inequations, $x \in R$:

a $(x+3)(x-3)>0$ *b* $(x-5)(x-8) \leqslant 0$ *c* $12-x-x^2>0$

10 Solve, $x \in R$: *a* $x+\dfrac{2}{x}-3=0$ *b* $\dfrac{3}{x+1}+\dfrac{1}{x-1}-2=0$

11 Which of the following are true and which are false for the function f defined by $f(x)=16-x^2$?

a $f(0)=16$ *b* $f(16)=0$ *c* The image of -4 under f is 32.
d The graph of f has a maximum turning point $(0, 16)$.
e If $f(x)>0$, then $-4<x<4$.

Revision Topic 10 Matrices

Reminders

1 A *matrix* is a rectangular array of numbers arranged in rows and columns. With m rows and n columns the order of the matrix is $m \times n$.

2 a $\begin{pmatrix} a & c \\ b & d \end{pmatrix} \pm \begin{pmatrix} p & q \\ r & s \end{pmatrix} = \begin{pmatrix} a \pm p & c \pm q \\ b \pm r & d \pm s \end{pmatrix}$ *b* $I = \begin{pmatrix} 1 & 0 \\ 0 & 1 \end{pmatrix}$ is a unit matrix

c $k\begin{pmatrix} a & b \\ c & d \end{pmatrix} = \begin{pmatrix} ka & kb \\ kc & kd \end{pmatrix}$ *d* $\begin{pmatrix} a & b \\ c & d \end{pmatrix}\begin{pmatrix} p & q \\ r & s \end{pmatrix} = \begin{pmatrix} ap+br & aq+bs \\ cp+dr & cq+ds \end{pmatrix}$

3 The *inverse* of $A = \begin{pmatrix} a & b \\ c & d \end{pmatrix}$ is $A^{-1} = \dfrac{1}{ad-bc}\begin{pmatrix} d & -b \\ -c & a \end{pmatrix}$, $ad-bc \neq 0$.

$ad-bc$ is the *determinant* of matrix A. $A^{-1}A = AA^{-1} = I$.

Exercise 10

1 Express each of the following as a single matrix:

a $\begin{pmatrix} 1 & 1 & 1 \\ 1 & -1 & -1 \end{pmatrix} + \begin{pmatrix} 2 & 1 & 1 \\ 0 & 2 & 1 \end{pmatrix}$ *b* $\begin{pmatrix} 3a & 1 \\ 4b & 3 \end{pmatrix} + \begin{pmatrix} a & 1 \\ -b & 0 \end{pmatrix} - \begin{pmatrix} 3a & -2 \\ b & -3 \end{pmatrix}$

2 Find p and q in each of the following:

a $\begin{pmatrix} 4p & 3 \\ 1 & 10q \end{pmatrix} = \begin{pmatrix} p+9 & 3 \\ 1 & 100 \end{pmatrix}$ *b* $(p+5 \quad 2) = (40 \quad q+2)$

3 Solve each of the following equations for the 2×2 matrix X:

a $X + \begin{pmatrix} 1 & 1 \\ 2 & 1 \end{pmatrix} = \begin{pmatrix} 7 & 0 \\ 1 & 2 \end{pmatrix}$ *b* $\begin{pmatrix} 3 & -4 \\ 2 & 8 \end{pmatrix} - X = \begin{pmatrix} -1 & 2 \\ 0 & 1 \end{pmatrix}$

4 $P = \begin{pmatrix} 2 & 1 & 1 \\ 1 & 0 & 1 \end{pmatrix}$, $Q = \begin{pmatrix} 2 & 0 & 2 \\ 2 & 2 & -2 \end{pmatrix}$. Simplify: *a* $P+2Q$ *b* $2P-3Q$

5 Find the following matrix products:

a $(1 \quad 4)\begin{pmatrix} 6 \\ 7 \end{pmatrix}$ *b* $\begin{pmatrix} 2 & 1 \\ 0 & 1 \end{pmatrix}\begin{pmatrix} 4 \\ 6 \end{pmatrix}$ *c* $\begin{pmatrix} 1 & 1 \\ 2 & 1 \end{pmatrix}\begin{pmatrix} 1 & 0 \\ 0 & 1 \end{pmatrix}$

d $(1 \quad 2 \quad 3)\begin{pmatrix} 5 \\ 8 \\ 0 \end{pmatrix}$ *e* $\begin{pmatrix} 1 & 0 & 1 \\ 2 & -1 & 1 \end{pmatrix}\begin{pmatrix} 1 \\ 2 \\ 1 \end{pmatrix}$ *f* $\begin{pmatrix} 1 & 0 & 1 \\ 2 & -1 & 1 \end{pmatrix}\begin{pmatrix} 1 & 0 \\ 2 & 0 \\ 1 & 0 \end{pmatrix}$

6 Suppose $P(x, y) \to P'(x', y')$, where $x' = x + 2y$ and $y' = 2x - y$.

a Write down the matrix equation for the mapping.
b Hence find the images of A(1, 2), B(−2, 1), C(−1, −2), D(2, −1).
c Show the points and their images on squared paper. What kind of shapes do they form?

7 *a* Find the 2×2 matrix associated with an anticlockwise rotation of 90° about the origin.

b Find the images of A(0, 0), B(2, 0), C(2, 2), D(0, 2) under the mapping with matrix $\begin{pmatrix} 0 & -2 \\ 2 & 0 \end{pmatrix}$.

Draw ABCD and its image on squared paper. What composite transformation would map the square onto its image?

8 Find the inverse of each of the following, if it exists:

a $\begin{pmatrix} 6 & 5 \\ 7 & 6 \end{pmatrix}$ *b* $\begin{pmatrix} 4 & 2 \\ -4 & -1 \end{pmatrix}$ *c* $\begin{pmatrix} 8 & -4 \\ -4 & 2 \end{pmatrix}$ *d* $\begin{pmatrix} -2 & -2 \\ -2 & -1 \end{pmatrix}$

9 Use a matrix method to solve the following systems of equations:

a $3x + y = 2$, $5x + 2y = 3$ *b* $7x + 9y = 12$, $5x + 2y = 5$ *c* $8x - 7y = 8$, $5x + 3y = 6$ $(x, y \in R)$

10 Under a mapping, $(x, y) \to (x', y')$, where $x' = x + 2y$ and $y' = y$. Construct the matrix equation for the mapping, and hence find the images of the vertices of the parallelogram ABCD where A is (0, 0), B(4, 0), C(0, 4), D(−4, 4). Sketch ABCD and its image on the same diagram. Find the matrix which will map the image onto ABCD, and check that it does so.

11 $P = \begin{pmatrix} 2 & 1 \\ 1 & 2 \end{pmatrix}$, $Q = \begin{pmatrix} x & 0 \\ y & 2 \end{pmatrix}$ and $PQ = \begin{pmatrix} 1 & 2 \\ 2 & 4 \end{pmatrix}$

Find x and y. Hence find QP, P^{-1} and Q^{-1} if possible.

12*a* Write down the matrix A associated with the transformation of reflection in the x-axis.

b What geometrical transformation does $B = \begin{pmatrix} 2 & 0 \\ 0 & 2 \end{pmatrix}$ represent?

c Find AB, and the image set under AB of $\{(0, 0), (4, 2), (6, -1)\}$.
d Find B^{-1}, and explain what transformation it represents.

Cumulative Revision Exercises

Exercise 1A

1 Given that $x = 9$ and $y = 16$, simplify:

a $4x^0y^{-1}$ *b* $x^{1/2} + y^{3/4}$ *c* $6x^{-1/2}y^{1/2}$ *d* $(x+y)^{3/2}$

2 Simplify: *a* $\sqrt{18} + \sqrt{50} - \sqrt{32}$ *b* $\sqrt{3}(\sqrt{6} - 2\sqrt{3})$ *c* $(\sqrt{2} + 1)^2$

3 $E = \{x \in Z: -3 \leqslant x \leqslant 3\}; P = \{x \in Z: -1 \leqslant x \leqslant 2\}; Q = \{x \in Z: x \geqslant 2\}$.

List: *a* P *b* P' *c* $P \cap Q$ *d* $P \cup Q$

4 The operation $*$ on R is defined by $a * b = (a+b)^2 - a^2 - b^2$.

a Simplify the right-hand side, and calculate $-5 * (-1)$ and $\frac{1}{3} * 0$.
b Prove that the operation $*$ is commutative (simplify $b * a$).

5 y varies as the square root of x, and $y = 3$ when $x = 1{\cdot}44$.

Find: *a* y when $x = 0{\cdot}36$ *b* x when $y = 10$.

6 Find the roots of these quadratic equations, rounded off to 1 decimal place:

a $x^2 - 4x + 1 = 0, x \in R$ *b* $5x^2 + 2x - 2 = 0, x \in R$

7 Simplify the following, and express each as a product of factors:

a $12k^2 + 12$ *b* $12k^2 - 12$ *c* $2x(x - \frac{1}{2}) - 1$

8 Find the solution sets of:

a $2 - 3(t-1) = 4t - 2, t \in Q$ *b* $4(r-1) - 2r + 1 \geqslant 7r, r \in Q$

9 Simplify: *a* $\dfrac{2b+2}{2b^2-2}$ *b* $1 + \dfrac{r}{2} - \dfrac{2r}{3}$ *c* $1 - \dfrac{1}{3-x} + \dfrac{1}{9-x^2}$

10 The sides of a right-angled triangle have lengths x cm, $(x-y)$ cm and $(x+y)$ cm, where $x > y > 0$. Find the value of $x:y$.

11 The nth term of a sequence is given by $T_n = 2^{n-1} + 1$.

a Find T_1, T_2, T_3 and T_4.
b Find n when $T_n = 17$.

12 Show by shading on squared paper the solution set of the system of inequations $x \geqslant 0$, $-5 \leqslant y \leqslant 5$ and $y \geqslant x - 3$, where $x, y \in R$.

Exercise 1B

1 Simplify: *a* $(4x^2y^4)^{-1/2}$ *b* $(xy)^{-1} \times (2x)^2$ *c* $\left(\frac{4x}{y}\right)^{-1}$.

2 $a = \sqrt{2}+1$ and $b = \sqrt{8}-1$. Show that $3a^2 - b^2 \notin Q$.

3 *a* Given that p, q and r are consecutive even numbers, and that $p<q<r$, find the value of $r-p$.

b The operation $*$ on the integers a and b is defined by $a * b = ab + b$. Find x, given that $4 * x = x * 4$.

4 p varies directly as q and inversely as r^2. $p = 0{\cdot}3$ when $q = 2$ and $r = 8$. Find:

a p when $q = 5$ and $r = 6$ *b* r when $p = 300$ and $q = 5$.

5 Find the solution sets of the following, where $x \in R$:

a $x(x-2)^2 = 0$ *b* $x^4 - x^2 = 0$ *c* $(x-1)^3 - (x-1)^2 = 0$

6 Which of these statements are true, and which are false?

a $x^2 = 196 \Leftrightarrow x = 14$. *b* If $\frac{1}{t} = \frac{1}{r} - \frac{1}{s}$, then $r > s \Rightarrow t > 0$.

c If the hands of a clock are in a straight line, the time is 6 o'clock.

d The converse of statement *c*.

7 Simplify: *a* $\frac{x^2-4}{x^2} \times \frac{x}{4-2x}$ *b* $\frac{2}{x^2+x-2} - \frac{1}{x-1}$

8 A number system comprising the set $S = \{2, 4, 6, 8\}$ and an operation $*$ is defined by the table shown.

$*$	2	4	6	8
2	4	8	2	6
4	8	6	4	2
6	2	4	6	8
8	6	2	8	4

a Name the identity element.

b Is the operation $*$ commutative?

c If $b \in S$, and $b * b = 4$, find b.

9 By pairing the first and last numbers, the second and second last, and so on, show that $1+2+3+\ldots+98+99+100 = 5050$.
Use the same method to find a formula for the sum of $1+2+3+\ldots+(n-2)+(n-1)+n$.

10 Simplify: $(8a-2b)(2a+b)-(4a-b)^2$

Exercise 2A

1 Make x the subject of each of the following formulae:

a $px+qy+r=0$ *b* $x^2+y^2=r^2$ *c* $\frac{x+a}{b}=\frac{c}{d}$

2 $P=\{(x, y):2x+3y+5=0\}$ and $Q=\{(x, y):5x-4y+1=0\}$, $x, y \in R$. Find $P \cap Q$.

3 A function f is defined by $f(x)=\frac{6}{x^2-1}$, $x \in R$ but $x \neq \pm 1$.

a Calculate the values of f at 2, -3 and $\frac{1}{2}$.
b What elements of the domain have image 2?

4 a The sum of a number and its square is 210. Form an equation, and solve it to find the number.
b $(x+2)$ twopence coins and $(x+5)$ fiftypence coins make up £8·78. Form an equation, and find x.

5 Simplify: *a* $1+\frac{1}{2x}+\frac{3}{x^2}$ *b* $\frac{a}{a-b}-\frac{a}{a+b}$ *c* $\frac{1}{x+1}-\frac{1}{x^2+3x+2}$

6 Draw the graphs of $y=4$ and $x+y=6$. Show, by shading, the regions of the plane defined by $x \geqslant 0$, $y \geqslant 0$, $y \leqslant 4$, $x+y \leqslant 6$. Subject to these restrictions, find the maximum value of $2x+3y$ and the corresponding values of x and y.

7 Simplify:

a $2(x+1)^2+(x-2)^2$ *b* $4-3(x-2)^2$ *c* $2(x-2)(x-3)-(x-1)^2$

8 Find the 2×2 matrix X such that $2X+\begin{pmatrix}1 & 2\\ 1 & 1\end{pmatrix}\begin{pmatrix}4 & 1\\ 0 & -1\end{pmatrix}=\begin{pmatrix}1 & 0\\ 2 & 0\end{pmatrix}$

9 If $a, b \in Z$, find a and b such that $x^2+12x+a=(x+b)^2$.

10 The side of a square is $(2x-1)$ mm long.

a Given that the perimeter is 94 mm, find x.
b Given that the area is 50 mm^2, find x (to 1 decimal place).

11 Simplify: *a* $1^0+1^{1/2}+1^{-2}$ *b* $(8^{1/3}+16^{1/4})^{-1}$ *c* $\left(\frac{4}{9}\right)^{-1/2}$

12 The nth term of a sequence is given by $T=\frac{1}{2}n(n+1)$. Find T when $n=15$, and n when $T=6$.

Exercise 2B

1 a Make n the subject of the formula $nE = I(r+Rn)$.

b If $r = \dfrac{\sqrt{(1-a^2)}}{a}$, express a in terms of r.

2 $P = \{(x, y): 4x+3y = -7\}$, $Q = \{(x, y): 5x-9y = -13\}$ and $R = \{(x, y): 14x-15y = -33\}$, $x, y \in R$. Prove that $P \cap Q \subset R$.

3 The domain of a function g given by the formula $g(x) = p(x-1)^2+4$, $p>0$, is $0 \leqslant x \leqslant 3$, $x \in R$.

a Find the least value of g.
b If the greatest value of g is 24, find p.
c Sketch the graph of g.

4 $P = (a^2+b^2)^2-(a^2-b^2)^2$, $a, b \in Z$.

a Prove that P is a positive even integer.
b If the point (a, b) lies on the curve $xy = 2$, find the value of P.

5 Show, by shading, the region for which $x \geqslant 0$, $y \geqslant 2$ and $2x+y \geqslant 8$. Subject to these restrictions find the minimum value of $4x+5y$, and the corresponding values of x and y, where $x, y \in R$.

6 a A certain rational number is halved, and 23 is added to the result. The same number is doubled, and 41 is added to the result. If the two numbers obtained are equal, find the rational number.
b There were 200 people at a concert. Some paid 30p each, the rest paid 20p each, and the total takings were £50·80. How many paid 20p?

7 Find in each case the matrix X such that:

a $\begin{pmatrix} 1 & 1 \\ 1 & 2 \end{pmatrix} X = \begin{pmatrix} 1 \\ 2 \end{pmatrix}$ *b* $X \begin{pmatrix} 1 & 1 \\ 1 & 2 \end{pmatrix} = (2 \ \ 1)$

8 Simplify:

a $(3p+4q)(3p-4q)-(p+q)(9p-16q)$ *b* $\dfrac{3a^2-5a-12}{a^2-3a}$

9 Simplify: *a* $\sqrt{18}-2\sqrt{8}$ *b* $3\sqrt{12} \times 2\sqrt{8}$ *c* $\sqrt{10}(\sqrt{5}-\sqrt{2})$

10 Find the roots of the quadratic equation $7x^2+11x-6=0$ to 1 decimal place by:

a using the quadratic formula *b* drawing a graph.

Exercise 3A

1 The breaking weight, W tonnes, of a rope of diameter d mm is proportional to the square of its diameter. Given that $W = 6$ when $d = 40$, find W when $d = 24$.

2 Sketch the graph of the function given by the formula $f(x) = (x-1)(x-2)$, $x \in R$. Hence find the interval of x over which the values of f are negative.

3 The sum of two real numbers, one of which is x, is 100. Prove that the product P of the numbers is given by $P = 2500 - (x-50)^2$. Write down the greatest value of P and the corresponding value of x.

4 Which of the following are true and which are false?

a $x^{1/2} \times x^2 = x$ *b* $(x^{-2})^{1/2} = x^{-5/2}$ *c* $\dfrac{1}{4x} = 4x^{-1}$

5 Figure 7 shows the regions into which the axes and the lines $y = 1$ and $y = 4x$ divide the plane. Which regions show the following sets?

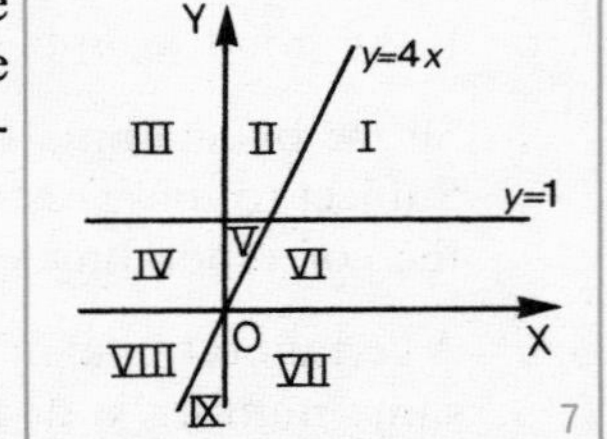

7

a $P = \{(x, y): y \geqslant 1\}$
b $S = \{(x, y): y \leqslant 4x\}$
c $P \cap S$ *d* $(P \cup S)'$

6 *a* Simplify: (1) $\sqrt{27} + \sqrt{12} - \sqrt{75}$ (2) $2\sqrt{2}(\sqrt{2} + \sqrt{8})$

b Taking $\sqrt{2} = 1{\cdot}414$, calculate $\dfrac{1}{2\sqrt{2}}$ to 2 decimal places.

7 Given $2x - y < 20 - 3x$, show that $y > 5(x-4)$. Hence find x for which $y > 20$.

8 Factorise: *a* $(p+q)^2 - r^2$ *b* $2x^2 - x - 10$ *c* $a^2b - 2ab^2 - 3b^3$

9 Solve, $x \in R$: *a* $2x^2 - x - 6 = 0$ *b* $2x^2 - x - 5 = 0$.

10 A photograph is mounted on a rectangular sheet of card of area 200 cm^2 and width x cm. If there are margins 2 cm wide at each side and 5 cm wide at the top and bottom, show that the area of the photograph is $\left(240 - 10x - \dfrac{800}{x}\right)\text{cm}^2$.

If this area is 60 cm^2, find the two possible values of x.

Exercise 3B

1 Simplify:

a $(y^{1/2})^2 \times (a^{-2})^0$ *b* $(x^{1/2}+x^{-1/2})^2$ *c* $(3x)^{-1}-\dfrac{1}{3x}$

2 Find the solution sets of the following, $x \in R$:

a $(x+5)(x-5)<0$ *b* $(x-4)(2-x)\geqslant 0$ *c* $6+x-x^2<0$

3 ABCD is a square of side 10 cm. E is a point on AB such that BE = x cm, and F is on BC such that CF = $2x$ cm. D, E, F are joined.

a Write down expressions for the lengths of AE and BF.
b If the area of $\triangle$DEF is A cm^2, prove that $A = 25+(x-5)^2$.
c Find the minimum value of A and the corresponding value of x.

4 Find the roots of $\dfrac{x}{x+1}-\dfrac{x+1}{x}=1$ to 1 decimal place.

5 $p=3+2\sqrt{2}$ and $q=3-2\sqrt{2}$. Find in their simplest forms:

a $p+q$ *b* $p-q$ *c* p^2-q^2 *d* pq

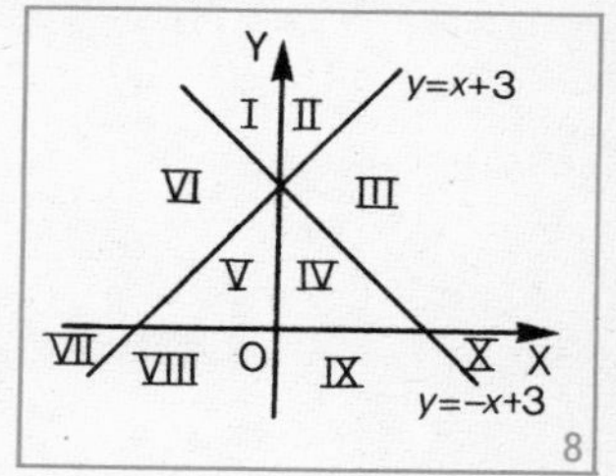

6 a Which regions in Figure 8 show
(*1*) $P=\{(x, y): y\geqslant x+3\}$
(2) $P\cap Q$, if $Q=\{(x, y): y\geqslant -x+3\}$?

b Give a set definition for the region containing VIII and IX only.

7 The stopping distances, d metres, of a car travelling at v m/s, are shown in the table.

v	5	15	30	60
d	5	25	80	280

a Assuming that d and v are connected by the formula $d = av^2+bv$, where a and b are constant, find a and b, and state the formula.
b Use the formula to calculate d when $v=20$, and v when $d=40$.

8 Matrix P represents the geometrical transformation of a rotation of 90° anticlockwise about O, and matrix Q reflection in the line $y=x$.

a Find P, Q, PQ and QP.
b What transformations are represented by PQ and QP?
c Find matrix R such that $PQR = QP$, and describe the transformation given by R.

Geometry

Circle 3: Tangents and Angles

1 *Tangents to a circle*

Exercise 1

1 a Draw a circle and one of its axes of symmetry.

b Draw a chord in the circle so that the whole figure is symmetrical about the axis in ***a***.

c Draw several more chords preserving symmetry about the same axis. What property do all the chords have in common?

d On tracing paper draw a long straight line XY with a point M near the middle of it. Place this line over one of your longer chords with M on the perpendicular diameter and slowly move the tracing so that the line moves away from the centre of the circle. Study its position where it has stopped covering any chords but is still in contact with the circle. Draw a line in this position.

What happens to the two points of intersection with the circle as XY approaches its final position?

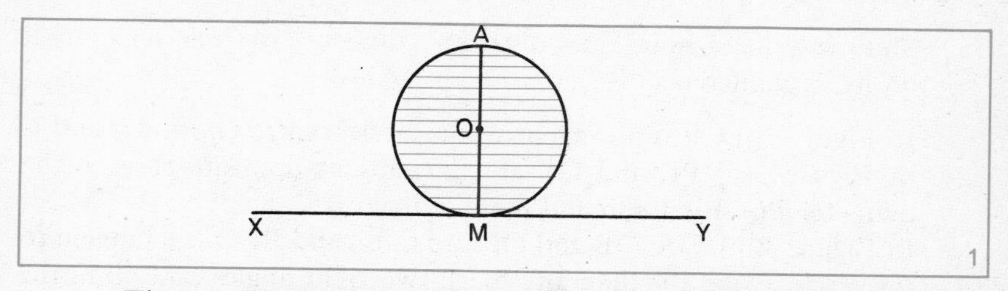

Figure 1 illustrates question *1*. AM is the diameter which is the axis of symmetry, and O is the centre of the circle. XY is the line parallel to the chords, at the end of the diameter.

Since all the chords are perpendicular to AM, XY is also perpendicular to AM and is called a *tangent* to the circle. M is called the *point of contact*.

A tangent to a circle is:

(i) a straight line which meets the circumference at one point only

(ii) perpendicular to the diameter or radius drawn to the point of contact.

From the symmetry of the figure we also see that:

(iii) a perpendicular to a diameter at its end is a tangent to the circle

(iv) a perpendicular to a tangent at its point of contact passes through the centre.

2 Which of the diagrams in Figure 2 shows a circle with a tangent XY?

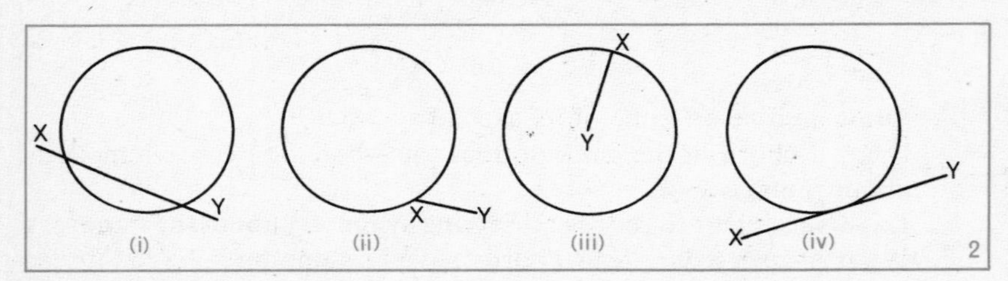

3 Draw a circle and lines which:

a cut the circle in two distinct points P and Q

b touch the circle at one point R

c do not meet the circle at all.

4 If S is the set of points forming the circumference of your circle in question ***3*** and L the set of points forming the straight line, list the members of $S \cap L$ in each of the three cases.

5 Show in a sketch how you could draw a tangent to a circle at a point P on its circumference.

6 a In Figure 3(i) CP is a tangent to the circle, centre O. Find x and y.

b In Figure 3(ii) PR and QS are tangents at opposite ends of the diameter PQ. Find a, b and c.

c In Figure 3(iii), OA, OB and OC are radii, and PCQ is a tangent to the circle. Copy the diagram, mark two right angles and fill in the sizes of all the other angles.

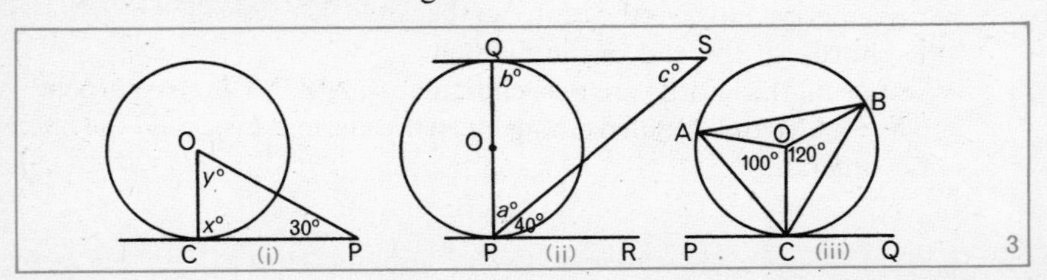

7 In Figure 4, GE is a diameter, and DEF is a tangent to the circle. DE = 8 cm, GE = 6 cm and GF = 6·5 cm. Calculate the lengths of DG and EF.

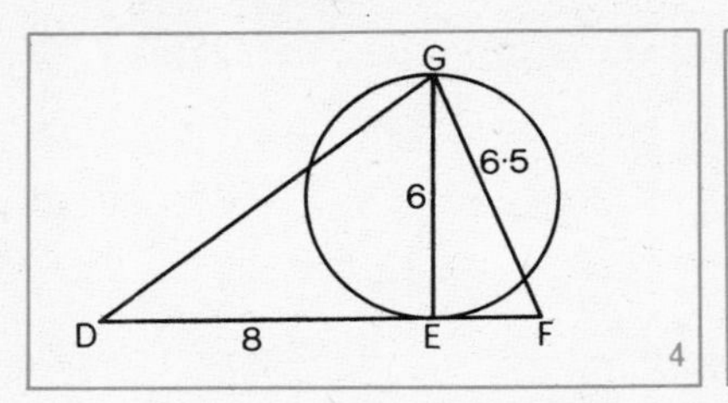

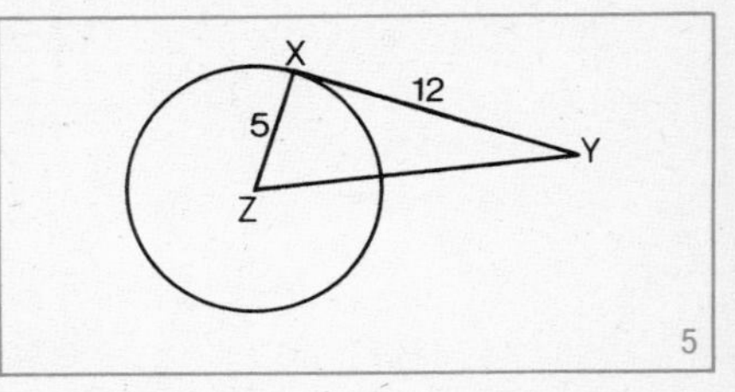

8 Figure 5 shows a circle, centre Z, with radius 5 cm long and a tangent XY 12 cm long.

a How far is Y from the centre of the circle?

b If XY is a rigid rod fixed at X to the rim of a wheel with centre Z, describe the path of Y as the wheel rotates.

c If YX is produced its own length to W, what is the locus of W as the wheel rotates?

9 Draw a circle with diameter UV. Draw tangents to the circle at U and V. Explain why these tangents are parallel.

10 In Figure 6, AB and CD are diameters and PQ, QR, RS and SP are tangents as shown. Explain why PQRS is a parallelogram.

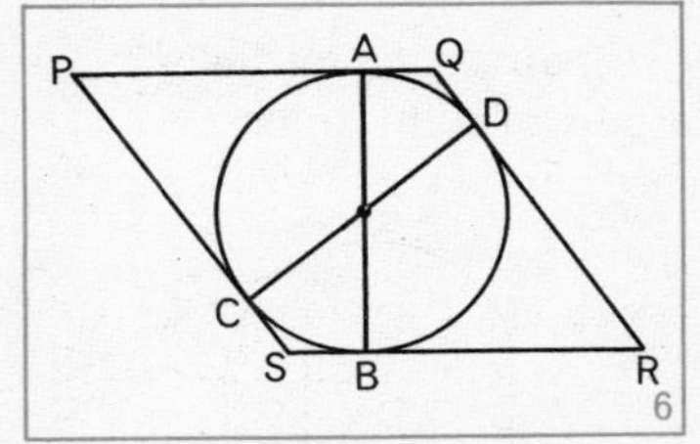

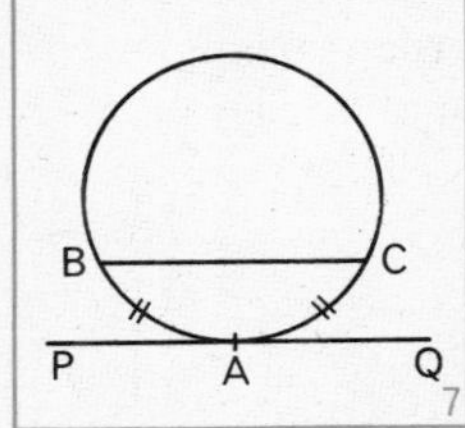

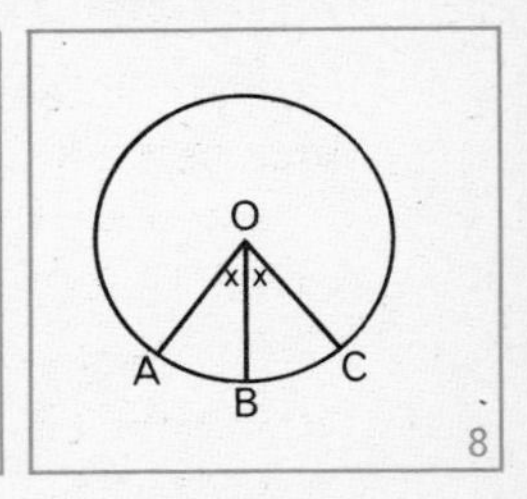

11 In Figure 7, PAQ is a tangent to the circle and arc AB = arc AC. Copy the diagram and draw an axis of symmetry. Explain why chord BC is parallel to PQ.

12 In Figure 8, O is the centre of the circle, and $\angle AOB = \angle BOC$. Copy the diagram and draw an axis of symmetry. Explain why chord AC is parallel to the tangent at B.

2 *The tangent-kite*

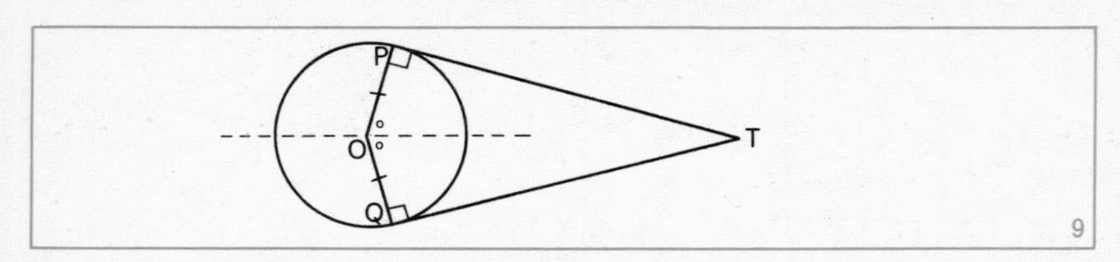

In Figure 9, tangents from T touch the circle, centre O, at P and Q. We can show that OT is an axis of symmetry as follows.

Under reflection in the bisector of ∠POQ,

O↔O

P↔Q (OP = OQ)

The line of PT↔ the line of QT (∠OPT = ∠OQT)

So T↔T

It follows that OT is an axis of symmetry and hence TP = TQ.

OPTQ is called a *tangent-kite*, and has axis of symmetry OT. Two tangents to a circle from a point outside the circle are equal in length.

Exercise 2

1 In Figure 10, ORTS is a tangent-kite. Name:

a one pair of congruent triangles *b* two pairs of equal lines
c three pairs of equal angles *d* a pair of right angles.

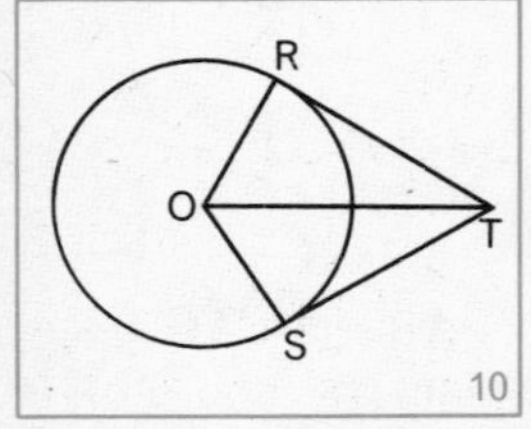

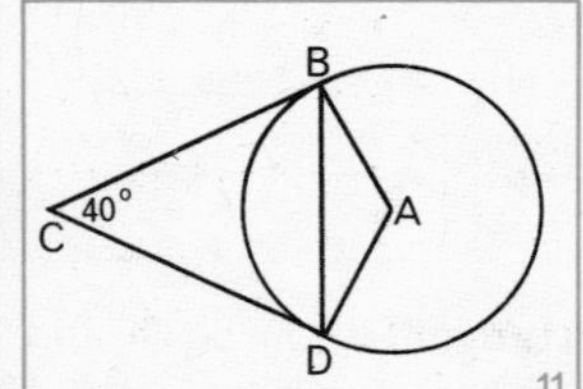

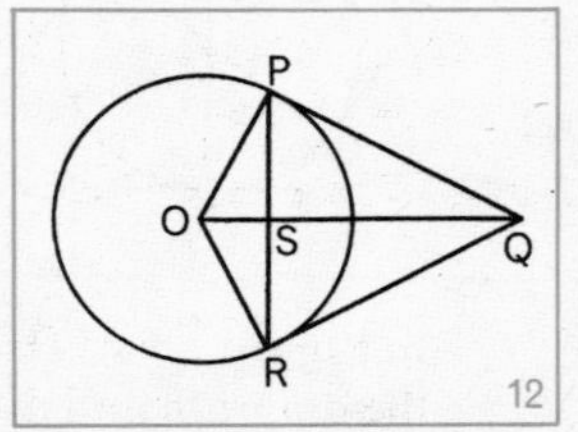

2 If, in Figure 10, OR = 3 cm and OT = 8 cm, calculate the length of the tangents RT and ST to one decimal place.

3 In Figure 11, ABCD is a tangent-kite and ∠BCD = 40°. Name two isosceles triangles and calculate the size of each angle in the diagram.

4 In Figure 12, OPQR is a tangent-kite.

a Name the axis of symmetry and explain why PR is perpendicular to OQ.

b If $\angle$SPQ = 70°, calculate the size of each angle in the diagram.

c If OQ = 7·2 cm and OP = 2·7 cm, calculate the length of the tangents to two significant figures.

5 Look back at Figure 6. We showed that PQRS was a parallelogram.

a This figure has two axes of symmetry. Name them.

b What can you now say about parallelogram PQRS?

6 In Figure 13, tangents BC, CA and AB to the circle touch it at X, Y and Z.

a Name three pairs of lines of equal length in the figure.

b If BZ = 4 cm, XC = 5 cm and AY = 3 cm, calculate the lengths of AB, BC and CA.

c If the perimeter of $\triangle$ABC is 42 cm, BX = 6 cm and CX = 7 cm, calculate the lengths of AB and AC.

d If the perimeter of $\triangle$ABC is 50 cm, AZ = 10 cm and CX = 12 cm, calculate the length of BC.

e If the perimeter of $\triangle$ABC is 48 cm and AZ = 6 cm, calculate the length of BC.

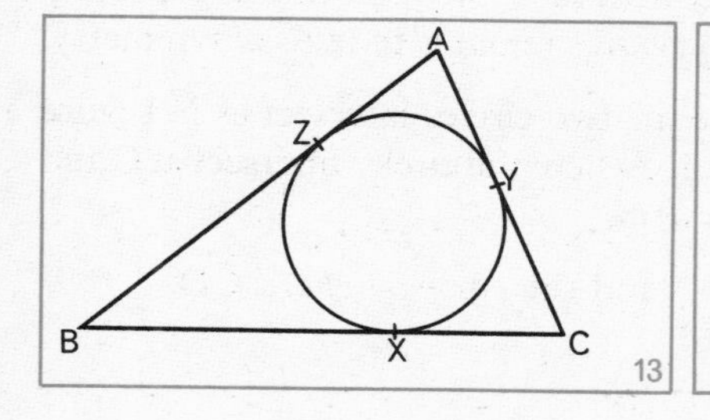

13

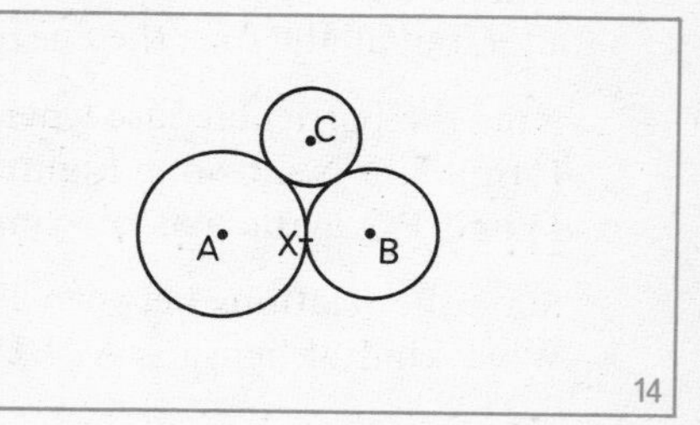

14

7 Three circular discs, with centres A, B and C and radii 5 cm, 4 cm and 3 cm respectively, are in contact as shown in Figure 14.

a If the first two circles touch at X, why are A, X and B collinear? (*Hint*. Think of the common tangent at X.)

b Calculate the lengths of BC, CA and AB, and hence make an accurate drawing of the arrangement.

c If $\triangle$ABC in Figure 14 has to have BC = 6 cm and AB = AC = 8 cm, calculate the radii of the three touching circles.

3 *Symmetry of two circles Common tangents*

Exercise 3

1 Figure 15 shows five pairs of circles with centres A and B.

a Describe the relative positions of the circles in each pair.
b How many axes of symmetry has each diagram?

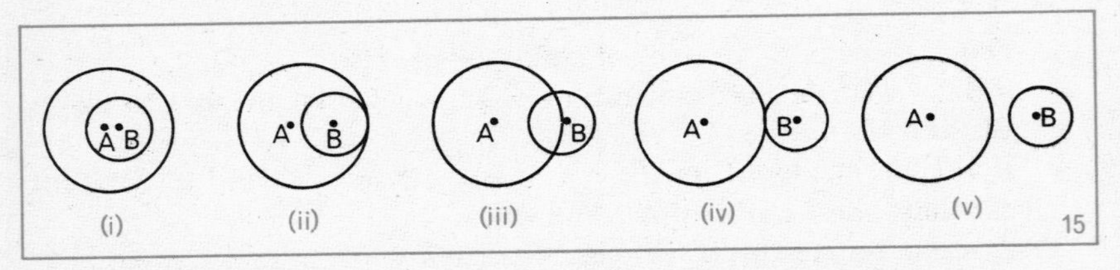

2 Draw a circle in your notebook and a smaller circle on tracing paper; mark their centres A and B. Place the tracing paper over your page with the smaller circle inside the larger, and gradually move the smaller outside the larger. Note how the relative positions gradually alter, but all the time the *line of centres* remains the axis of symmetry.

3 Note the particular case where the two circles intersect as in Figure 15(iii). This is shown in Figure 16, where the circles intersect at C and D and PQ is the axis of symmetry.

a State the relations between PQ and the *common chord* CD.
b What kind of figure is ACBD?

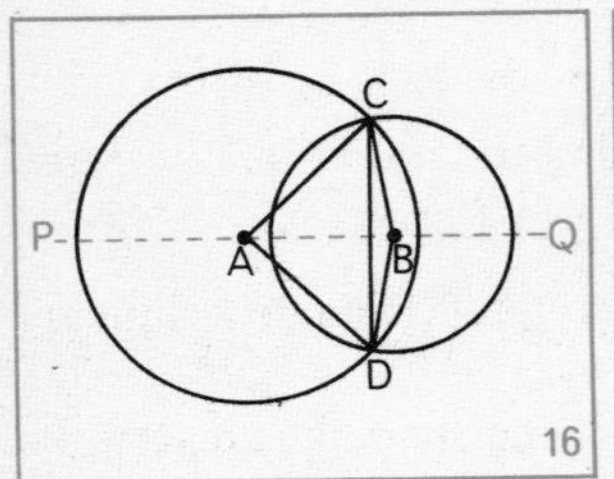

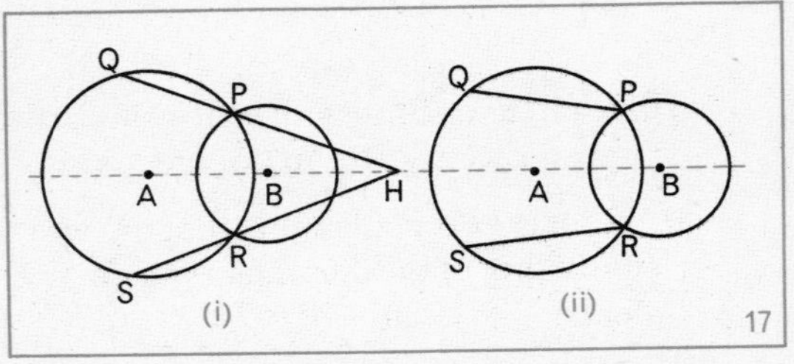

The *line of centres* of two circles is their axis of symmetry; it bisects the common chord of two intersecting circles at right angles.

4 In Figure 17(i), H is a point on the axis of symmetry of two circles, centres A and B. HP and HR cut the circle, centre A, again in Q and S respectively.

a Under reflection in AB, write down the images of H, P and Q.
b What can you say about (*1*) PR and QS (*2*) PQ and RS?

5 In Figure 17(ii) two circles, centres A and B, intersect at P and R. QS is parallel to PR.

a Write down the images of P and Q under reflection in AB.
b If QP and SR are produced to meet at T, what can you say about T?

6 In Figure 18(i) AB is a common tangent to two circles with centres P and Q, and meets the line of centres at T. C and D are images of A and B under reflection in PQ.

a Why are T, D and C collinear?
b How is TDC related to the two circles?
c What can you say about AB and CD?

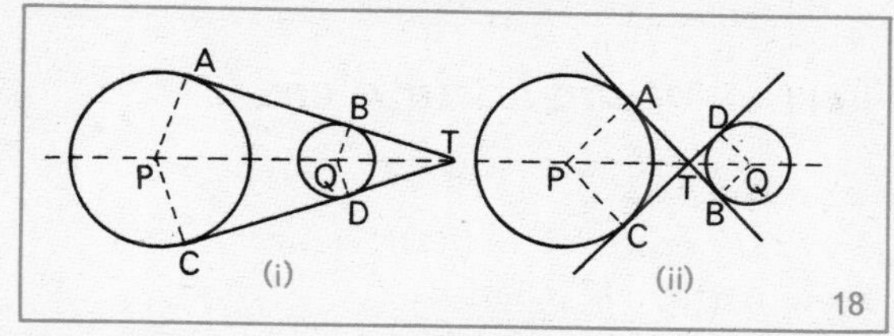

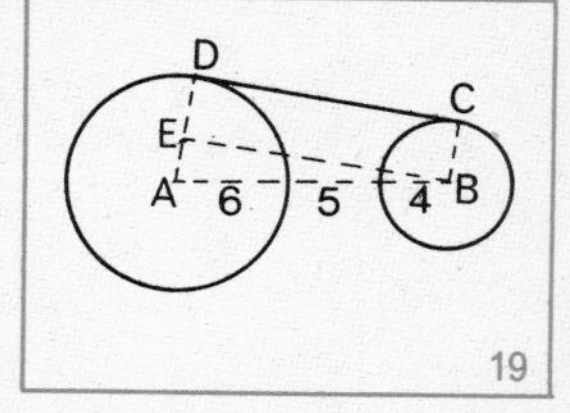

7 Repeat question ***6*** for Figure 18(ii).

Common tangents to two circles touch both circles. Figure 18(i) shows *direct* common tangents and Figure 18(ii) *transverse* common tangents. Direct common tangents are equal in length and transverse common tangents are equal in length.

8 a If, in Figure 18(i), the radii are 5 cm and 2 cm, state the centre and scale factor of the dilatation which maps the smaller circle onto the greater. If PQ = 6 cm, calculate the length of TP.
b Repeat for Figure 18(ii).

9 Copy Figure 15, parts (ii)–(v) and draw in all the common tangents. Make sure that you find ten common tangents altogether.

10 In Figure 19 we can calculate the length of the common tangent CD by drawing BE perpendicular to AD as shown. The circles have radii 6 cm and 4 cm long, and AB is 15 cm long.

a Explain why BCDE is a rectangle.
b Write down the lengths of BC, DE, AD and AE.
c Use the theorem of Pythagoras to calculate the length of BE and hence CD.

11 Repeat question ***10*** for circles with radii 11 cm and 4 cm long and AB 25 cm long.

12 Repeat question ***10*** for circles with radii 2 cm and 12 cm long and AB 26 cm long.

13 Two circles with centres A and B touch each other externally (as in Figure 15(iv)) at C. XCY is a common tangent and DCE is a *double chord* with D on one circumference and E on the other.

a Prove that: (*1*) ACB is a straight angle (*2*) ∠ADC = ∠BEC.
b Why is DA parallel to BE?

4 *Properties of angles in a circle*

(*i*) *Angles at the centre and circumference subtended by the same arc*

Exercise 4

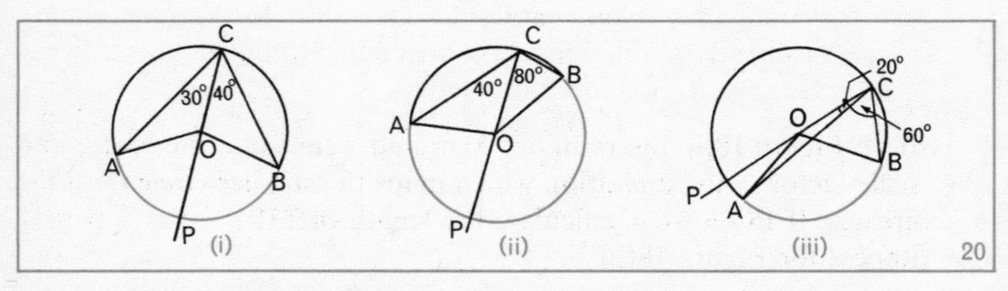

20

1 In Figure 20, O is the centre of each circle.

a Copy Figure 20(i) and fill in the sizes of all the angles.
b Repeat for Figure 20(ii). *c* Repeat for Figure 20(iii).
State any connection you see between angles AOB and ACB 'standing on' the coloured arc AB in all three cases.

2 In Figure 21(i), (ii) and (iii), calculate the sizes of all the angles in terms of x and y. Is it still true that $\angle AOB = 2\angle ACB$?

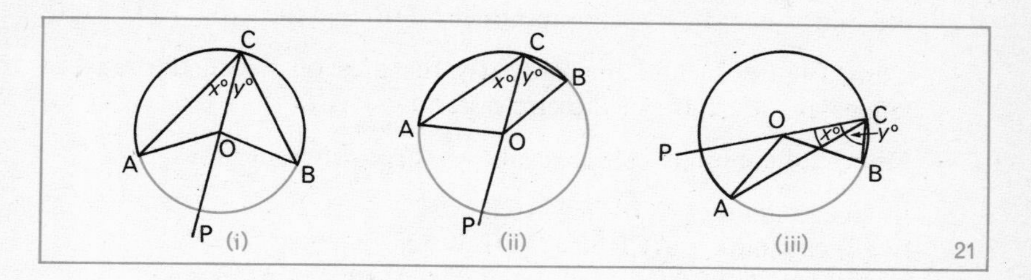

We can amplify question *2* as follows.

$\angle ACO = x°$ and $\angle CAO = x°$, so
$\angle AOC = (180-2x)°$ and $\angle AOP = 180° - (180-2x)° = 2x°$.
Similarly $\angle POB = 2y°$.

In Figures 21(i) and (ii),

$$\begin{aligned}\angle AOB &= \angle AOP + \angle POB\\ &= 2x° + 2y°\\ &= 2(x+y)°\\ &= 2\angle ACB\end{aligned}$$

In Figure 21(iii)

$$\begin{aligned}\angle AOB &= \angle POB - \angle POA\\ &= 2y° - 2x°\\ &= 2(y-x)°\\ &= 2\angle ACB\end{aligned}$$

An angle at the centre of a circle is double any angle at the circumference subtended by (*or* standing on) the same arc.

Note. In Figure 21(ii) the angles are subtended by the *major arc* AB; in Figures 21(i) and (iii) the angles are subtended by the *minor arc* AB. If we wish to name the angle at the centre in Figure 21(ii), we call it the *reflex angle* AOB.

A *reflex angle* is greater than two right angles and less than four right angles.

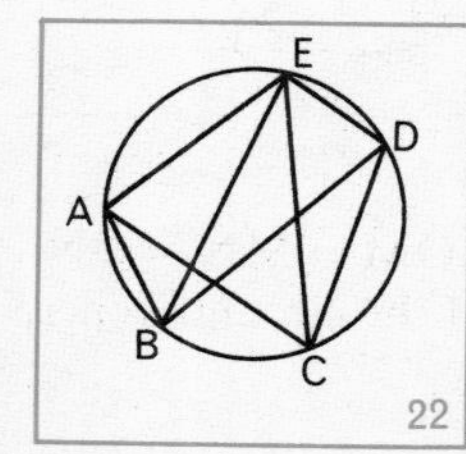

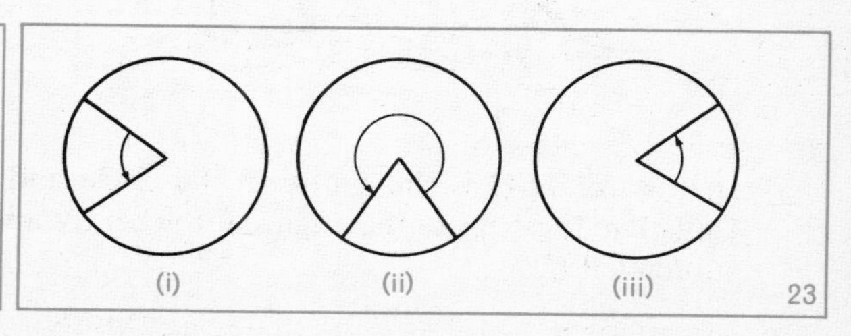

3 In Figure 22 name two angles at the circumference which stand on (or are subtended by):

a the minor arc BC *b* the minor arc DE *c* the arc AED.

4 Copy Figure 23 and mark clearly the arcs on which the marked angles at the centre of the circle stand.

5 Figure 24 summarises the result for three different cases.

a In (i) calculate $\angle AOB$ if $\angle ACB = 38°$.
b In (ii) calculate $\angle ACB$ if $\angle AOB = 52°$.
c In (iii) calculate reflex $\angle AOB$ and $\angle AOB$ if $\angle ACB = 115°$.

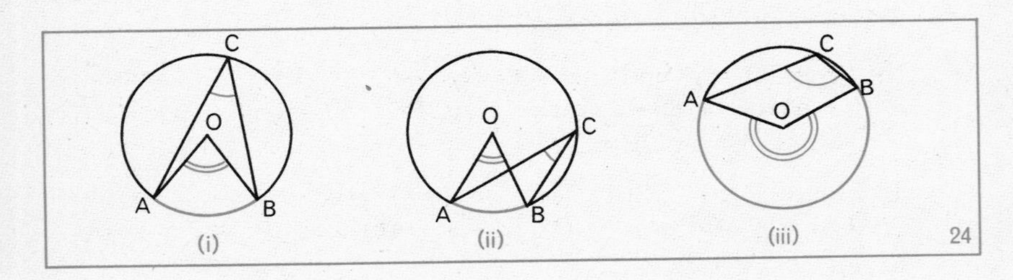

24

6 In Figure 25, O is the centre of the circle. Calculate the size of:

a $\angle MKN$ (standing on the minor arc MJN)
b $\angle MJN$ (standing on the major arc MKN; think of the size of the corresponding angle at the centre of the circle).

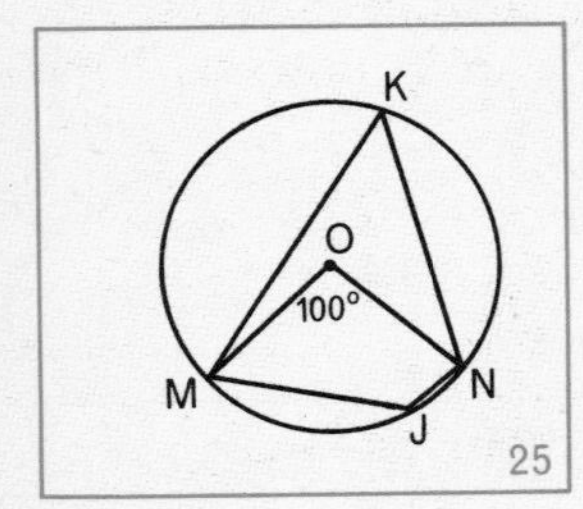

25

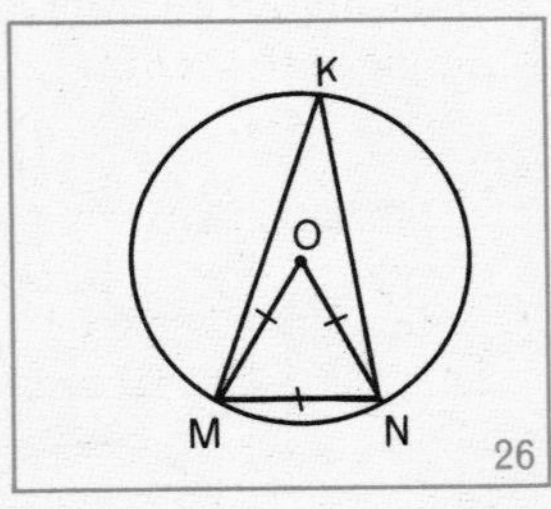

26

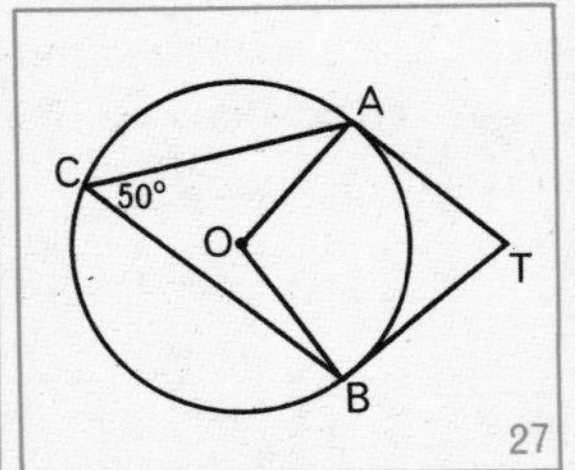

27

7 In Figure 26, O is the centre of the circle and OM = MN = ON. Calculate the sizes of the angles at the centre and the circumference subtended by:

a the minor arc MN *b* the major arc MKN.

8 In Figure 27, TA and TB are tangents, and O is the centre of the circle.

a If $\angle ACB = 50°$, calculate the size of $\angle ATB$.
b If $\angle ACB = x°$, calculate the size of $\angle ATB$ (in terms of x).

9 The vertices of a regular pentagon inscribed in a circle divide the circumference into five equal arcs. Calculate the sizes of the angles at the centre and at the circumference subtended by one of these arcs.

10 Repeat question ***9*** for a regular octagon (8 sides), decagon (10 sides) and n-gon (n sides).

11 $\triangle XYZ$ is inscribed in a circle with centre O, and OY and OZ are joined. If $\angle YXZ = 60°$, calculate the sizes of the angles of $\triangle YOZ$.

12 $\triangle PQR$ is inscribed in a circle with centre O, and OP, OQ and OR are joined. If $\angle POQ = 110°$ and $\angle QOR = 130°$, calculate the sizes of the angles of $\triangle PQR$.

(ii) Angles in a semicircle

Exercise 5

1 In Figure 28, O is the centre of the circle and AOB is a diameter, which makes arc ACB a semicircle.
a What is the size of $\angle AOB$? *b* What is the size of $\angle ACB$?

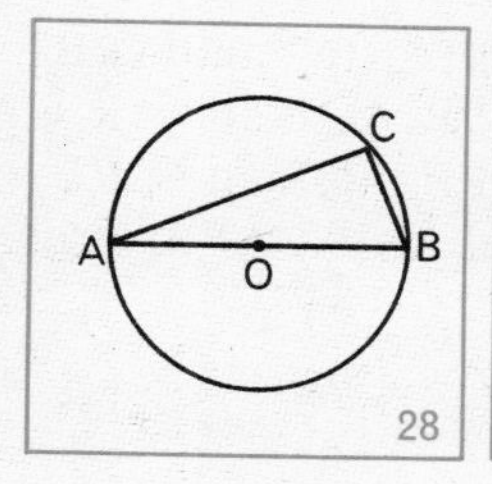

28

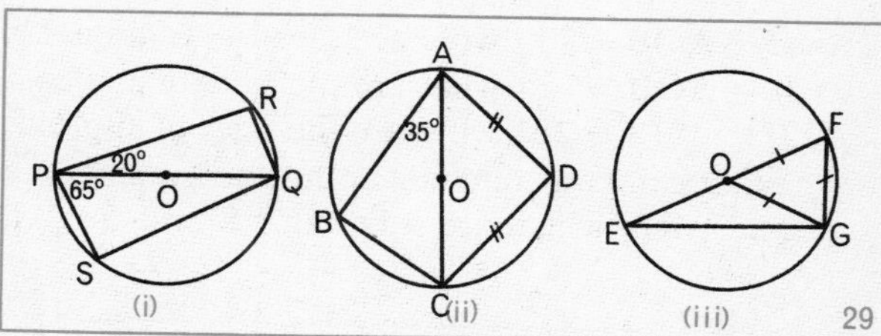

(i) (ii) (iii) 29

Every angle in a semicircle is a right angle.

2 Copy the diagrams in Figure 29, where O is the centre of the circle in each case, and mark in the sizes of all the angles.

3 Draw two diameters AOB and COD in a circle. Explain why ACBD is a rectangle.

4 Using compasses, draw a circle with centre O and radius 2 cm. Mark a point T 6 cm from O. Join OT and with your compasses draw a circle on OT as diameter to cut the first circle at A and B. Join TA and TB, and explain why these are tangents from T to the first circle. (This gives a method of constructing tangents to a circle from an *external* point.)

(iii) Angles in the same segment

A *segment* of a circle is a region of the circle bounded by a chord and an arc. Every chord divides a circle into two segments.

Angles at the circumference of a circle standing on the same arc are often described as *angles in the same segment*. Note that it is always possible to name angles subtended by the same arc beginning and ending with the same letter, e.g. in Figure 30, angles **APB, AQB, ARB** at the circumference and angle **AOB** at the centre.

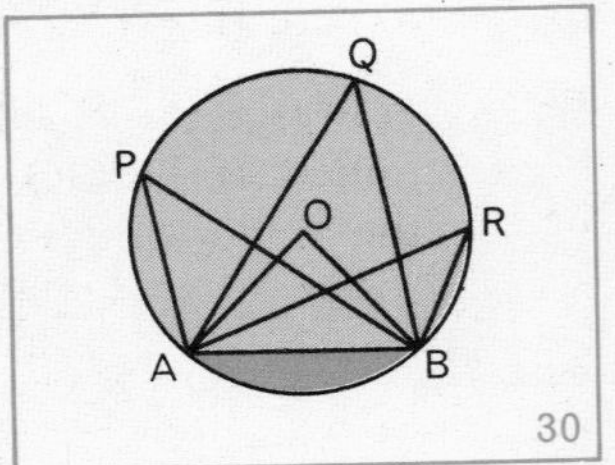

30

Exercise 6

1 In Figure 31, O is the centre of the circle.

a If $\angle AOB = 80°$, what are the sizes of angles APB, AQB and ARB?
b If $\angle AOB = 72°$, what are the sizes of angles APB, AQB and ARB?
c If $\angle AOB = 2x°$, what are the sizes of angles APB, AQB and ARB?
d What can you conclude about the sizes of *all* angles like APB, AQB and ARB, which are subtended by the arc AB?

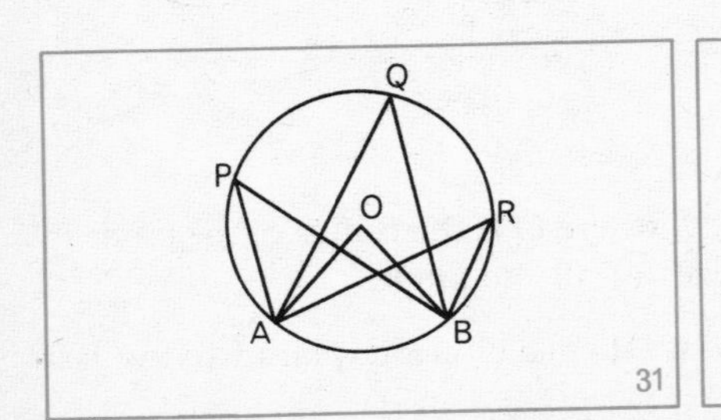

31

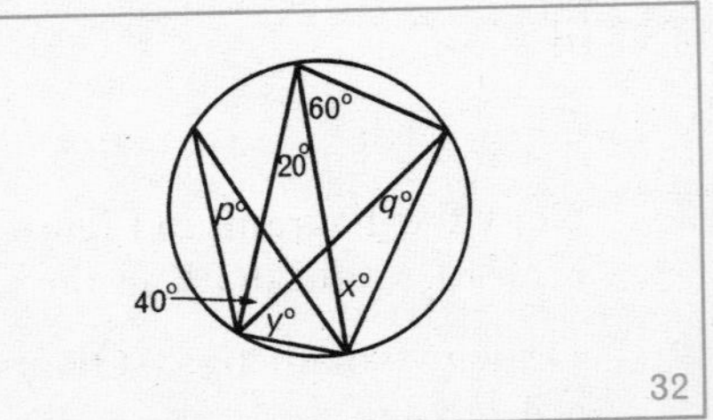

32

Angles in the same segment of a circle are equal.

2 In Figure 32, find the values of x, y, p and q.

3 Copy Figure 33(i) and mark four pairs of equal angles at the circumference.

4 Copy Figure 33(ii) and fill in the sizes of all the angles in the diagram.

5 Copy Figure 33(iii) and fill in the sizes of all the angles in the two triangles. Why are the triangles similar?

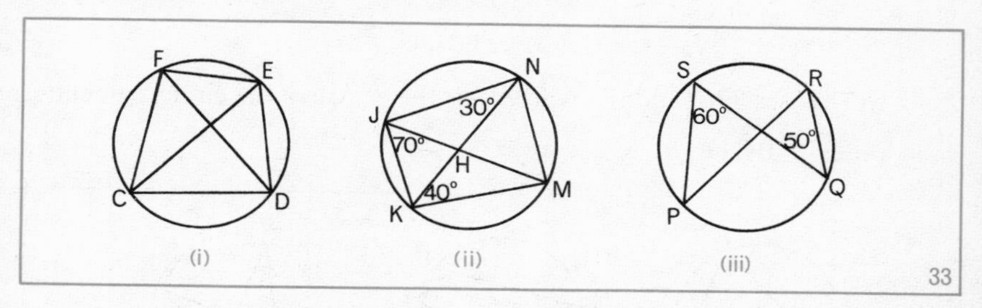

6 In Figure 34, XY is parallel to AB. Calculate the sizes of as many angles as possible.

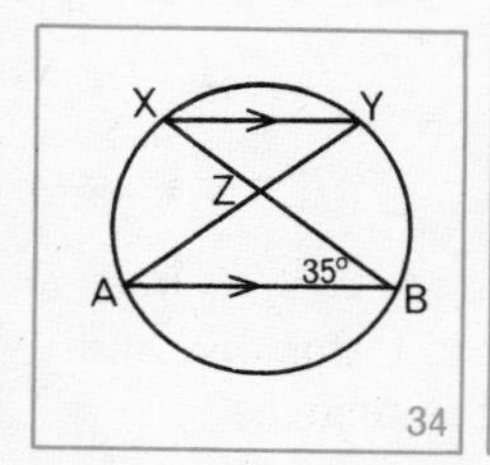

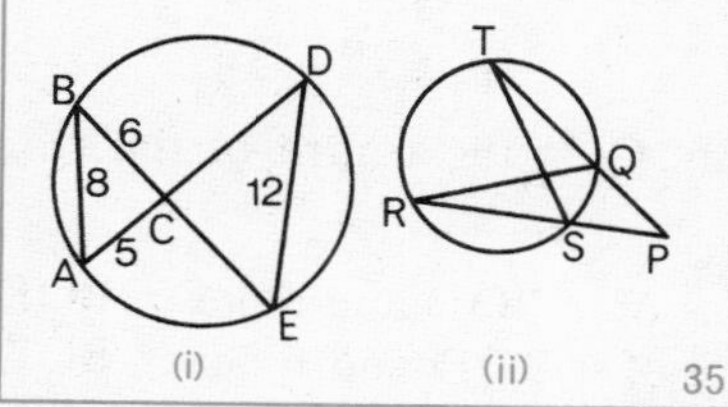

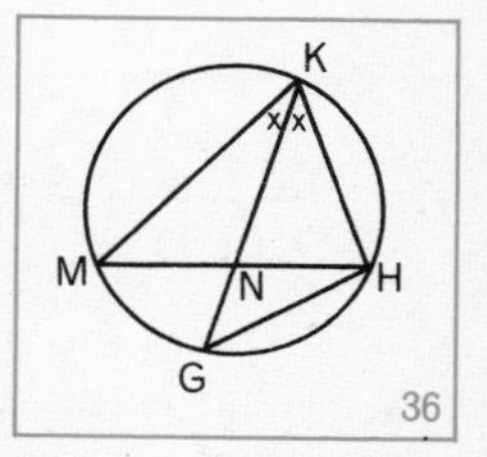

7 Prove that triangles ABC and EDC in Figure 35(i) are similar. If AB = 8 cm, BC = 6 cm, AC = 5 cm and DE = 12 cm, calculate the lengths of CD and CE.

8 Prove that triangles PQR and PST in Figure 35(ii) are similar. If TQ = 6 cm, QP = 4 cm and PR = 8 cm, calculate the length of PS.

9 In Figure 36, $\angle$MKN = $\angle$GKH. Prove that triangles MKN and GKH are similar. Name another pair of similar triangles in the figure, and prove that they are similar.

5 *Cyclic quadrilaterals*

A quadrilateral whose vertices lie on the circumference of a circle is called a *cyclic quadrilateral.* The vertices are said to be *concyclic points.* (*Cyclic* comes from the Greek word for circle; compare bicycle.) It will be seen from some of the examples later in this section that the actual circle is not always drawn.

Exercise 7

1 In Figures 37(i) and (ii), where O is the centre of the circle, calculate:

a reflex $\angle$BOD *b* x *c* y *d* $x+y$.

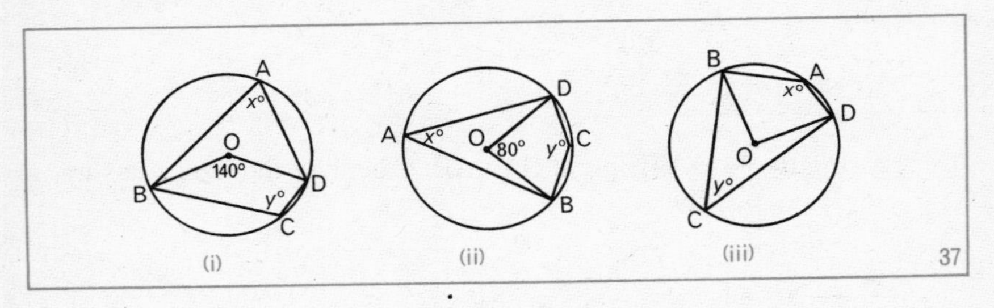

2 In Figure 37(iii), O is the centre of the circle, $\angle$BAD $= x°$ and $\angle$BCD $= y°$.

a Express the sizes of $\angle$BOD and reflex $\angle$BOD in terms of x and y.
b What is the sum of $\angle$BOD and reflex $\angle$BOD in degrees?
c What is the sum of $\angle$BOD and reflex $\angle$BOD in terms of x and y?
d What is the sum of x and y?

The answers to question **2** demonstrate that in all cases $x+y=180$. Hence $\angle$BAD $+ \angle$BCD $= 180°$ and $\angle$ABC $+ \angle$ADC $= 180°$, since the sum of the angles of a quadrilateral is 360°.

The opposite angles of a cyclic quadrilateral are supplementary.

3 Calculate the sizes of as many angles as possible in each part of Figure 38. In (ii), O is the centre of the circle.

4 Calculate the sizes of all the angles in Figure 39, in which ABC and DEF are double chords.

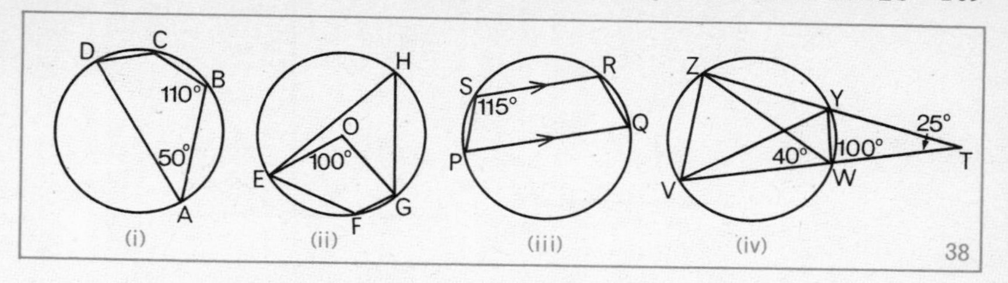

5 In Figure 40, calculate the size of ∠ABC in two different ways.

6 In Figure 41, ABCD is a parallelogram. Prove that △ABK is isosceles by showing that two of its angles are equal.

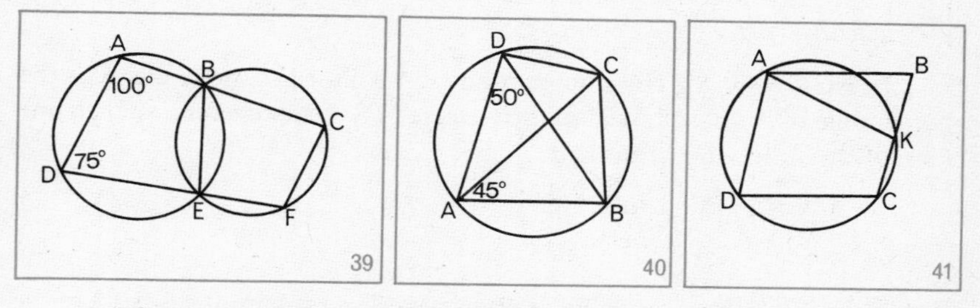

The dependence of the results of Section 5 on those of Section 4 may be seen by arrangement in a kind of family tree, as follows.

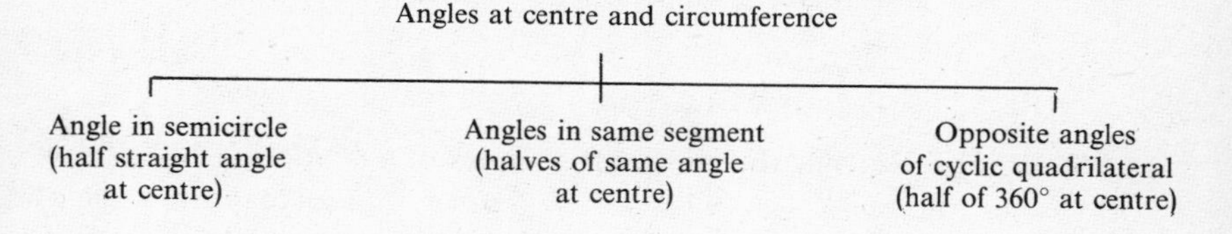

Exercise 7B

1 In Figure 42, PQRS is a quadrilateral inscribed in a circle, and PQ is produced to T. What is the relation between:

a ∠RQT and ∠RQP *b* ∠RSP and ∠RQP

c ∠RQT and ∠RSP?

We could state this result as a theorem.

An exterior angle of a cyclic quadrilateral is equal to the interior opposite angle.

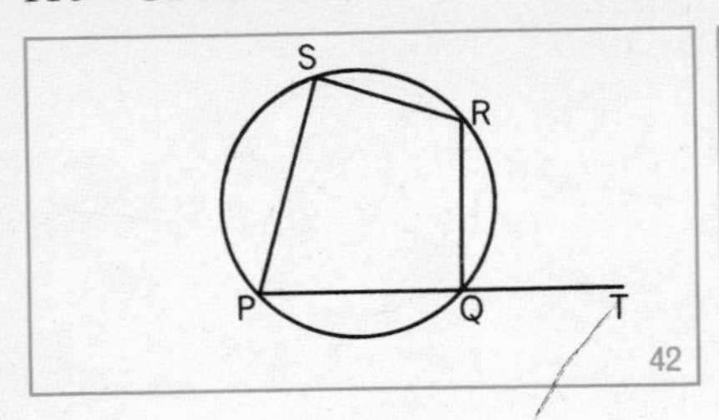

42

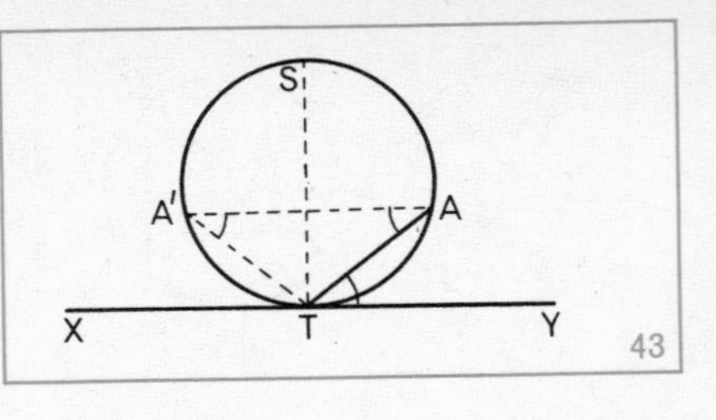

43

2 Suppose that in question *1*, Q moves round the circle to P. What does the line PQT become? Name two angles you think are equal in the resulting diagram. Suggest a further angle property of the circle arising from this special case.

3 In Figure 43, XTY is a tangent and TA a chord of the circle. A′ is the image of A under reflection in diameter TS. Give reasons for each line of the following argument.

A′A is parallel to XTY.
$\angle ATY = \angle A'AT$
$= \angle AA'T$
= any angle in segment AA′T on the minor arc TA.

4 Use the result of question *3* and supplementary angles to show that $\angle XTA$ = any angle in the opposite segment.

An angle between a tangent and a chord through the point of contact is equal to any angle in the opposite segment.

5 Remembering that the sum of the angles of a triangle is 180°, show that in Figure 44, $\angle ACD = \angle A + \angle B$; i.e. that an exterior angle of the triangle is equal to the sum of the two interior opposite angles.

It follows that $\angle ACD > \angle A$ and $\angle ACD > \angle B$; i.e. an exterior angle of a triangle is greater than an interior opposite angle.

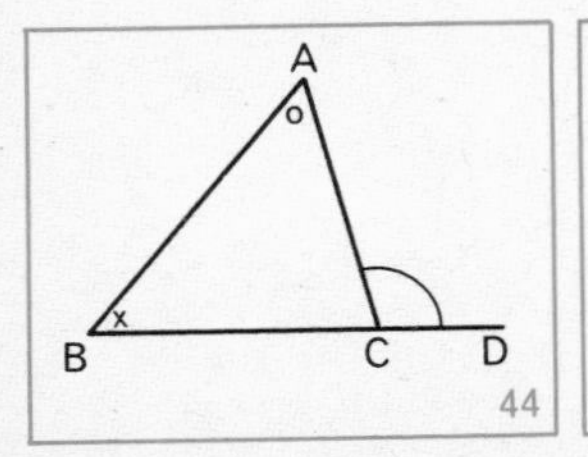

44

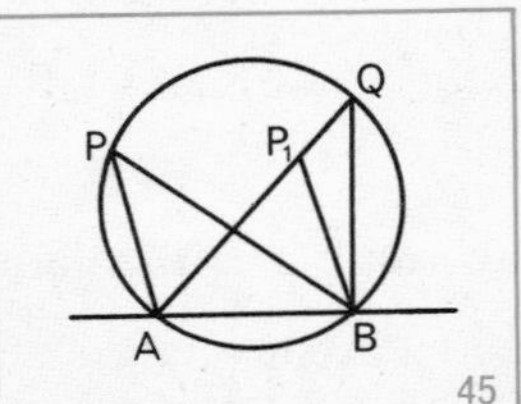

45

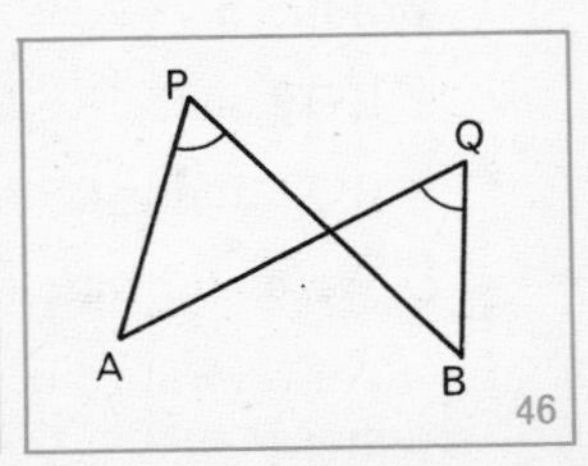

46

6 Using the result of question ***5*** prove that, in Figure 45, $\angle AP_1B > \angle APB$.

Is it possible to find a point P_1 inside the segment APB of the circle for which $\angle AP_1B$ is not greater than $\angle APB$?

Describe the set $\{P_1 : \angle AP_1B > \angle APB; P_1$ on the same side of AB as P$\}$ in another way.

7 If, in Figure 45, P_2 is a point above AB and outside the circle (on AQ produced, for example), prove that $\angle AP_2B < \angle APB$.

Describe $\{P_2 : P_2$ lies outside the circle on the same side of AB as P$\}$ in another way.

8 In Figure 46, if A, P, Q and B are concyclic points, then $\angle APB = \angle AQB$. State the converse of this theorem. Is it true? (Think of questions ***6*** and *7*.)

9 Use the converse you stated in question ***8*** to prove that:

a In Figure 47(i), if $\angle B + \angle D = 180°$, then ABCD is cyclic (*Hint*. Make AECD cyclic and compare angles ABC and AEC.)

b In Figure 47(ii), if $\angle ACB$ is a right angle, then C lies on the circle with AB as diameter. (*Hint*. Make ADB an angle in a semicircle and compare angles ACB and ADB.)

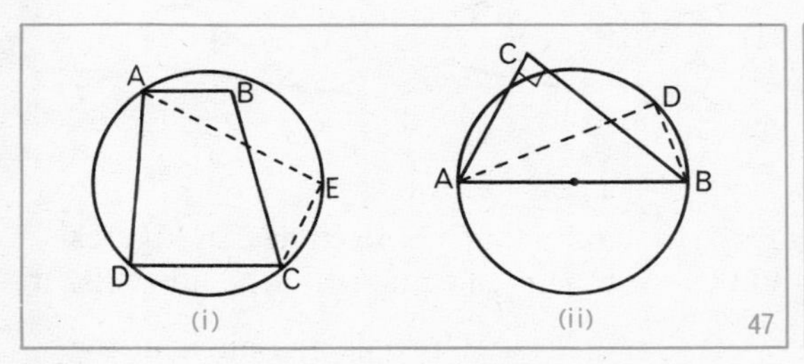

(i) (ii) 47

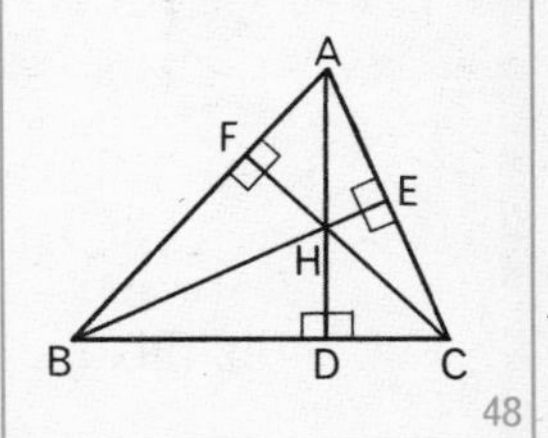

48

10 Figure 48 shows a triangle with its three altitudes intersecting at H.

a Explain why AEHF is a cyclic quadrilateral. Find two others like AEHF.

b Explain why B, F, E, C are concyclic points. Find two other sets like B, F, E, C.

c Where are the centres of the circles that can be drawn through these six sets of concyclic points?

Summary

1 *A tangent to a circle* is a straight line which
(i) meets the circle at one point only
(ii) is perpendicular to the diameter or radius to the point of contact.

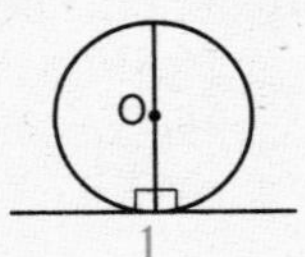

1

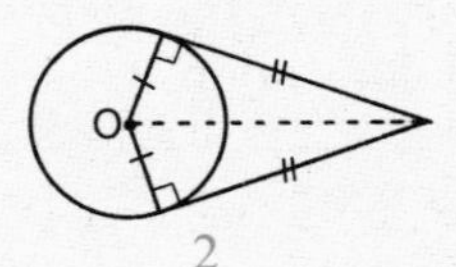

2

2 *Two tangents to a circle* from a point outside the circle are equal; with two radii they form a *tangent-kite*.

3 *The line of centres of two circles* is an axis of symmetry, and bisects the common chord at right angles.

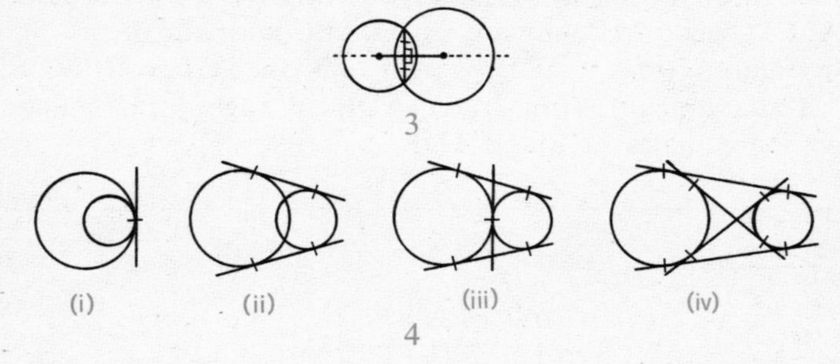
3

(i) (ii) (iii) (iv)
4

4 *Common tangents* to two circles touch both circles. In Figures 4(ii), (iii) and (iv) there are pairs of equal common tangents.

5 *An angle at the centre of a circle* is double any angle at the circumference standing on the same arc.

6 *Every angle in a semicircle* is a right angle.

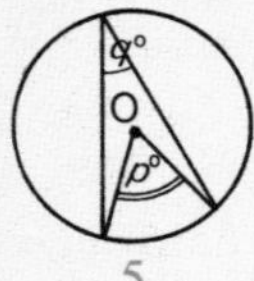

5

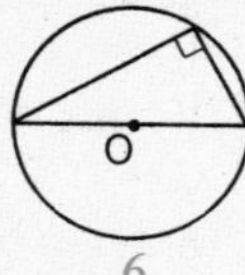

6

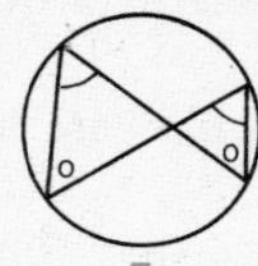
7

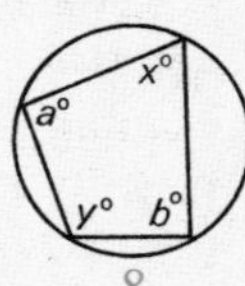

8

7 *Angles in the same segment* **are equal.**

8 *Opposite angles of a cyclic quadrilateral* **are supplementary.**

9 *An exterior angle of a cyclic quadrilateral* **is equal to the interior opposite angle.**

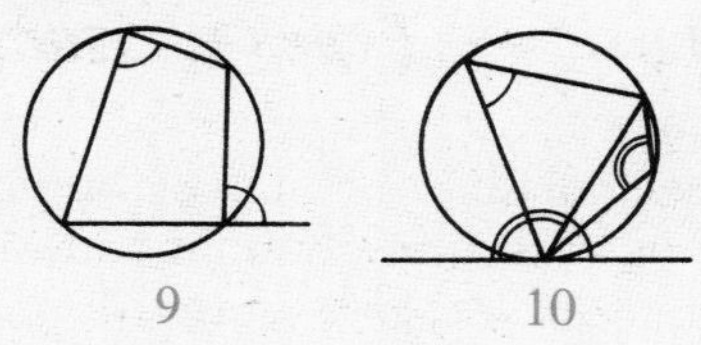

10 *An angle between a tangent and a chord through the point of contact* **is equal to any angle in the opposite segment.**

Composition of Transformations 1

1 Revision—Types of transformation

We have now met four different kinds of transformation of points in a plane: translation, reflection, rotation and dilatation. Translation, reflection and rotation are *isometric* transformations, i.e. under these transformations a figure and its image are *congruent*. Under dilatation a figure and its image are *similar* (see Figure 1).

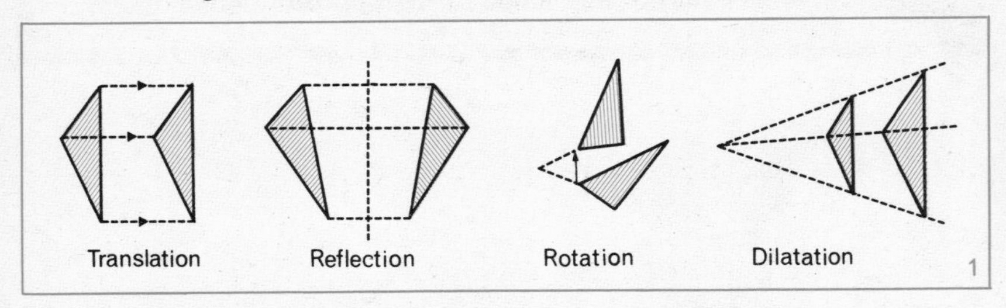

1

A *translation* is defined in *magnitude* and *direction* by a directed line segment, e.g. $\overrightarrow{PQ}$, or by a number pair, e.g. $\begin{pmatrix} a \\ b \end{pmatrix}$.

A *reflection* is defined if we know the *axis of reflection*.

A *rotation* is defined by its *centre*, *magnitude* and *sense* (e.g. clockwise or anticlockwise).

A *dilatation* is defined by its *centre* and *scale factor*, e.g. [P, 3].

Exercise 1

1 Copy Figure 2 on squared paper and draw the image of triangle ABC under the translation represented by $\overrightarrow{XY}$.

2 Copy Figure 3 on squared paper and draw the image of triangle DEF under reflection in ST.

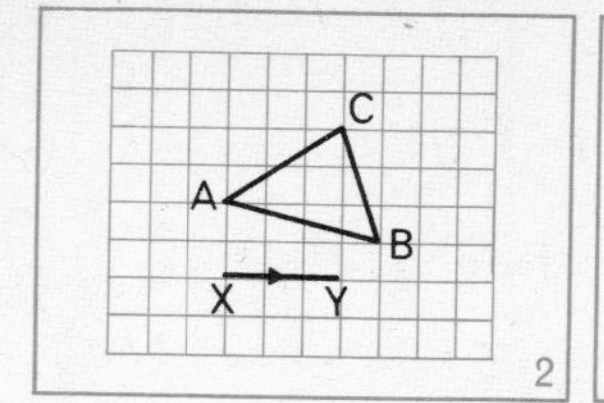

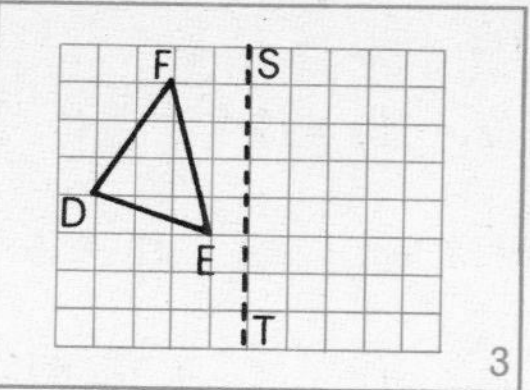

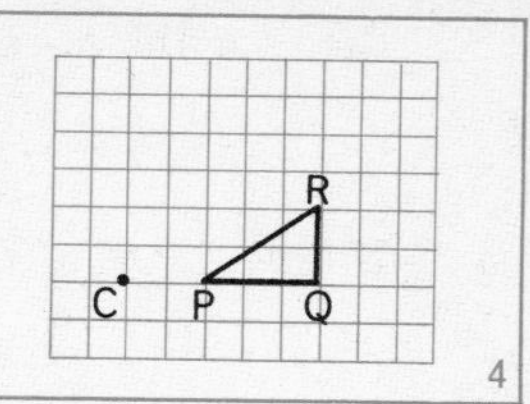

3 Copy Figure 4 on squared paper and draw the image of triangle PQR under:

a an anticlockwise rotation of 90° about C
b the dilatation [C, 3].

4 A triangle has vertices at O (0, 0), A(4, 2) and B(2, −1). On squared paper draw the image of triangle OAB under each of the following transformations. Write down the coordinates of the images of O, A and B in each case.

a Translation $\begin{pmatrix}3\\1\end{pmatrix}$ *b* Dilatations [O, 3] and [O, $\frac{1}{2}$]
c Reflection in the y-axis *d* Reflection in the x-axis
e Reflection in $y = x$ *f* Clockwise rotation of 90° about O.

2 *Two translations*

In the chapter on Translation in Book 4, the composition of two successive translations was carried out as shown in Figure 5:
(i) by a head-to-tail arrangement of directed line segments representing the translations, and
(ii) by addition of components in the number pairs which define the translations.

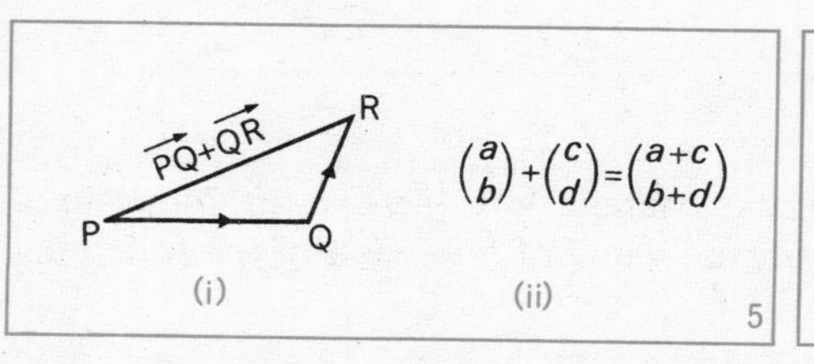

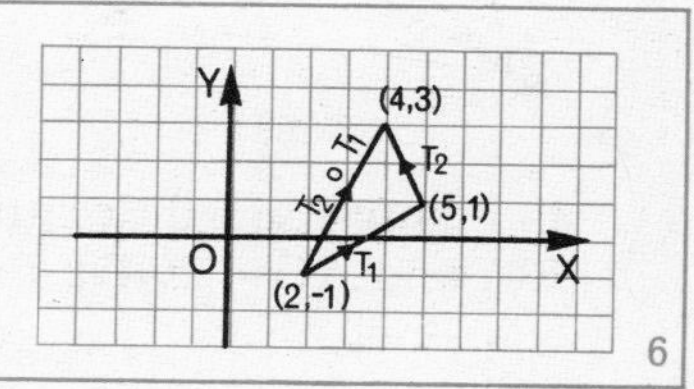

A new notation. If T_1 is a translation $\begin{pmatrix}3\\2\end{pmatrix}$ in component form, then for the point $(2, -1)$,

$$T_1(2, -1) = (2+3, -1+2) = (5, 1).$$

If now T_2 is a translation $\begin{pmatrix} -1 \\ 2 \end{pmatrix}$, then for the point $(5, 1)$

$$T_2(5, 1) = (5-1, 1+2) = (4, 3).$$

From the above, T_1 followed by T_2 maps the point $(2, -1)$ to $(4, 3)$ as shown in Figure 6. The composition of transformations 'T_1, then T_2' is often written $T_2 \circ T_1$.

$$T_2 \circ T_1 = \begin{pmatrix} -1 \\ 2 \end{pmatrix} + \begin{pmatrix} 3 \\ 2 \end{pmatrix} = \begin{pmatrix} 2 \\ 4 \end{pmatrix} \text{ so that}$$

$$(T_2 \circ T_1)(2, -1) = (2+2, -1+4) = (4, 3).$$

Combining two translations gives the same result as performing the translations one after the other.

Note. When using this notation you should remember that the transformation written second, *next to the point to be transformed*, is performed first, i.e. $T_2 \circ T_1(a, b)$ means $T_2[T_1(a, b)]$

Exercise 2

1 In Figure 7, PQRS is a parallelogram. Find single representatives of the translations represented by:

a $\vec{PQ} + \vec{QR}$ *b* $\vec{PQ} + \vec{PS}$ *c* $\vec{PQ} + \frac{1}{2}\vec{QS}$ *d* $\frac{1}{2}\vec{PR} + \frac{1}{2}\vec{SQ}$.

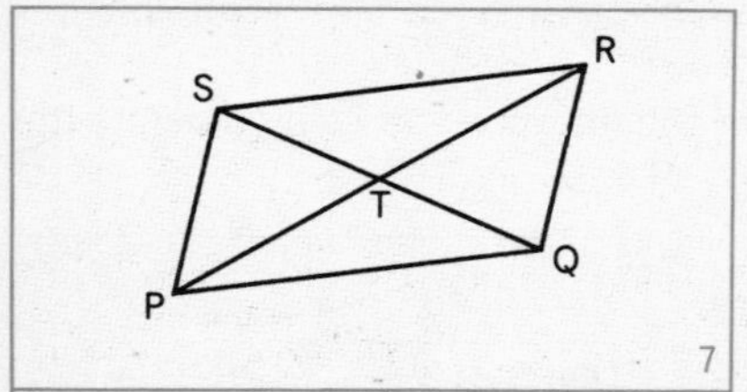

7

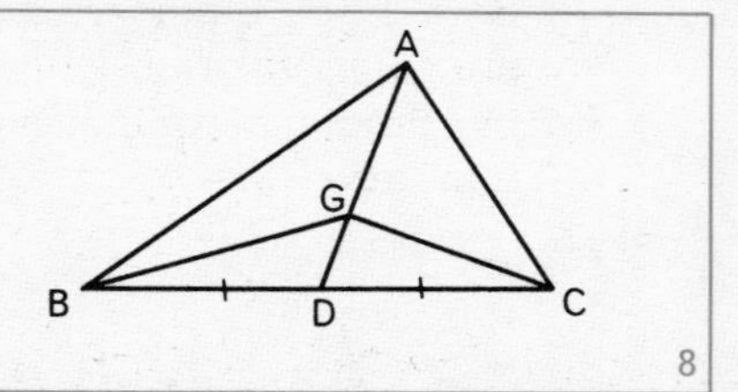

8

2 In Figure 8, G is the centroid of triangle ABC (i.e. D is the midpoint of BC, and AG = 2GD). Find the value of k in each of the following equations.

a $\vec{AG} = k\vec{AD}$ *b* $\vec{AD} = \vec{AB} + k\vec{BC}$

c $\vec{BG} = \frac{1}{2}\vec{BC} + k\vec{DA}$ *d* $\vec{GC} = k\vec{DA} + \vec{AC}$

3 If, in Figure 7, $\overrightarrow{PQ}$ represents the translation $\begin{pmatrix}4\\1\end{pmatrix}$ and $\overrightarrow{QS}$ the translation $\begin{pmatrix}-3\\2\end{pmatrix}$, express in component form the translations represented by: *a* $\overrightarrow{PS}$ *b* $\overrightarrow{PR}$ *c* $\overrightarrow{PT}$.

4 Translation $T_1 = \begin{pmatrix}1\\1\end{pmatrix}$ and translation $T_2 = \begin{pmatrix}2\\2\end{pmatrix}$.

(3, −1) is a given point.

a Find $T_1(3, -1)$ and *hence* $T_2[T_1(3, -1)]$.

b Express $T_2 \circ T_1$ as a single transformation and *hence* find $(T_2 \circ T_1)(3, -1)$.

5 Repeat question *4* for the points (2, 0), and then (*a*, *b*), when

$T_1 = \begin{pmatrix}4\\-1\end{pmatrix}$ and $T_2 = \begin{pmatrix}0\\3\end{pmatrix}$.

6 $T_1 = \begin{pmatrix}3\\0\end{pmatrix}$ and $T_2 = \begin{pmatrix}1\\-2\end{pmatrix}$.

a Find the images of the points (4, 4) and (1, −3) under the translations $T_2 \circ T_1$ and $T_1 \circ T_2$.

b Illustrate your results in a diagram.

c Which law does this suggest that composition of translations obeys?

3 *Reflections in two parallel axes*

Exercise 3

1 Copy Figure 9 on squared paper (width required, about 20 units). Draw the image of the capital F under reflection in the axis AB. Then draw the image of this *image* under reflection in the axis CD. Call the first image F_1 and the second image F_2.

Find out if F can be mapped to F_1 or F_2 by sliding the plane (or a tracing of the plane) or by turning the plane over.

Try to describe the simplest transformation that would map F to F_2. How far is F_2 from the original position? How far apart are the axes?

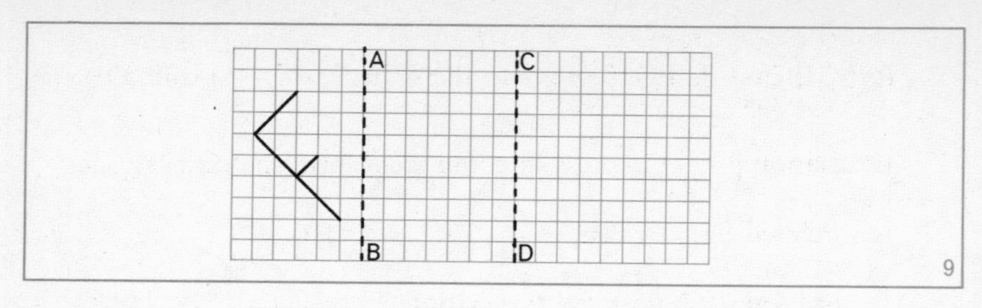

Note. A transformation which maps a figure onto its image by sliding the plane is called a *direct* transformation.

A transformation which maps a figure onto its image by turning the plane over is called an *opposite* transformation.

For example, a translation and a rotation are direct transformations while a reflection in an axis is an opposite transformation.

2 To explore further the situation described in question *1*, cut out three congruent F-shapes to represent F, F_1 and F_2. (You should keep these for use later in this chapter.) Draw parallel axes AB and CD on a sheet of squared paper. Remember that in each case the first reflection is in AB and the second in CD.

What transformation maps F to F_2 if:

a F is placed upside down or on its side
b F is placed so that it overlaps AB
c F is placed between AB and CD
d F is placed to the right of CD?

3 Explore further what happens, for various positions of F, if the *first* reflection is in CD and the *second* in AB.

Describe the transformation which maps F to F_2 in this case.

4 OX and OY are rectangular axes. We shall denote by M_1 (M for mirror) the operation 'reflection in the line $x = 6$' and by M_2 the operation 'reflection in the line $x = 11$'. P is a given point, P_1 its image under M_1 and P_2 the image of P_1 under M_2.

With the help of a diagram on squared paper complete the following table.

P	(4, 2)	(2, 3)	(0, 0)	(6, −1)	(a, b)
P_1					
P_2					

From your table, state what you think is the effect of the composition $M_2 \circ M_1$ (M_1, then M_2) on the point P.

5 a Repeat question ***4*** for the same set of points, but using $x = 0$ as the first axis and $x = 5$ as the second.

b Repeat question ***4*** for the same set of points, but using $x = 2$ and $x = 7$ as axes of reflection.

c Compare your answers to questions ***4***, ***5a*** and ***5b***.

d State two more pairs of axes of reflection which have the same combined effect as those in question ***4***.

6 For the set of points in question ***4*** complete the table when P_1 is the image of P under reflection in $x = 11$ (M_2) and P_2 is the image of P_1 under reflection in $x = 6$ (M_1).
Write down an equation connecting $M_2 \circ M_1$ with $M_1 \circ M_2$.

We can generalise the results noticed above (see Figure 10). m_1 is the first axis of reflection and m_2 the second. PP_1P_2 must cut m_1 and m_2 at right angles. Let PP_1P_2 cut m_1 at X_1 and m_2 at X_2.

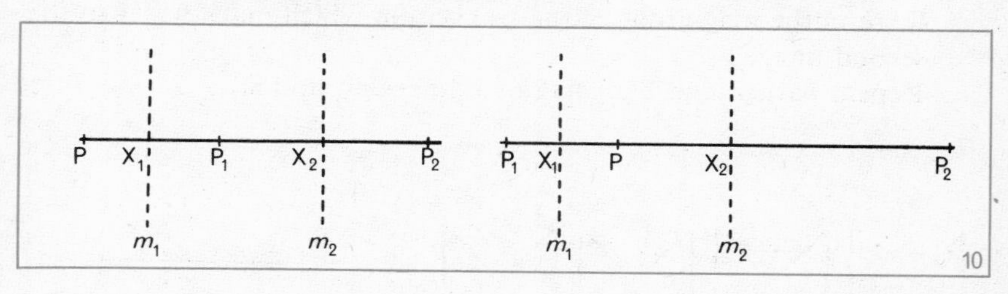

10

Then in every possible figure (we have shown only two),

$$\overrightarrow{PX_1} = \overrightarrow{X_1P_1} \text{ and } \overrightarrow{P_1X_2} = \overrightarrow{X_2P_2}$$

$$\Leftrightarrow \overrightarrow{PP_1} = 2\overrightarrow{X_1P_1} \text{ and } \overrightarrow{P_1P_2} = 2\overrightarrow{P_1X_2}$$

$$\text{So } \overrightarrow{PP_2} = \overrightarrow{PP_1} + \overrightarrow{P_1P_2}$$

$$= 2\overrightarrow{X_1P_1} + 2\overrightarrow{P_1X_2}$$

$$= 2(\overrightarrow{X_1P_1} + \overrightarrow{P_1X_2})$$

$$= 2\overrightarrow{X_1X_2}$$

Hence after the two reflections, *every* point P is translated twice the distance between the axes in a direction at right angles to them, from the first axis to the second.

The composition of two reflections in parallel axes gives a translation of:

(i) magnitude equal to twice the distance between the two axes
(ii) direction from the first axis to the second and perpendicular to the two axes.

Note. The actual position of the two axes does not alter the final result. Only their direction and the distance between them are important.

Exercise 3B

1 Figure 11 shows part of an infinite strip pattern with axes of symmetry $m_1, m_2, \ldots, m_7$.

a Make a copy of the diagram on squared paper.
b What is the image of flag 1 in m_1? What is the image of this *image* in m_3?
c How many units are there between m_1 and m_3? How many units are there in the magnitude of the translation which maps flag 1 to the second image?
d Repeat parts ***b*** and ***c*** for flag 6 and axes m_4 and m_2.

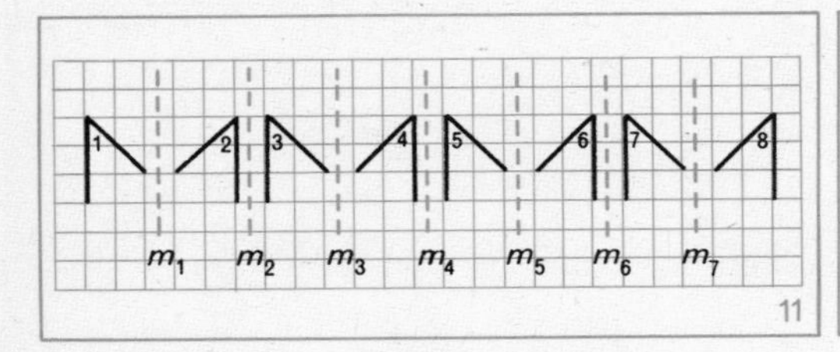

11

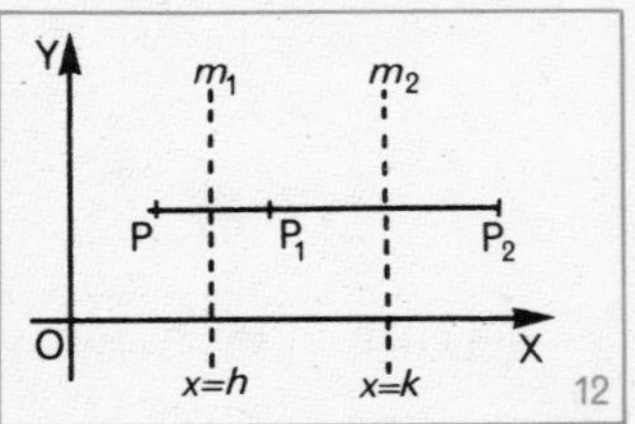

12

2 In Figure 12, P_1 is the image of P(a, b) under reflection in m_1 ($x=h$), and P_2 is the image of P_1 under reflection in m_2 ($x=k$).

a What is the y-coordinate of P_1 and P_2?
b What is the distance between m_1 and m_2?
c What translation maps P to P_2?
d What are the coordinates of P_2?

3 Use the result of question ***2d*** to calculate the coordinates of the images of the given points under successive reflections in the given axes.

	Point (a, b)	First axis $(x=h)$	Second axis $(x=k)$
a	(3, 4)	$x=1$	$x=4$
b	(0, 5)	$x=2$	$x=7$
c	(5, 0)	$x=-1$	$x=6$
d	(−1, −2)	$x=5$	$x=2$
e	(1, 4)	$x=3$	$x=-2$

4 What are the images of the point (2, 1) under successive reflections in:

a $y=3$ followed by $y=5$ *b* $y=3$ followed by $y=1$?

5 *a* Sketch a part of the graph of the function $f: x \to \sin x°$. The line $x=90$ is an axis of symmetry of the whole sine curve. State any other axis of symmetry. Does a translation equal to twice the distance between the axes conserve the pattern?

b Repeat for the function $g: x \to \cos x°$. Choose your own axes of symmetry.

6 Write out a proof that $\overrightarrow{PP_2} = 2\overrightarrow{X_2X_1}$ (similar to that after Figure 10) where the first reflection is in m_2 and the second in m_1.

4 *Reflections in two perpendicular axes*

Exercise 4

1 Copy Figure 13 on squared paper. Draw the image F_1 of the letter F under reflection in the axis AB and then the image F_2 of F_1 under reflection in the axis CD. Describe a single transformation that would take F to the position F_2.

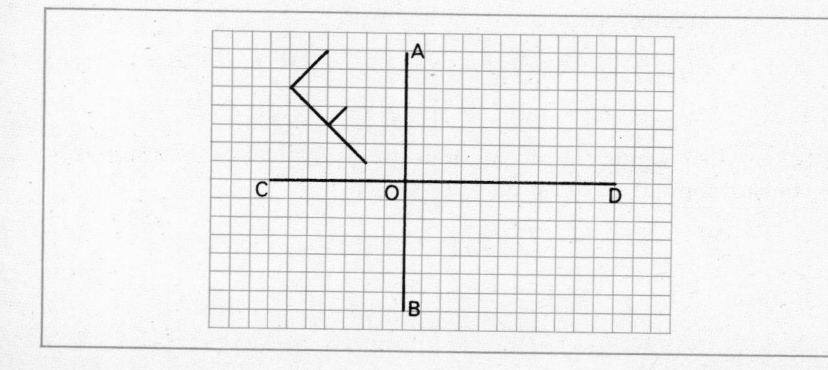

13

2 On a sheet of squared paper draw perpendicular axes AB and CD intersecting at O. Use your cut-out F-shapes (see Exercise 3, question 2) to discover what happens after successive reflections in AB and CD if F lies:

a in different positions between OA and OC
b across OA or across OC
c across both OA and OC
d in any of the other three quadrants formed by AB and CD.

3 Repeat question **2**, reflecting first in CD and then in AB.

Note. Let the two axes intersect at right angles at O. If M_1 is the operation of reflection in the first axis, M_2 the operation of reflection in the second axis and H the operation of a half turn about O, we could write $M_2 \circ M_1$ (M_1 then M_2) $= H = M_1 \circ M_2$.

In coordinates it will be convenient to represent the operation of reflection in the x-axis by X, reflection in the y-axis by Y and a half turn about the origin by H;

e.g. we could write $X(3,2) = (3,-2)$, $Y(3,2) = (-3,2)$ and $H(3,2) = (-3,-2)$.

As usual $X \circ Y$ will mean Y, then X;

e.g. $X \circ Y(3,2) = X[Y(3,2)] = X(-3,2) = (-3,-2)$

We can also use I to represent the identity transformation, i.e. $I(a,b) = (a,b)$.

4 State which points are indicated by the following.

a $X(-2,1)$ *b* $Y(4,-3)$ *c* $H(-3,2)$

d $X(-5,-6)$ *e* $Y(-5,-6)$ *f* $H(-5,-6)$

5 Find the points indicated by the following.

a $X \circ Y(-3,2)$ *b* $Y \circ X(-3,2)$ *c* $H \circ Y(-3,2)$

d $Y \circ H(-3,2)$ *e* $H^2(-3,2)$, $(H^2 = H \circ H)$ *f* $X^2(-3,2)$

6 State which single transformations are equivalent to the composite transformations in question **5**.

Exercise 4B

1 (a,b) is a given point. Find:

a $X(a,b)$ *b* $Y(a,b)$ *c* $H(a,b)$.

2 Which point is given by $X \circ Y(a, b)$? Why is it reasonable to write $X \circ Y = H$?

3 Express the following as single transformations.

a $X \circ H$ *b* $H \circ X$ *c* $H \circ Y$ *d* X^2

4 Complete this table. (I is the identity transformation.)

	$\circ$	Transformation acting first I	H	X	Y
Transformation acting second	I				
	H				
	X				
	Y				

For example, the first row is

$I \circ I$, $I \circ H$, $I \circ X$, $I \circ Y$,

i.e. I H X Y

5 By the use of the above table, or otherwise, simplify the following.

a $H \circ X \circ Y$ *b* $X \circ H \circ Y$ *c* $H \circ X^2$ *d* $X^2 \circ Y^2$

6 Let T be the operation of reflection in the line $y = 3$, S the operation of reflection in the line $x = 6$ and H the operation of a half turn about the point $(6, 3)$.

Copy and complete the following table with the help of a diagram on squared paper.

(x, y)	$(2, 1)$	$(0, 0)$	$(7, 4)$	$(-1, 2)$
$T(x, y)$				
$S(x, y)$				
$H(x, y)$				
$T \circ S(x, y)$				
$H \circ T(x, y)$				

Comment on any interesting features of your answers.

5 *Rotations about the same centre*

In Figure 14, $OP = OP_1 = OP_2$. $\angle POP_1 = 30°$ and $\angle P_1OP_2 = 40°$. An anticlockwise rotation of 30° about O maps P to P_1. An anticlockwise rotation of 40° maps P_1 to P_2. Hence

P is mapped to P_2 under an anticlockwise rotation about O of $40° + 30°$, i.e. 70°.

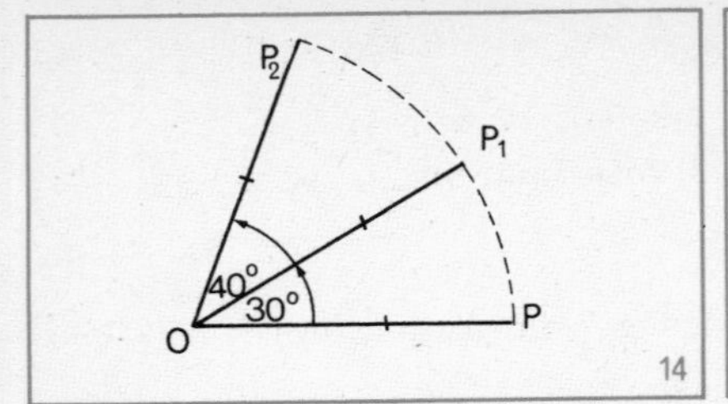

14

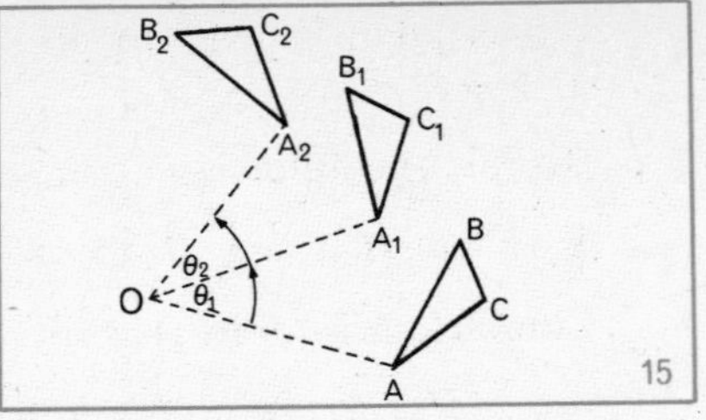

15

In Figure 15, $\triangle A_1B_1C_1$ is the image of $\triangle ABC$ under an anticlockwise rotation of $\theta_1°$ about O. $\triangle A_2B_2C_2$ is the image of $\triangle A_1B_1C_1$ under an anticlockwise rotation of $\theta_2°$ about O.

Then $\quad \angle AOA_1 = \angle BOB_1 = \angle COC_1 = \theta_1°$

$\quad\quad \angle A_1OA_2 = \angle B_1OB_2 = \angle C_1OC_2 = \theta_2°$

and hence $\quad \angle AOA_2 = \angle BOB_2 = \angle COC_2 = (\theta_1 + \theta_2)°.$

The composition of two rotations $\theta_1°$ and $\theta_2°$ about O with the same sense is a rotation $(\theta_1 + \theta_2)°$ about O with the same sense.

Care must be taken with the senses of the rotations. For opposite senses we shall have to subtract instead of add.

Remember that this simple result holds *only* when both rotations have the *same* centre.

Exercise 5

1 Copy Figure 16 and make sketches of the image F_1 of F under an anticlockwise rotation of 20° about O and the image F_2 of F_1 under an anticlockwise rotation of 40° about O. State the composite transformation which maps F to F_2.

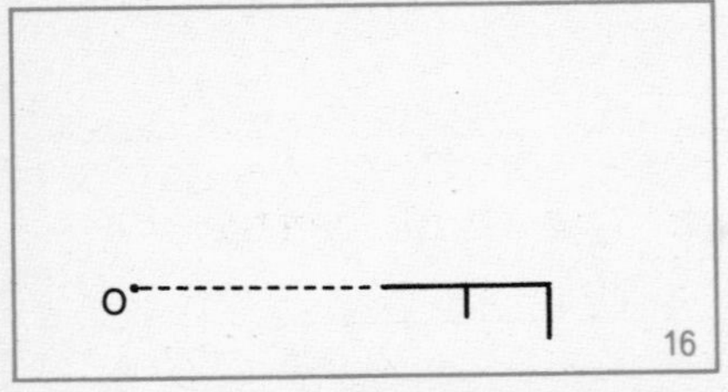

16

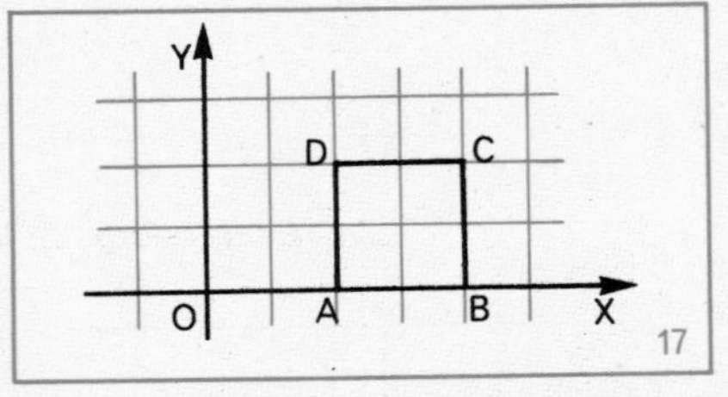

17

2 In Figure 17, ABCD is a square. Copy the figure and sketch the image $A_1B_1C_1D_1$ of ABCD under an *anticlockwise* rotation of 90° about O, and the image $A_2B_2C_2D_2$ of $A_1B_1C_1D_1$ under a *clockwise* rotation of 45° about O. State the composite transformation which maps ABCD to $A_2B_2C_2D_2$.

3 In Figure 18, ABCDEF is a regular hexagon with centre O. R_1, R_2, R_3, R_4 and R_5 represent anticlockwise rotations of 60°, 120°, 180°, 240° and 300° respectively about O. Let I represent the identity rotation (i.e. the rotation which leaves the figure in its original position).

Sketch the images of the hexagon under the following rotations. (You need not put in the letters every time as long as you show the black arrow in the right place.)

a R_2 *b* $R_1 \circ R_2$ *c* $R_2 \circ R_2 \circ R_2$ *d* $R_2 \circ R_5$.

In ***b***, $R_1 \circ R_2 = R_3$. Write similar equations for ***c*** and ***d***. (Use only the symbols mentioned in the question.)

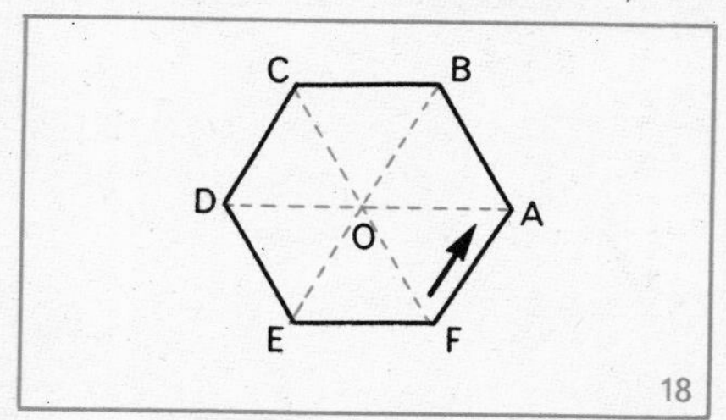

18

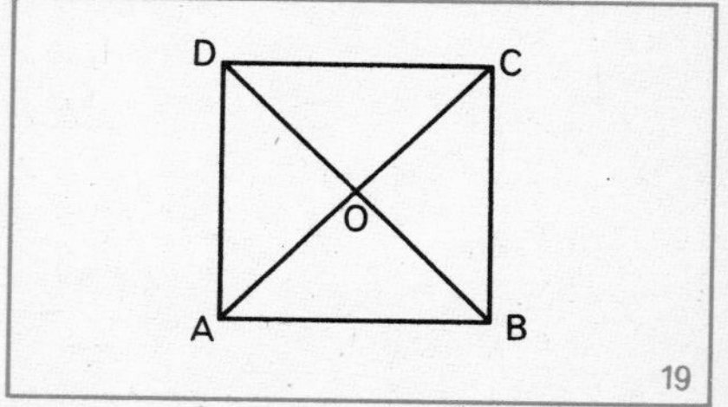

19

4 In Figure 18, state the images of the following parts of the figure under the given rotations.

a OA under $R_1 \circ R_3$ *b* OA under $R_3 \circ R_1$ *c* BC under $R_2 \circ R_2$ (i.e. $R_2{}^2$)

d F under $R_3 \circ R_4$ *e* $\triangle$OCD under $R_2 \circ R_1$ *f* OABC under $R_1{}^2$.

5 Simplify:

a $R_1 \circ R_4$ *b* $R_5 \circ R_1$ *c* $R_3{}^2$ *d* $(R_1 \circ R_3) \circ R_4$.

6 In Figure 19, ABCD is a square. Sketch the square (with letters) as it would look after successive rotations about O of:

a +90° and +90° *b* +90° and −45°

c +90° and +270° *d* +90° and +45°.

7 Use squared paper to find the coordinates of the images of the following points under the given rotations about the origin.

a (1, 0) under successive rotations of +90° and +90°
b (2, 1) under successive rotations of +180° and −90°
c (2, 0) under successive rotations of +90° and +45°

6 *Reflections in two intersecting axes*

Figure 20 represents a regular hexagon and arrow drawn on tracing paper. The six axes of symmetry are shown as fixed dotted lines drawn, not on the tracing paper, but on paper underneath it. You should prepare tracing paper and a diagram like this.

Let A represent the operation of reflection in AA_1 (or turning over about AA_1), X represent reflection in XX_1, etc.

$X \circ A$ means 'operation A, then operation X'.

R_1, R_2, R_3, R_4 and R_5 represent the operations of rotation of 60°, 120°, 180°, 240° and 300° respectively anticlockwise about O.

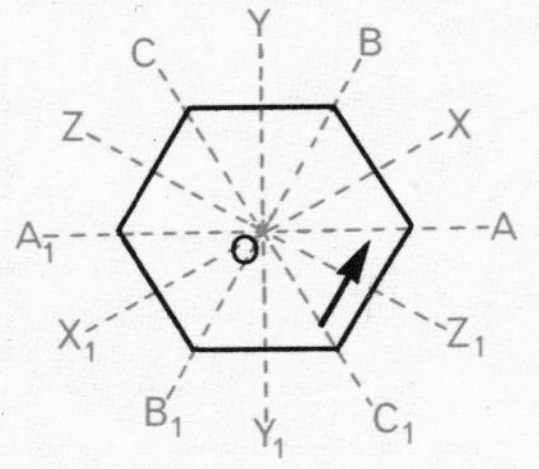

20

Exercise 6

1 Sketch images of the hexagon (showing the arrow) under the operations:

a A *b* X *c* B *d* R_1 *e* R_2.

2 Use your tracing to verify that:

a $X \circ A = R_1$ *b* $B \circ X = R_1$ *c* $Y \circ B = R_1$.

Note that two successive reflections in axes inclined at 30° combine to produce a rotation through 60°.

3 a What rotation is equivalent to $B \circ A$?

b Write down two more compositions of reflections giving the same rotation.

4 Repeat question *3* for the composition $Y \circ A$.

The work in questions *1–4* suggests that reflections in two intersecting axes result in a rotation

(i) about the point of intersection of the axes
(ii) of magnitude equal to twice the angle between the axes
(iii) of sense from the first axis to the second.

Note that Section 4, reflection in perpendicular axes, is a particular case of this more general result.

5 Draw two axes at an angle of 45°. Use your cut out F-shapes (see Exercise 3, question *2*) to illustrate the truth of the above suggestions about composition of reflections. (Use a variety of positions of the first F, and vary the order of the two axes.)

6 Find, with the help of a diagram on squared paper, the coordinates of the images A_2, B_2 and C_2 of A(3, −1), B(5, 0) and C(4, 2) under reflection in the x-axis followed by reflection in the line $y = x$.

Write down the coordinates of the images of A, B and C under a rotation of +90° about the origin.

7 Find the coordinates of the images P_2 and Q_2 of P(1, −3) and Q(2, 4) under reflection in the y-axis followed by reflection in the line $y = x$.

Which clockwise rotation is the result of combining these two reflections? State this transformation as an anticlockwise rotation.

8 Repeat question *7* but reverse the order of the reflections.

Do the answers to questions *6*, *7* and *8* bear out the statements in colour after question *4* above?

In Figure 21, under reflection in OA, $P \to P_1$, and under reflection in OB, $P_1 \to P_2$. Let PP_1 cut OA in M and P_1P_2 cut OB in N.

It follows that, under the same reflections

$\angle POM \to \angle P_1OM$ and $\angle P_1ON \to \angle P_2ON$

Also $OP = OP_1 = OP_2$.

Hence

$$\begin{aligned}\angle POP_2 &= \angle POP_1 + \angle P_1OP_2\\ &= 2\angle MOP_1 + 2\angle P_1ON\\ &= 2(\angle MOP_1 + \angle P_1ON)\\ &= 2\angle AOB\end{aligned}$$

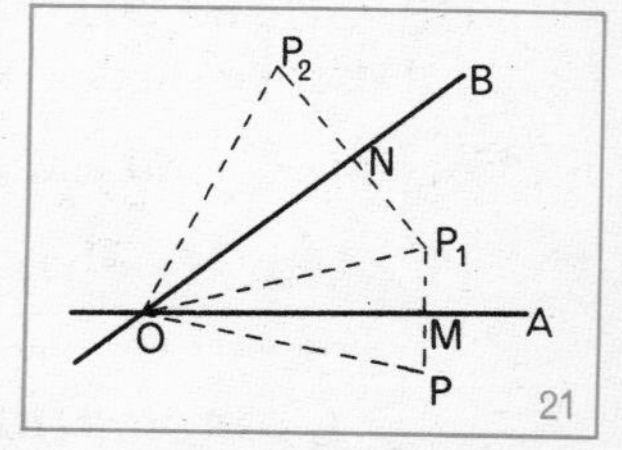

21

This suggests that for every position of P, $\angle POP_2 = 2\angle AOB$, and $OP = OP_2$. Other positions of P will be considered in Exercise 6B.

Under successive reflections in OA and OB, every point P is rotated about O (i) through an angle equal to $2\angle AOB$ (ii) in the sense from OA to OB.

Exercise 6B

1 Draw two axes OA and OB as in Figure 21 with the angle AOB = 35°. Draw accurate diagrams to show what happens under successive reflections in OA and OB when P lies:

a on OA *b* on OB *c* between OA and OB.

2 Write out a proof in question ***1c*** that $\angle POP_2 = 2\angle AOB$.

3 In $\triangle ABC$, $\angle BAC = 40°$. $\triangle AB_1C$ is the image of $\triangle ABC$ under reflection in the line of AC and $\triangle AB_1C_1$ is the image of $\triangle AB_1C$ under reflection in the line of AB_1.

a Describe exactly the single transformation which maps $\triangle ABC$ to $\triangle AB_1C_1$.

b How many such pairs of reflections will map $\triangle ABC$ onto itself?

4 $\triangle ABC$ is right-angled at B. The bisector of $\angle BAC$ meets BC at X. In sketches show the image of the triangle after successive reflections in: *a* AX and AB *b* AB and AX.

5 In polar coordinates P_2 is the image of $P(r, \theta°)$ under successive reflections in $\theta = 20$ and $\theta = 45$. State the rotation produced by the composition of the two reflections. Complete the following table.

P	(2, 10°)	(10, 20°)	(5, 30°)	(3, 45°)	(4, 60°)
P_2					

6 OA, OB, OC and OD are four lines in that order intersecting at O. Let M_1, M_2, M_3 and M_4 represent the operations of reflection in OA, OB, OC and OD.

Describe single transformations equivalent to:

a $M_1 \circ M_2$ *b* $M_3 \circ M_4$ *c* $(M_1 \circ M_2) \circ (M_3 \circ M_4)$

d $(M_4 \circ M_3) \circ (M_2 \circ M_1)$ *e* $(M_3 \circ M_4) \circ (M_1 \circ M_2)$.

7 *Composition of transformations by matrices*

Figure 22(i) represents a rectangle with an arrow drawn near one side. A tracing of the rectangle will help you. Corresponding to the four ways of placing this shape in its outline, we can think of four operations:

I, 'replace it as you found it', the identity operation;
X, 'turn it over about XX_1';
Y, 'turn it over about YY_1';
H, 'turn it round', a half turn about the point marked H.
These are illustrated in Figure 22.

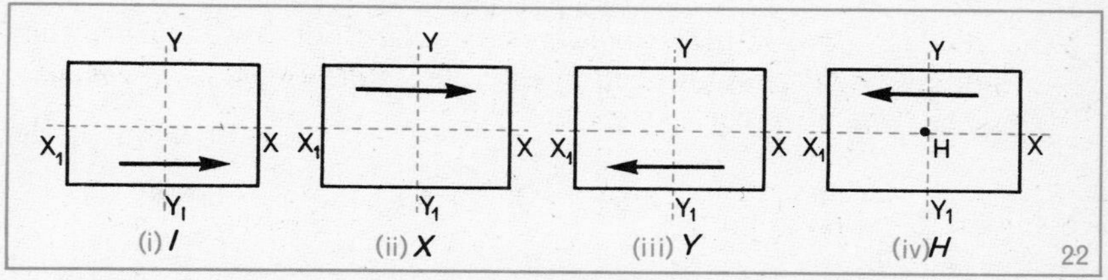

Figure 23 shows the result of combining two operations.

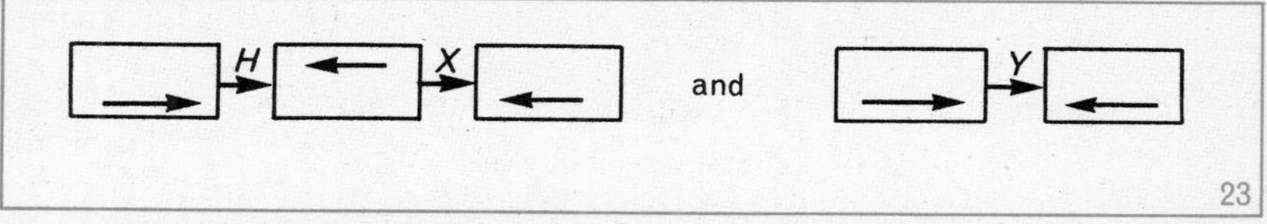

H then X yields the same result as Y. This could be written in symbols $X \circ H = Y$.

Exercise 7B

1 Interpret the diagrams in Figure 24 as illustrated under Figure 23.

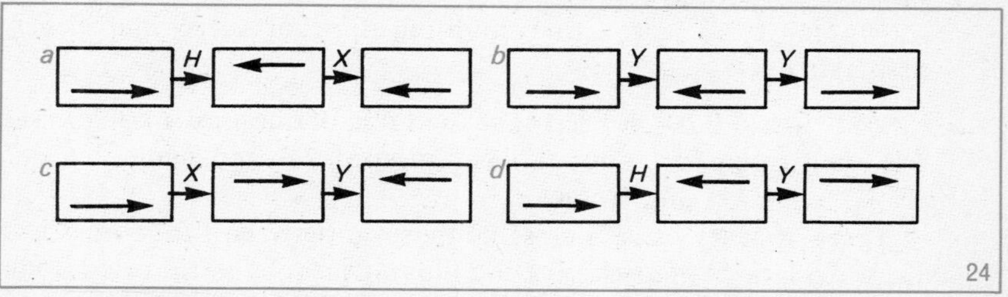

2 Use the above and similar results to complete this table.

		First operation			
	$\circ$	I	H	X	Y
Second operation	I				
	H				
	X				
	Y				

3 *a* With the help of the above table state why you are sure that the composition denoted by $\circ$ is commutative for I, H, X and Y.

b Use the table to verify that $\circ$ is associative (e.g. $X\circ(Y\circ X) = (X\circ Y)\circ X$) by examining some particular cases of your own choice. Can you find an exception?

Reminder about matrix multiplication. In Book 5 we learned that matrices could be multiplied by the rule '*Multiply rows into columns, and add the resulting products*'.

$$\text{e.g.} \begin{pmatrix} 2 & 3 \\ 1 & 4 \end{pmatrix}\begin{pmatrix} 1 & 1 \\ 3 & 2 \end{pmatrix} = \begin{pmatrix} 2\times1+3\times3 & 2\times1+3\times2 \\ 1\times1+4\times3 & 1\times1+4\times2 \end{pmatrix} = \begin{pmatrix} 11 & 8 \\ 13 & 9 \end{pmatrix}.$$

4 Let $I_1 = \begin{pmatrix} 1 & 0 \\ 0 & 1 \end{pmatrix}$, $H_1 = \begin{pmatrix} -1 & 0 \\ 0 & -1 \end{pmatrix}$, $X_1 = \begin{pmatrix} 1 & 0 \\ 0 & -1 \end{pmatrix}$, $Y_1 = \begin{pmatrix} -1 & 0 \\ 0 & 1 \end{pmatrix}$.

Complete this table by multiplying pairs of matrices.

	$\times$	*Right-hand matrix* I_1	H_1	X_1	Y_1
Left-hand matrix	I_1				
	H_1				
	X_1				
	Y_1				

5 *a* Compare the tables in questions *2* and *4*. Do you see anything interesting about them?

b Is multiplication for this set of matrices commutative? Is this true for all matrix multiplication?

c Verify in a few cases that multiplication of these matrices is associative.

You should have noticed that the table of question ***4*** could have been constructed from the table of question ***2*** by replacing $\circ$ by $\times$, I by I_1, X by X_1, and so on.

There appears to be a close connection between the geometrical transformations of question ***2*** and the matrices of question ***4***. In Book 5 we saw that certain matrices could be associated with geometrical transformations. Here is a summary of the results.

Matrix product	*Mapping*	*Transformation*
$\begin{pmatrix}1 & 0\\0 & 1\end{pmatrix}\begin{pmatrix}a\\b\end{pmatrix}=\begin{pmatrix}a\\b\end{pmatrix}$	$(a,b)\to(a,b)$	Identity (I)
$\begin{pmatrix}-1 & 0\\0 & -1\end{pmatrix}\begin{pmatrix}a\\b\end{pmatrix}=\begin{pmatrix}-a\\-b\end{pmatrix}$	$(a,b)\to(-a,-b)$	Half turn about O (H)
$\begin{pmatrix}1 & 0\\0 & -1\end{pmatrix}\begin{pmatrix}a\\b\end{pmatrix}=\begin{pmatrix}a\\-b\end{pmatrix}$	$(a,b)\to(a,-b)$	Reflection in x-axis (M_x)
$\begin{pmatrix}-1 & 0\\0 & 1\end{pmatrix}\begin{pmatrix}a\\b\end{pmatrix}=\begin{pmatrix}-a\\b\end{pmatrix}$	$(a,b)\to(-a,b)$	Reflection in y-axis (M_y)
$\begin{pmatrix}0 & 1\\1 & 0\end{pmatrix}\begin{pmatrix}a\\b\end{pmatrix}=\begin{pmatrix}b\\a\end{pmatrix}$	$(a,b)\to(b,a)$	Reflection in $y=x$ $(M_{y=x})$
$\begin{pmatrix}0 & -1\\-1 & 0\end{pmatrix}\begin{pmatrix}a\\b\end{pmatrix}=\begin{pmatrix}-b\\-a\end{pmatrix}$	$(a,b)\to(-b,-a)$	Reflection in $y=-x$ $(M_{y=-x})$
$\begin{pmatrix}0 & -1\\1 & 0\end{pmatrix}\begin{pmatrix}a\\b\end{pmatrix}=\begin{pmatrix}-b\\a\end{pmatrix}$	$(a,b)\to(-b,a)$	Rotation of $+90°$ about O (R_{90})
$\begin{pmatrix}0 & 1\\-1 & 0\end{pmatrix}\begin{pmatrix}a\\b\end{pmatrix}=\begin{pmatrix}b\\-a\end{pmatrix}$	$(a,b)\to(b,-a)$	Rotation of $-90°$ about O (R_{-90})

We can regard the matrix as operating on the coordinates of all points in the plane and so producing a geometrical transformation.

6 Under $M_{y=x}$ use the associated matrix to find the image of the parallelogram OABC where A is (2,0), B is (4,4) and C is (2,4).

7 Under R_{90}, use the associated matrix to find the image of the isosceles triangle OAB where A is (4,0) and B is (2,6).

Exercise 8B

1 Carry out the necessary multiplications and state the geometrical transformations associated with each of the following.

a $\begin{pmatrix}1 & 0\\0 & -1\end{pmatrix}$, $\begin{pmatrix}-1 & 0\\0 & 1\end{pmatrix}$ and $\begin{pmatrix}1 & 0\\0 & -1\end{pmatrix}\begin{pmatrix}-1 & 0\\0 & 1\end{pmatrix}$

b $\begin{pmatrix}-1 & 0\\0 & -1\end{pmatrix}$, $\begin{pmatrix}1 & 0\\0 & -1\end{pmatrix}$ and $\begin{pmatrix}-1 & 0\\0 & -1\end{pmatrix}\begin{pmatrix}1 & 0\\0 & -1\end{pmatrix}$

c $\begin{pmatrix}-1 & 0\\0 & 1\end{pmatrix}$ and $\begin{pmatrix}-1 & 0\\0 & 1\end{pmatrix}\begin{pmatrix}-1 & 0\\0 & 1\end{pmatrix}$

d $\begin{pmatrix}0 & -1\\1 & 0\end{pmatrix}$, $\begin{pmatrix}0 & 1\\-1 & 0\end{pmatrix}$ and $\begin{pmatrix}0 & -1\\1 & 0\end{pmatrix}\begin{pmatrix}0 & 1\\-1 & 0\end{pmatrix}$

e $\begin{pmatrix} 0 & -1 \\ 1 & 0 \end{pmatrix}$ and $\begin{pmatrix} 0 & -1 \\ 1 & 0 \end{pmatrix}\begin{pmatrix} 0 & -1 \\ 1 & 0 \end{pmatrix}$

In ***a*** we see that $M_x \circ M_y = H$. Make similar statements for ***b***, ***c***, ***d*** and ***e***.

Can multiplication of matrices be interpreted as composition of transformations?

2 Carry out the following matrix multiplications.

a $\begin{pmatrix} 1 & 0 \\ 0 & -1 \end{pmatrix}\left[\begin{pmatrix} -1 & 0 \\ 0 & -1 \end{pmatrix}\begin{pmatrix} 3 \\ 2 \end{pmatrix}\right]$ and $\left[\begin{pmatrix} 1 & 0 \\ 0 & -1 \end{pmatrix}\begin{pmatrix} -1 & 0 \\ 0 & -1 \end{pmatrix}\right]\begin{pmatrix} 3 \\ 2 \end{pmatrix}$

b $\begin{pmatrix} -1 & 0 \\ 0 & -1 \end{pmatrix}\left[\begin{pmatrix} -1 & 0 \\ 0 & 1 \end{pmatrix}\begin{pmatrix} 5 \\ -3 \end{pmatrix}\right]$ and $\left[\begin{pmatrix} -1 & 0 \\ 0 & -1 \end{pmatrix}\begin{pmatrix} -1 & 0 \\ 0 & 1 \end{pmatrix}\right]\begin{pmatrix} 5 \\ -3 \end{pmatrix}$

c $\begin{pmatrix} -1 & 0 \\ 0 & 1 \end{pmatrix}\left[\begin{pmatrix} 0 & 1 \\ 1 & 0 \end{pmatrix}\begin{pmatrix} p \\ q \end{pmatrix}\right]$ and $\left[\begin{pmatrix} -1 & 0 \\ 0 & 1 \end{pmatrix}\begin{pmatrix} 0 & 1 \\ 1 & 0 \end{pmatrix}\right]\begin{pmatrix} p \\ q \end{pmatrix}$

If you use the product of two matrices do you get the same result as if you use one matrix after the other?

3 Use the matrix $\begin{pmatrix} \frac{1}{\sqrt{2}} & -\frac{1}{\sqrt{2}} \\ \frac{1}{\sqrt{2}} & \frac{1}{\sqrt{2}} \end{pmatrix}$ to transform the points $(2, 0)$, $(\sqrt{2}, \sqrt{2})$, $(0, 2)$ and $(-\sqrt{2}, \sqrt{2})$. Make a sketch showing the positions of points and images. Suggest a rotation associated with the given matrix.

4 Simplify

$$\begin{pmatrix} \frac{1}{\sqrt{2}} & -\frac{1}{\sqrt{2}} \\ \frac{1}{\sqrt{2}} & \frac{1}{\sqrt{2}} \end{pmatrix}^2$$

With which geometrical transformation is the answer associated? Does this agree with the rotation suggested in question ***3***?

Matrices can be used for many more general transformations than the simple isometric ones we have been discussing. The composition of these transformations can be carried out by multiplication of the associated matrices. We can choose any 2×2 matrix and investigate the result of transforming points in the x–y plane.

Example. O, A, B and C are the corners of a square with A(1, 0), B(1, 1) and C(0, 1). Draw the square and its image under the transformation associated with the matrix $\begin{pmatrix} 0 & -2 \\ 2 & 0 \end{pmatrix}$.

$$\begin{pmatrix} 0 & -2 \\ 2 & 0 \end{pmatrix}\begin{pmatrix} 0 & 1 & 1 & 0 \\ 0 & 0 & 1 & 1 \end{pmatrix}$$

$$= \begin{pmatrix} 0 & 0 & -2 & -2 \\ 0 & 2 & 2 & 0 \end{pmatrix}$$

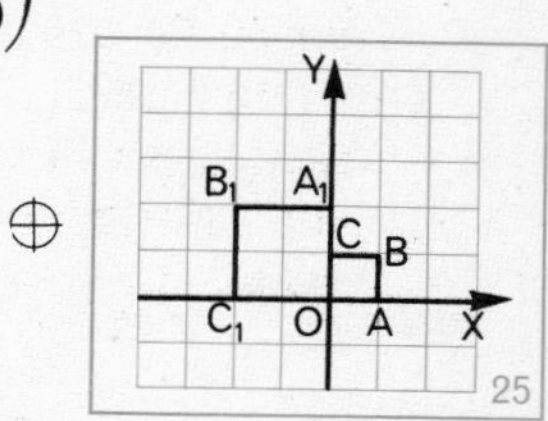

So the image square has vertices O(0, 0), $A_1(0, 2)$, $B_1(-2, 2)$ and $C_1(-2, 0)$.

When we plot these points on a diagram (see Figure 25) we can see that the square OABC has undergone a rotation of 90° about O and a dilatation [O, 2].

5 Show that the same composite transformation takes place, as in the worked example, by transforming first by the matrix $\begin{pmatrix} 2 & 0 \\ 0 & 2 \end{pmatrix}$ and then by the matrix $\begin{pmatrix} 0 & -1 \\ 1 & 0 \end{pmatrix}$.

Do you get the same answer by first using $\begin{pmatrix} 0 & -1 \\ 1 & 0 \end{pmatrix}$ and then $\begin{pmatrix} 2 & 0 \\ 0 & 2 \end{pmatrix}$?

What are the products of the matrices in the two cases? Illustrate the separate transformations and the composite transformation in a sketch.

Note. The matrix $\begin{pmatrix} 2 & 0 \\ 0 & 2 \end{pmatrix}$ is associated with the dilatation [O, 2].

Similarly $\begin{pmatrix} k & 0 \\ 0 & k \end{pmatrix}$ is associated with the dilatation [O, k].

6 Use the matrix $\begin{pmatrix} 0 & -2 \\ 2 & 0 \end{pmatrix}$ to transform the square with vertices P(2, 1), Q(4, 1), R(4, 3) and S(2, 3). Does your answer agree with the conclusion in the worked example above?

7 Investigate the geometrical effect of transforming the square PQRS of question *6* by the matrix $\begin{pmatrix} 1 & 1 \\ 0 & 1 \end{pmatrix}$.

Repeat for two other squares of your own choice. Do you see any common pattern in the images? Is this an isometric transformation? In each case calculate the area of the square and its image.

Note that this matrix has a *shearing* effect on the square (it is twisted sideways).

8 Repeat question *7* for the matrix $\begin{pmatrix} 2 & 1 \\ 1 & 2 \end{pmatrix}$.

Summary

1 *Two successive translations* produce a translation which can be found:

(i) by a head-to-tail addition

(ii) by addition of components in number pairs.

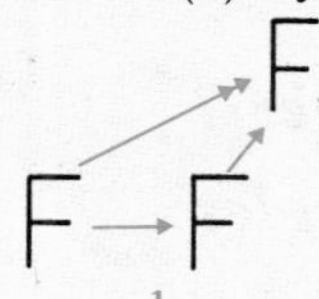
1

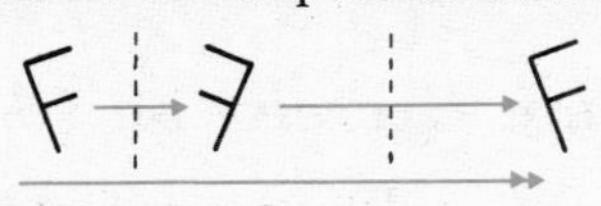
2

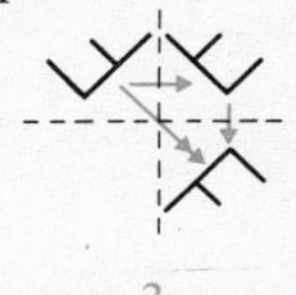
3

2 *Two successive reflections in parallel axes* produce a translation of:

(i) magnitude twice the distance between the axes

(ii) direction at right angles to the axes from the first axis to the second.

3 *Two successive reflections in perpendicular axes* produce a half turn about the point of intersection of the axes.

4 *Two successive rotations about the same centre* produce a rotation about that centre with magnitude equal to the sum of the two original magnitudes.

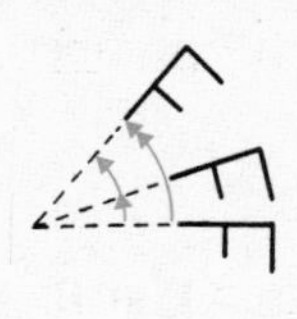
4

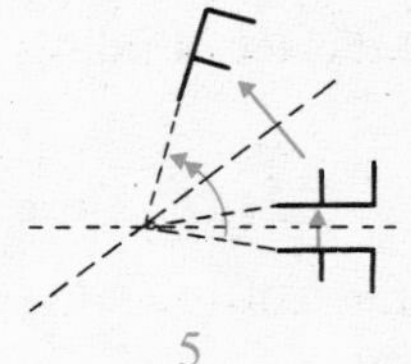
5

5 *Two successive reflections in intersecting axes* produce a rotation about the point of intersection of the axes of:

(i) magnitude twice the angle between the axes

(ii) sense from the first axis to the second.

6 *2 × 2 matrices* are associated with transformations of the plane, e.g.

$$I=\begin{pmatrix}1 & 0\\ 0 & 1\end{pmatrix} \quad H=\begin{pmatrix}-1 & 0\\ 0 & -1\end{pmatrix} \quad M_x=\begin{pmatrix}1 & 0\\ 0 & -1\end{pmatrix} \quad M_y=\begin{pmatrix}-1 & 0\\ 0 & 1\end{pmatrix}$$

$$M_{y=x}=\begin{pmatrix}0 & 1\\ 1 & 0\end{pmatrix} \quad M_{y=-x}=\begin{pmatrix}0 & -1\\ -1 & 0\end{pmatrix} \quad R_{90}=\begin{pmatrix}0 & -1\\ 1 & 0\end{pmatrix} \quad R_{-90}=\begin{pmatrix}0 & 1\\ -1 & 0\end{pmatrix}$$

7 *Products of matrices* are associated with compositions of transformations.

Revision Exercises

Revision Exercise on Chapter 1
Circle 3. Tangents and angles

Revision Exercise 1

1 A point A is 2·5 cm from the centre O of a circle of radius 1·5 cm. Calculate the length of a tangent from A to the circle.

2 PQ is a diameter of a circle. Tangents to the circle are drawn at P and Q.

a Mark right angles in the figure.
b Explain why the two tangents are parallel.
c Indicate on a diagram any axes of symmetry of the figure.

3 Describe the set of points from which the number of tangents to a circle is:

a 0 *b* 1 *c* 2 *d* more than 2.

4 AB is a tangent touching a circle, centre O, at B. AO = 17 cm and OB = 8 cm.

a Calculate the length of AB.
b If AB is produced 12 cm to C, calculate how far C is from the centre O. (Answer to one decimal place.)
c Use your tables to calculate the angles of triangle AOC.

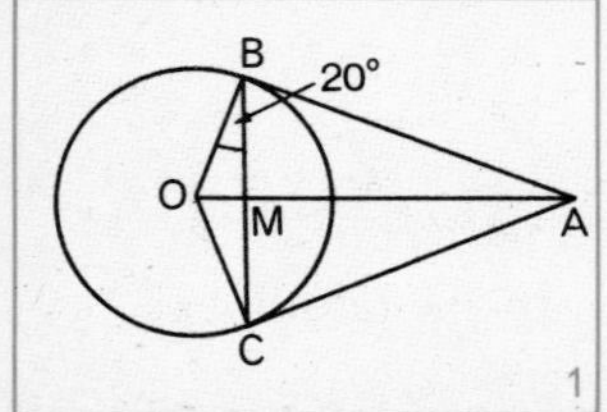

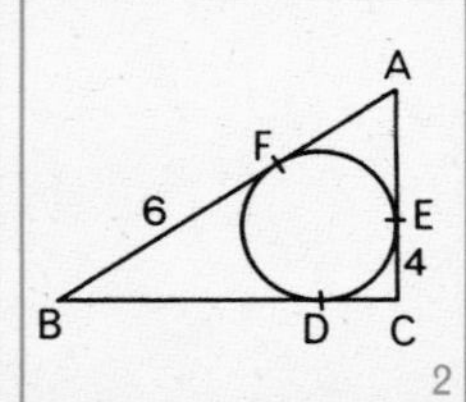

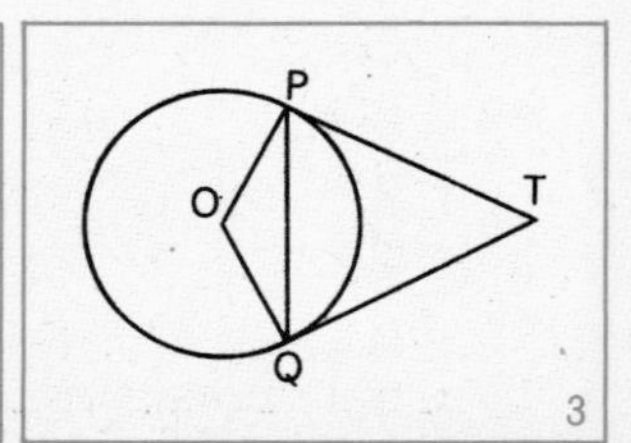

5 In Figure 1, AB and AC are tangents to the circle, centre O. If ∠OBM = 20°, copy the diagram and fill in the sizes of all the angles in the figure.

6 In Figure 2 a circle touches the sides of △ABC at D, E and F.

a Write down three pairs of equal lines in the figure.
b If BF = 6 cm and CE = 4 cm, calculate the length of BC.
c If the perimeter of the triangle is 30 cm, calculate the length of AF.

7 In Figure 3, PT and QT are tangents to the circle centre O.

a If ∠PQT = 63°, calculate the sizes of angles PTQ and POQ.
b If ∠PQT = x°, calculate the sizes of angles PTQ and POQ in terms of x.
c Draw a circle of radius 3 cm. Draw accurately a pair of tangents to the circle intersecting at an angle of 40°

8 A point P is 22 cm from the centre O of a circle. The angle between the tangents from P is 66°. Use your tables to calculate the radius of the circle.

9 Draw sketches of two circles to illustrate the cases where there can be:

a no common tangents to the two circles
b exactly two common tangents to the two circles.

10 A and B are two points 50 mm apart. A is the centre of a circle of radius 22 mm. B is the centre of a circle of radius 8 mm. Use Pythagoras' theorem to calculate the length of the direct common tangents to the two circles.

11 AB is a diameter of a circle and C is a point on the circumference. ∠BAC = 45°. What size is ∠ACB? Calculate angle ABC.

12 AB is a diameter of a circle of radius 10 cm. BC is a chord of this circle of length 16 cm.

a Calculate the length of AC.
b If D is the image of B under reflection in AC, calculate the area of △ABD.
c If E is the image of C under reflection in AB, calculate the area of quadrilateral ACBE.

13 In Figure 4, O is the centre of the circle and the figure is symmetrical about AO. If ∠BOC = 120°, show that △ABC is equilateral.

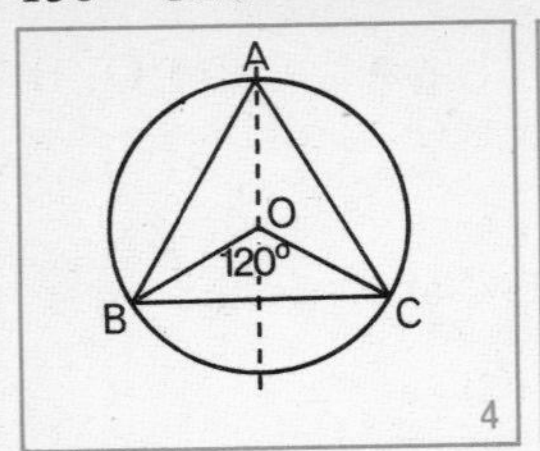

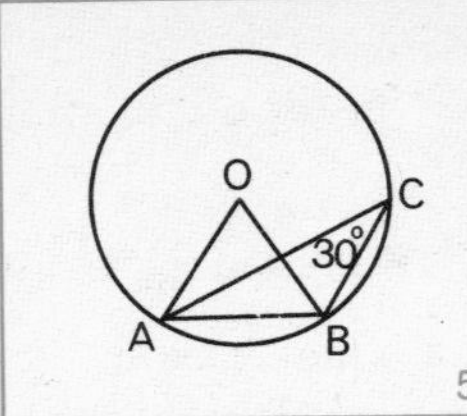

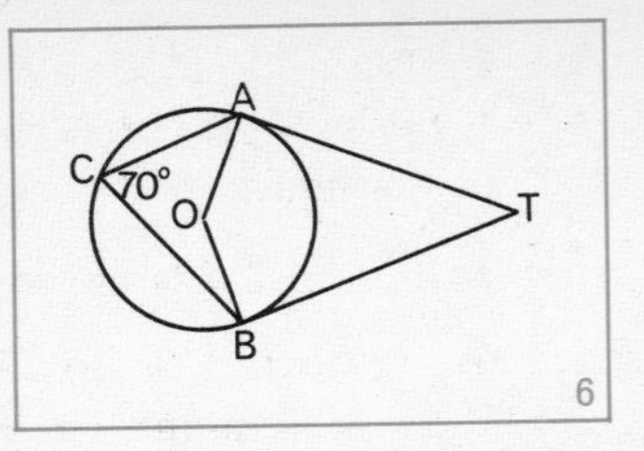

14 In Figure 5, O is the centre of the circle and $\angle ACB = 30°$. Show that AB is equal in length to a radius of the circle.

15 In Figure 6, TA and TB are tangents to a circle centre O.

a If $\angle ACB = 70°$, calculate angle ATB.
b Repeat when $\angle ACB = p°$.

16 Copy Figure 7 and mark pairs of equal angles standing on the minor arcs AB, BC, CD and DA respectively.

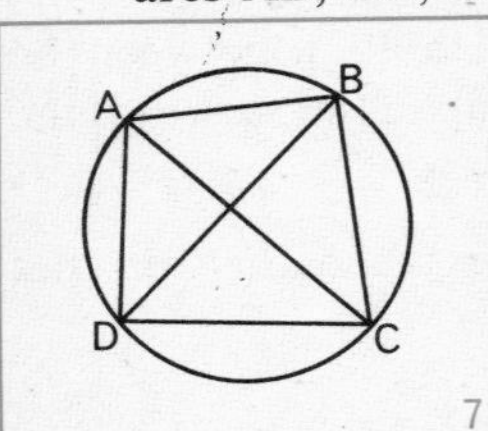

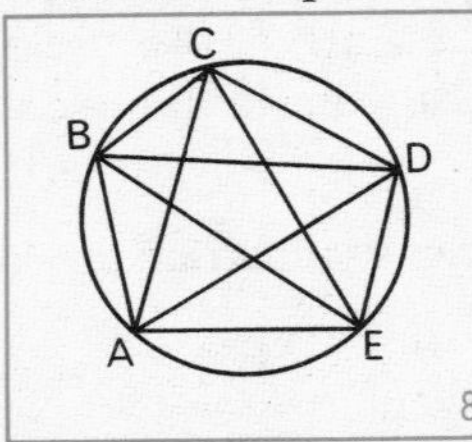

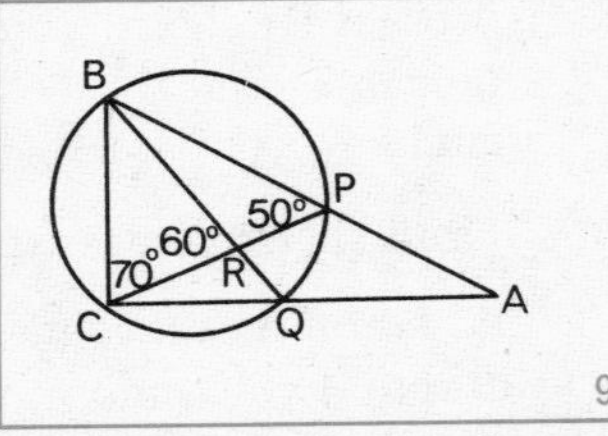

17 Copy Figure 8 and mark (by means of equal numbers) sets of equal angles standing on:

a the minor arcs AB, BC, CD, DE and EA respectively
b the arcs ABC, BCD, CDE, DEA and EAB respectively.

18 In Figure 9, $\angle BPC = 50°$, $\angle BRC = 60°$ and $\angle BCR = 70°$. Calculate the angles of $\triangle ABC$.

19 In Figure 10, PQ and RS are common tangents to two touching circles. Explain why:

a $SP = SR = SQ$ *b* $\angle PRQ$ is a right angle.

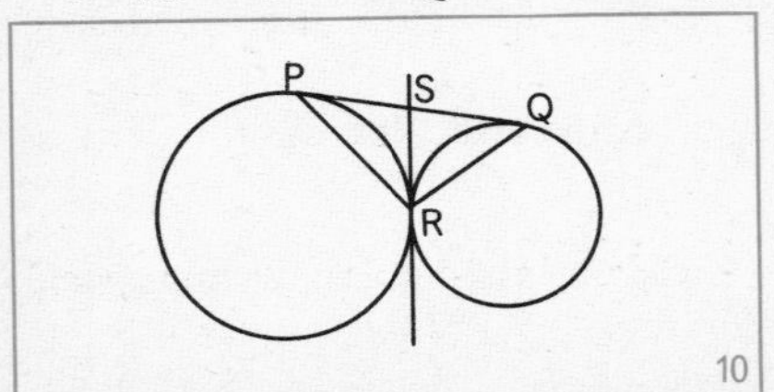

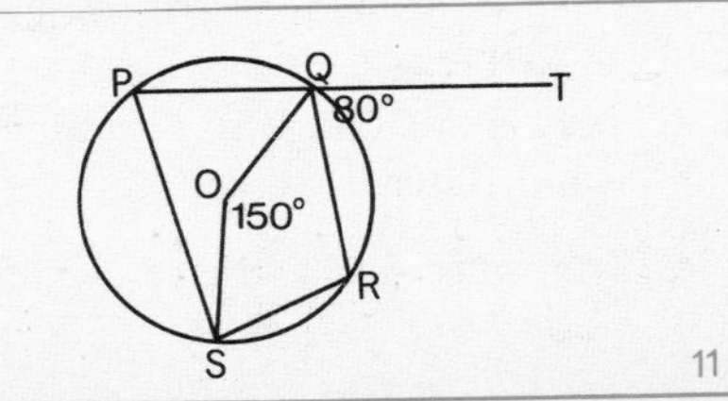

20 In Figure 11, O is the centre of the circle. Calculate the angles of quadrilateral PQRS.

21a In Figure 12, explain why triangles AEB and DEC are similar.

b If AB = 9 cm, AE = 7·5 cm, BE = 6 cm and CE = 4 cm, calculate the lengths of DE and DC.

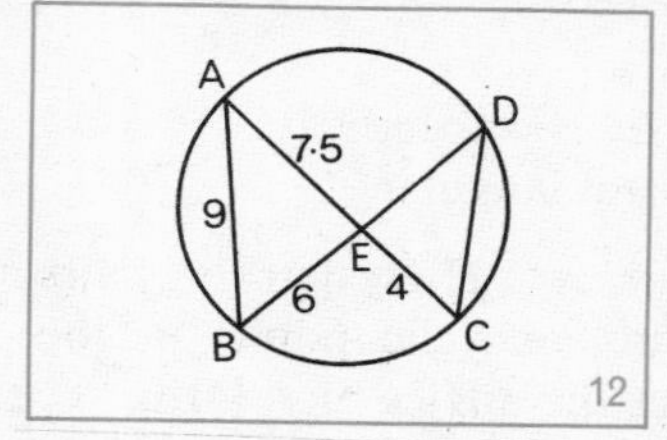

12

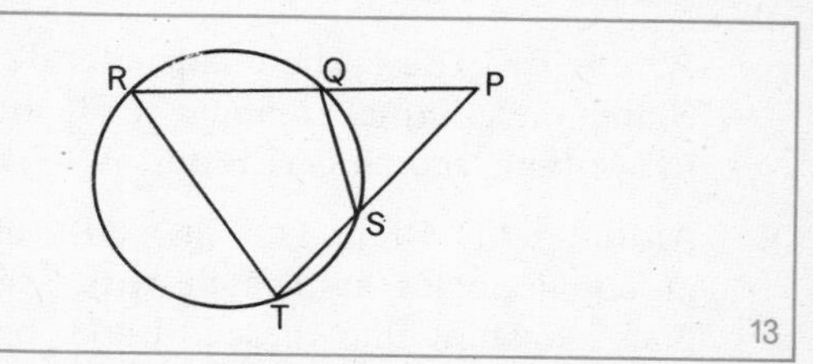

13

22a In Figure 13, show that triangles PQS and PTR are similar.

b If PQ = 10 cm, PR = 18 cm and PT = 15 cm, calculate the lengths of PS and TS.

23 In Figure 14, calculate ∠BCD, and explain why AE is parallel to DC.

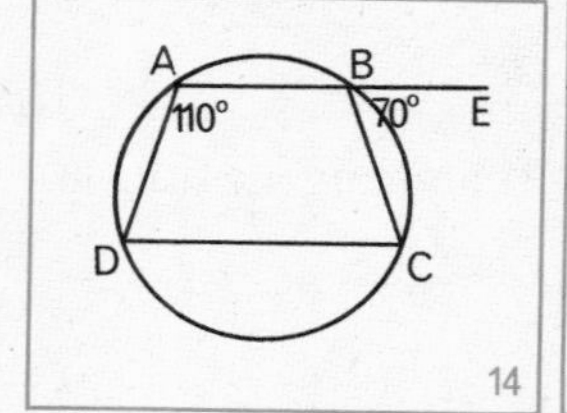

14

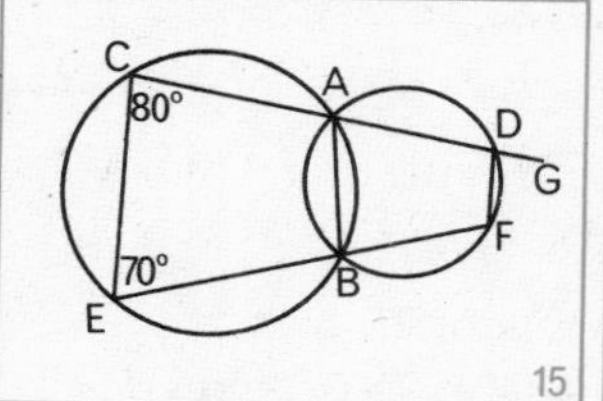

15

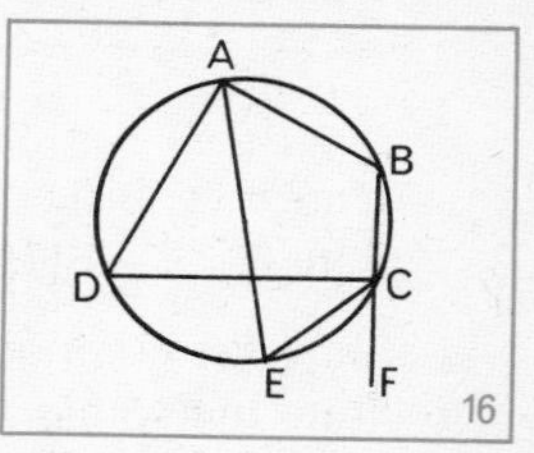

16

24a Copy Figure 15 and fill in the sizes of all the angles. Why is CE parallel to DF?

b Repeat when $\angle C = p^\circ$ and $\angle E = q^\circ$.

25a In Figure 16, use the properties of cyclic quadrilaterals to find an angle equal to ∠ECF and an angle equal to ∠DCF.

b If CE bisects ∠DCF, what can you say about AE?

26 In Figure 17, PT is a tangent to the circle.

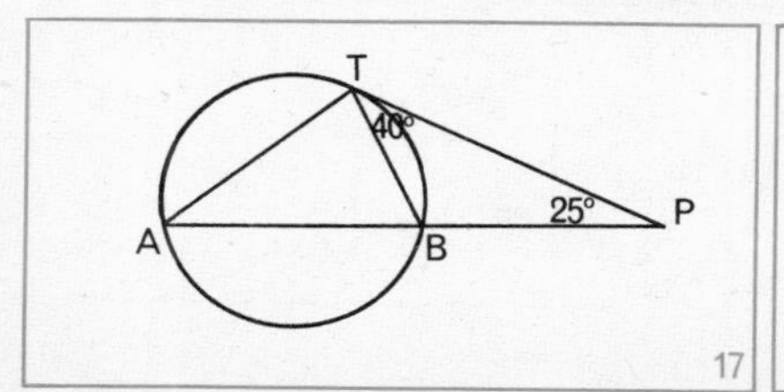

17

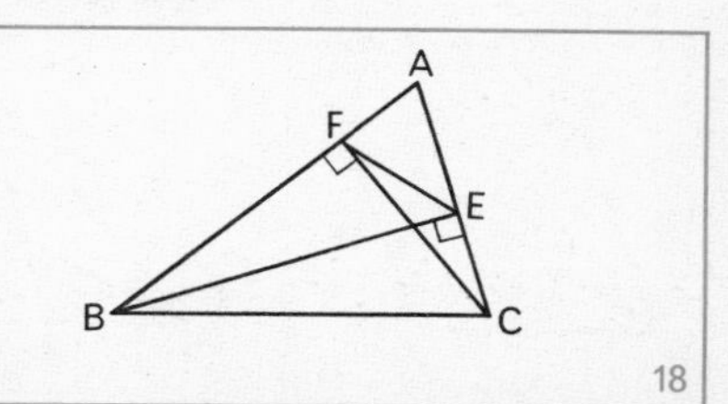

18

a Copy the figure and mark the sizes of all the angles.
b Repeat when $\angle PTB = x°$ and $\angle TPB = y°$.
c If PB = 4 cm and PA = 9 cm use a pair of similar triangles in the figure to calculate the length of the tangent PT.

27 In Figure 18, BE and CF are altitudes of $\triangle$ABC.

a Why is BFEC a cyclic quadrilateral?
b Name angles equal to angles EFC and ECF respectively.
c Prove that triangles AEF and ABC are equiangular.

28 Altitudes AD, BE and CF in $\triangle$ABC intersect at H. By marking pairs of equal angles associated with sets of concyclic points in the diagram, show that these altitudes bisect the angles of $\triangle$DEF.

Revision Exercise on Chapter 2
Composition of Transformations 1

Revision Exercise 2

1 Under a translation T_1, the origin O→A(5, −3), and under a translation T_2, A→B(4, 1).

a State the components of T_1 and T_2.
b State the components of $T_1 \circ T_2$ and $T_2 \circ T_1$.
c Find the coordinates of the images of A and B under $T_2 \circ T_1$.

2 In Figure 19, ABCD is a rhombus. State a transformation which will map:

a $\triangle$AEB to $\triangle$CED *b* $\triangle$AEB to $\triangle$CEB *c* $\triangle$AEB to $\triangle$AED.

3 In Figure 20, G is the centroid of $\triangle$ABC. Find single representatives of the composite translations represented by:

a $\vec{BA} + \frac{2}{3}\vec{AD}$ *b* $\frac{1}{3}\vec{DA} + \vec{GC}$ *c* $\vec{AD} + \frac{1}{2}\vec{CB}$.

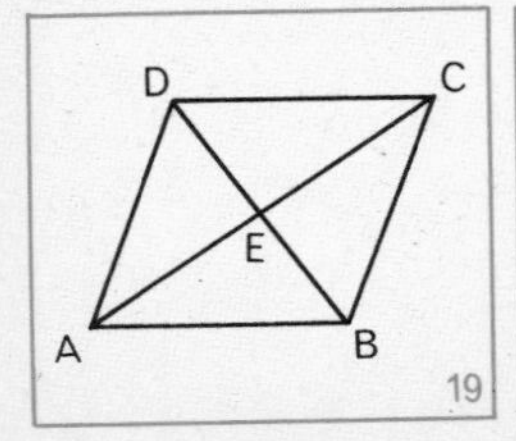

19

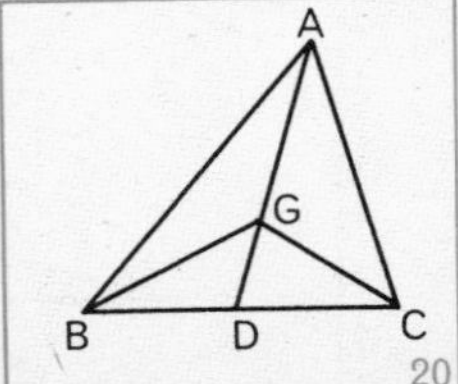

20

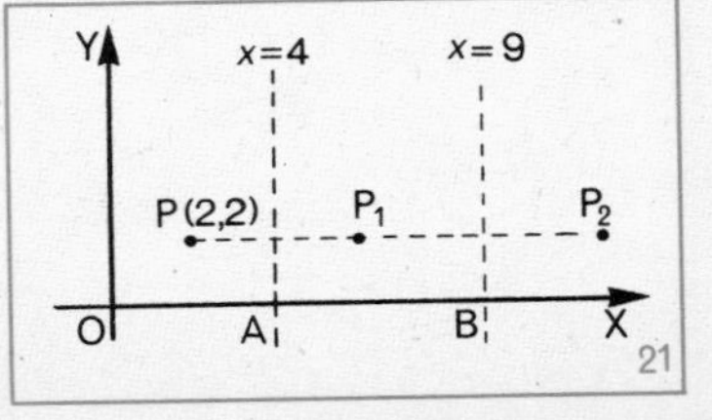

21

4 In Figure 21, P is the point (2, 2). P_1 is the image of P under reflection in the line $x = 4$ and P_2 is the image of P_1 under reflection in $x = 9$.

a Copy the diagram and write down the coordinates of P_1 and P_2.
b What single transformation maps P to P_2?
c How is the transformation related to the two axes $x = 4$ and $x = 9$?

5 A point A(2, 3) is translated to $A_2(12, 3)$ by means of successive reflections in parallel axes.

a If the axes are $x = 5$ and $x = k$, find k.
b If the axes are $x = h$ and $x = 6$, find h.
c If the axes are $x = h$ and $x = k$, write down an equation connecting h and k.

6 M_1 represents the operation of reflection in $y = 2$ and M_2 the operation of reflection in $y = 4$.

a Find the coordinates of the image of the point (4, 1) under the transformations $M_2 \circ M_1$ and $M_1 \circ M_2$.
b Explain why you could write $M_2 \circ M_1 = -M_1 \circ M_2$.

7 *a* Express $M_2 \circ M_1$ as a single transformation when M_1 represents reflection in the line $x = 2$ and M_2 reflection in the line $x = 5$.
b Verify that the points (0, 0) and $(-6, -12)$ lie on the line $y = 2x$.
c What are the images of (0, 0) and $(-6, -12)$ under $M_2 \circ M_1$?
d What is the equation of the image of the line $y = 2x$ under $M_2 \circ M_1$?

8 Repeat question 7 when M_1 represents reflection in $y = 0$ and M_2 reflection in $y = 6$.

9 Figure 22 shows part of a tiling of congruent triangles with AD at right angles to DG.

a State single transformations which will map:
(1) $\triangle$ABK to $\triangle$CDE (2) $\triangle$BKL to $\triangle$FEL
(3) $\triangle$KLH to $\triangle$LKB (4) $\triangle$ABK to $\triangle$ADG.

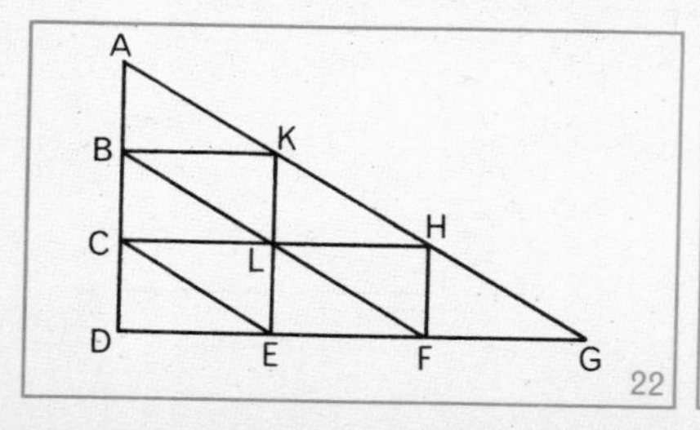

22

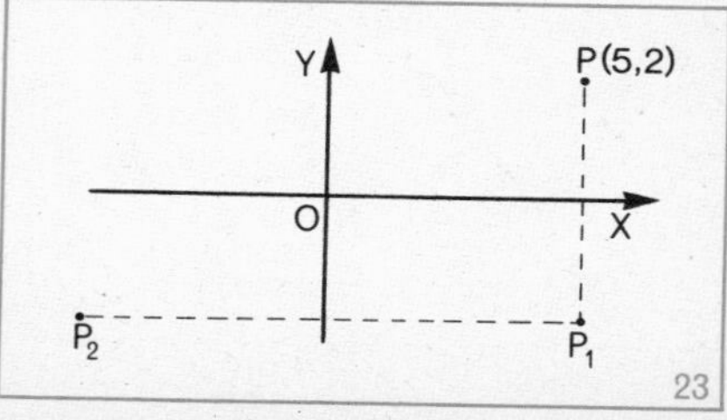

23

b Using lines in the figure as axes, state two successive reflections in the correct order which are equivalent to each of (*1*) and (*2*).

10 Figure 23 shows P_1 the image of P(5, 2) under reflection in the *x*-axis (the transformation X) and P_2 the image of P_1 under reflection in the *y*-axis (the transformation Y).

a Write down the coordinates of P_1 and P_2.
b Draw another diagram and write down the coordinates of P_1 and P_2 if P_1 is the image of P under Y and P_2 the image of P_1 under X.
c Write down an equation connecting $X \circ Y$ and $Y \circ X$.
d What single transformation will replace $X \circ Y$?

11 If H represents a half turn about the origin, draw diagrams on squared paper to illustrate the following:

a $X \circ Y(-2, 3) = H(-2, 3) = Y \circ X(-2, 3)$
b $H \circ Y(-4, -3) = X(-4, -3) = Y \circ H(-4, -3)$
c $X \circ H(5, -5) = Y(5, -5) = H \circ X(5, -5)$

12 What single transformation results from reflection in the *x*-axis followed by reflection in the *y*-axis?

Use diagrams on squared paper to find the equation of the images of the following lines after successive reflections in the *x* and *y*-axes.

a $y = \frac{1}{2}x$ *b* $y = x + 2$ *c* $y = 3x - 2$

13 In polar coordinates P is the point (2, 26°), (see Figure 24). P_1 is the image of P under a rotation of $+12°$ about O and P_2 is the image of P_1 under a rotation of $+15°$ about O.

a The equation of OP in polar coordinates is $\theta = 26$. Write down the equations of OP_1 and OP_2.
b Write down the polar coordinates of P_1 and P_2.
c What rotation maps P to P_2?
d How is this composite rotation related to the two original rotations?

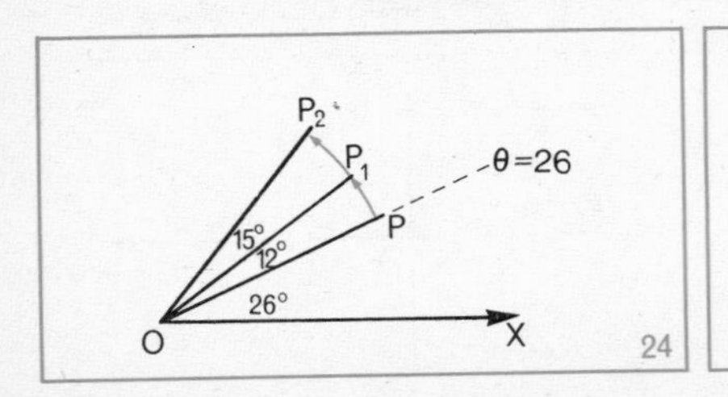

24

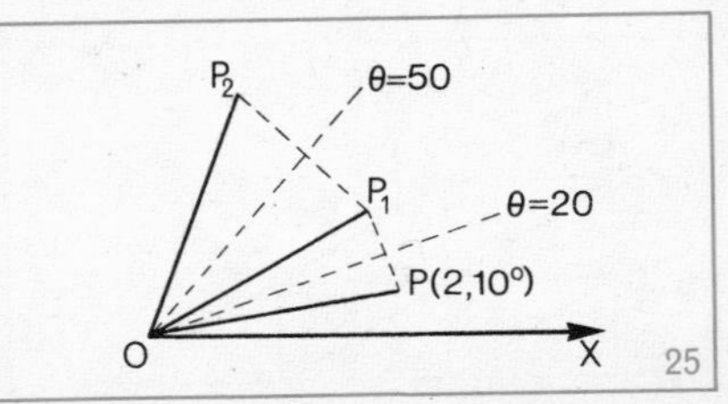

25

14 In polar coordinates P is the point (2, 10°), (see Figure 25). P_1 is the image of P under reflection in $\theta = 20$ and P_2 the image of P_1 under reflection in $\theta = 50$.

a Write down the polar coordinates of P_1 and P_2.
b State the size of the angle POP_2.
c If M_1 represents reflection in $\theta = 20$ and M_2 reflection in $\theta = 50$, which single transformation can replace $M_2 \circ M_1$?

15 Draw a diagram to illustrate what happens when P(2, 10°) is reflected first in $\theta = 50$ and its image reflected in $\theta = 20$. If M_1 and M_2 have the same meanings as in question ***14*** compare $M_2 \circ M_1$ and $M_1 \circ M_2$.

16a On squared paper draw the line L with equation $y = \frac{1}{2}x$.
b Draw L_1 the image of L in the line $y = x$.
c Draw L_2 the image of L_1 in the line $x = 0$.
d What single transformation maps L to L_2?
e Write down the coordinates of a point (not the origin) on the line $y = \frac{1}{2}x$ and of its image under the composite transformation.
f Write down the equation of L_2.

17a State as a single transformation the result of successive reflections in two parallel axes.
b State as a single transformation the result of successive reflections in two intersecting axes.
c A(2, 1), B(0, 3) and C(−1, 2) are vertices of a triangle. On squared paper draw the triangle and its image $\triangle A_1B_1C_1$ under reflection in the y-axis followed by reflection in the line $x = -2$.
d Draw also the image $\triangle A_2B_2C_2$ of $\triangle ABC$ under reflection in the y-axis followed by reflection in the line $y = x$.

18 $\triangle AOB$ has its vertices at O(0, 0), A(3, 0) and B(3, 3). On squared paper draw:

a $\triangle OA_1B_1$ the image of $\triangle OAB$ under a half turn about O
b $\triangle OA_2B_2$ the image of $\triangle OAB$ under reflection in OA followed by reflection in OB
c $\triangle OA_3B_3$ the image of $\triangle OAB$ under reflection in OB followed by reflection in OA.

What single rotation will map $\triangle OA_1B_1$ to $\triangle OA_3B_3$? State two successive reflections in the correct order equivalent to this rotation.

Does the diagram formed by $\triangle AOB$ and its three images have bilateral symmetry? Does it have point symmetry? What order of rotational symmetry does it have?

19 ABCD is a rectangle with AB = 1 unit and AD = 2 units. M is the midpoint of AD.

a Show that △ABM is similar to △MBC.
b △ABM can be mapped to △MBC by a rotation about B and a dilatation with centre B. State the magnitude and sense of the rotation and the scale factor of the dilatation.

20a What transformations correspond to the matrices

$$M=\begin{pmatrix}1 & 0\\0 & -1\end{pmatrix} \text{ and } N=\begin{pmatrix}0 & -1\\1 & 0\end{pmatrix}?$$

b If $P = MN$, find matrix P, and describe the transformation associated with it.

21 Under a transformation, $(a, b) \rightarrow (a', b')$ where $a' = a + 2b$ and $b' = a + b$.

a What is the matrix associated with this transformation?
b Use the matrix to find the images of O(0, 0), A(1, 0), B(1, 1) and C(0, 1).
c Is the transformation isometric?
d Has the transformation changed the area of the square?

22 By the use of matrices, find the images of (2, −4) and (5, −2) under the following transformations.

a Reflection in the y-axis
b Reflection in the line $y = x$
c Clockwise rotation of 90° about O
d Dilatation [O, 2]

23 Use products of matrices to find the images of (3, 4) and (−2, −3) under the following composite transformations.

a Reflection in the x-axis followed by reflection in the line $y = x$.
b Reflection in the line $y = x$ followed by reflection in the line $y = -x$.
c Clockwise rotation of 90° about O followed by the dilatation [O, 3].

Cumulative Revision Section

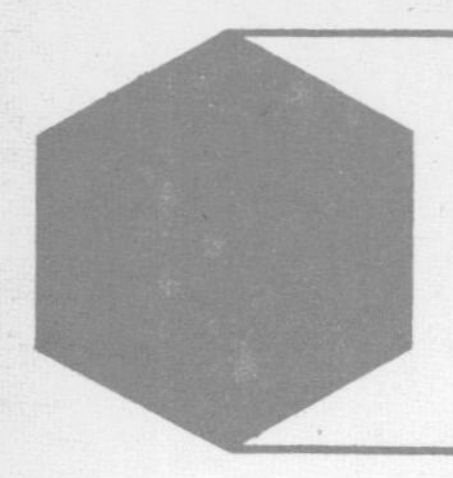

Cumulative Revision Section (Books 1–7)

Revision Topic 1. Rectangle and Square

Reminders

1 *a* Each face of a *cuboid* is a *rectangle*.
b Each face of a *cube* is a *square*.
c Cubes and cuboids have three sets of four *parallel edges*.
d They can be constructed from suitable nets.

2 *Axioms.* *a* A *rectangle* fits its outline in *four* ways as shown.

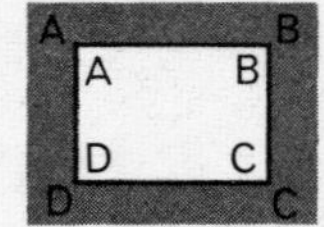

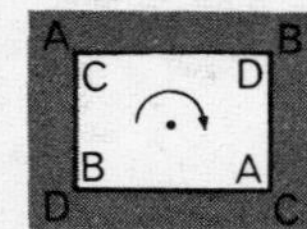

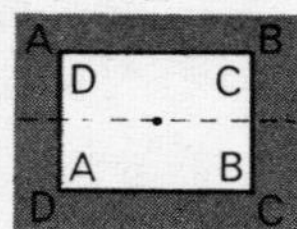

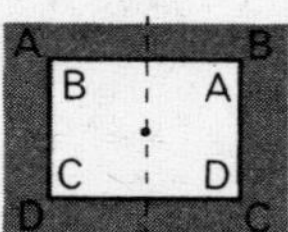

b Congruent rectangles can be fitted together to cover the plane exactly.

Deductions. *a* Opposite sides are equal and parallel.
b The diagonals are equal and bisect each other.
c Each angle is right.

3 A *square* fits its outline in *eight* ways.

Deductions.
a A square has all the properties of a rectangle.
b All the sides are equal.
c The diagonals are perpendicular and bisect the angles.

4 *Parallel lines* are straight lines which keep the same distance apart.

Exercise 1

1 Make a sketch of a net, with dimensions clearly marked, which can be folded to form an *open* box 10 cm by 6 cm by 4 cm.

2 If congruent cubes can be fitted exactly into the box above, find:

a the greatest size of cube that can be used
b the number of cubes of that size that can be fitted into the box.

3 A cuboid is 10 cm long, 3 cm broad and 3 cm high.

a How many edges does it have?
b How many different lengths of edge does it have?
c How many faces does it have?
d How many different sizes of face does it have?

4 Which of the following is (are) true about *both* rectangle and square?

a Diagonals are equal. *b* Diagonals bisect the angles.
c Diagonals are perpendicular. *d* Diagonals bisect each other.

5 Explain why it is always possible to draw a circle through the vertices of a rectangle.

6 Show how to construct accurately a rectangle with:

a a diagonal 8 cm long making an angle of 35° with one side.
b diagonals 8 cm long making an angle of 35° with each other.

7 PQRS is a rectangle with its diagonals intersecting at O. $\angle PQO = 36°$. Sketch the figure and mark in the sizes of as many angles as you can.

8 Construct a rectangle which has one side 9 cm long and an area equal to that of a square which has a side 6 cm long.

9 Find the coordinates of D and of the point of intersection of the diagonals of a rectangle ABCD when three of the vertices are:

a A(−1, 2), B(5, 2), C(5, 6) *b* A(2, 1), B(8, 4), C(6, 8).

10 Show how to construct accurately a square with a diagonal 10 cm long.

11 One vertex of a square is the point (2, 2) and the intersection of its diagonals is the point (5, 5). Find the coordinates of the other three vertices.

12 On squared paper construct a square which has two opposite vertices at the points (−1, 1) and (7, 3). Give the coordinates of the other vertices.

13 The straight line joining the points (1, 2) and (5, 3) is a side of a square. Find the coordinates of the remaining two vertices (two sets).

Revision Topic 2. Angles and Triangles

Reminders

1 $0^\circ <$ *an acute angle* $<$ *a right angle* $(90^\circ) <$ *an obtuse angle* $<$ *a straight angle* $(180^\circ) <$ *a reflex angle* $<$ *a complete turn* (360°).

2 *Directions and bearings*

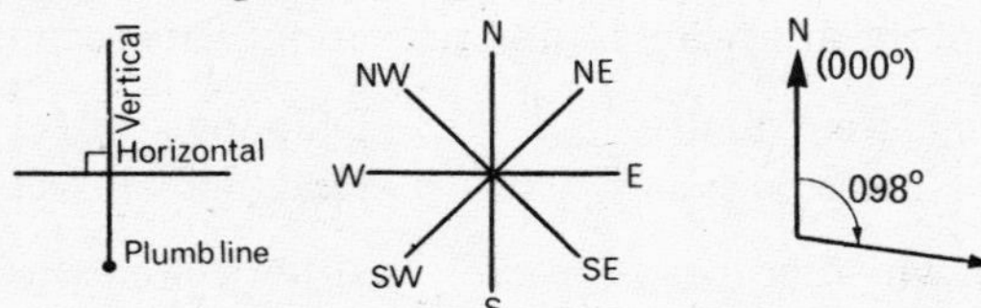

3 *Pairs of angles*

Supplementary angles Complementary angles Vertically opposite angles.

4 *Kinds of triangle*

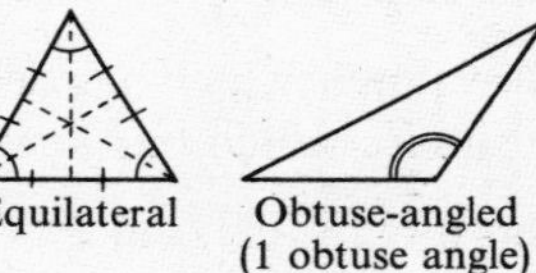
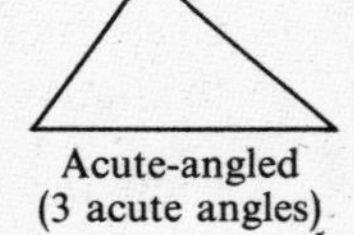

Right-angled Isosceles Equilateral Obtuse-angled (1 obtuse angle) Acute-angled (3 acute angles).

5 a The sum of the angles of a triangle is 180°.

b Area of a triangle $= \frac{1}{2}$ base $\times$ height.

6 Triangles are *congruent* if: *a* 3 sides = 3 sides

b 2 sides, included angle = 2 sides, included angle

c 2 angles, side = 2 angles, corresponding side.

7 *Concurrent lines in a triangle*

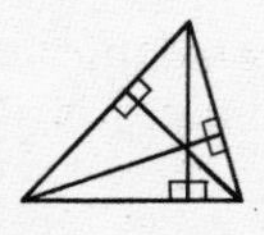

Altitudes

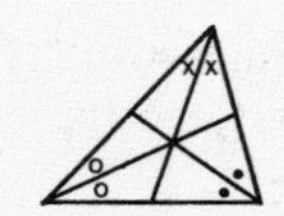

Angle bisectors

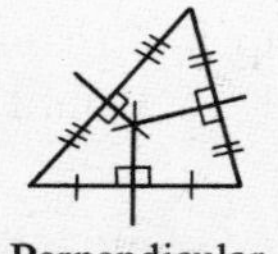

Perpendicular bisectors of sides

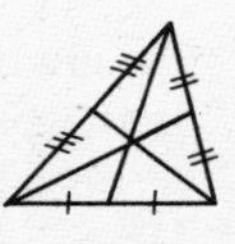

Medians

Exercise 2

1 The sizes of two angles are $(100-x)^\circ$ and $(3x+50)^\circ$. Calculate x if these two angles are: *a* supplementary *b* vertically opposite.

2 AB and CD intersect so that a pair of vertically opposite angles are complementary. Calculate all the angles between AB and CD.

3 Are the following true or false?

a Every triangle must have at least two acute angles.
b Each angle of an isosceles triangle must be 60°.
c The altitudes of a triangle are concurrent.

4 A rectangle which is 8 cm by 6 cm is divided into two right-angled triangles.

a What are the lengths of the bases and the altitudes of the two different isosceles triangles that can be formed from these triangles?
b Sketch and name other shapes that can be formed by fitting together equal sides of the two right-angled triangles.

5 An isosceles triangle ABC has AB = AC. Explain why the triangle must be equilateral if: *a* $\angle BAC = 60^\circ$ *b* $\angle ABC = 60^\circ$.

6 The vertices of a triangle are P(3, 2), Q(5, 6) and R(1, 8). Plot these points on squared paper and by the use of a circumscribing rectangle, or otherwise, calculate the area of $\triangle PQR$.

7 A ship sails on a bearing of 138° from a port P for a distance of 15 km to a point Q and then sails on a bearing of 012° to a point R. If the bearing of R from P is 053°, find from an accurate drawing to scale:

a the distance from Q to R *b* the ship's final distance from P.

8 Construct a triangle ABC whose area is 40 cm² and whose base AB is 8 cm. Do you have eough information to specify the triangle uniquely?

If also $\angle BAC = 50^\circ$, construct the triangle. Is there now a unique solution?

9 Construct the medians AD, BE, CF in a large $\triangle ABC$. If the medians are concurrent at M, check that AM : MD = 2 : 1.

Revision Topic 3. Parallelograms and Parallel Lines

Reminders

1 Under a *half turn* about O:

a O is an invariant point
b O is the midpoint of the line joining a point and its image
c A line and its image are parallel, or in the same line, but their directions are opposite.

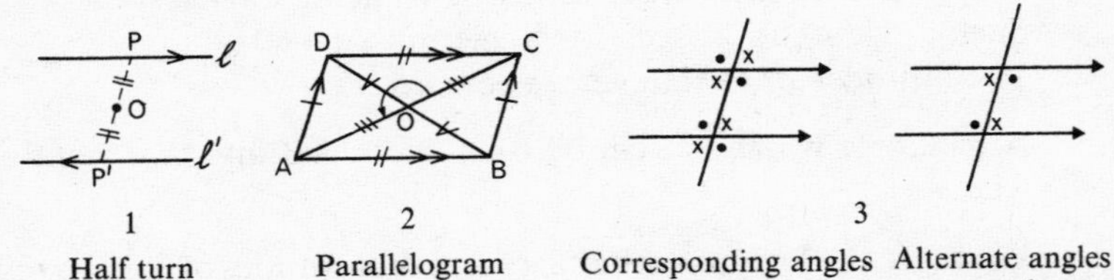

1 Half turn

2 Parallelogram

3 Corresponding angles equal — Alternate angles equal

Angles connected with parallel lines

2 A *parallelogram* is formed by a triangle and its image under a half turn about the midpoint of one side. Also:

a opposite sides are equal and parallel
b opposite angles are equal
c diagonals bisect each other
d it has half turn symmetry
e it can be formed by two pairs of parallel lines
f a plane can be covered by a tiling of congruent parallelograms containing a tiling of congruent triangles
g area of a parallelogram = base × perpendicular height.

Exercise 3

1 ABCD is a parallelogram in which $\angle A = 72°$ and AB is equal in length to diagonal BD. Calculate the sizes of all the angles in the figure.

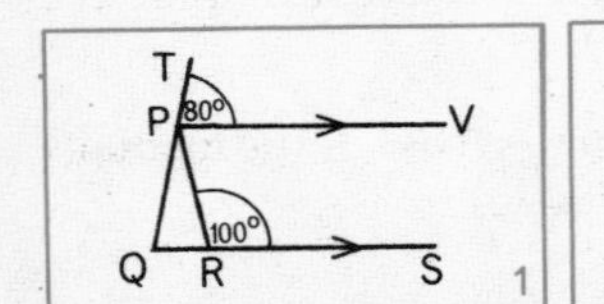

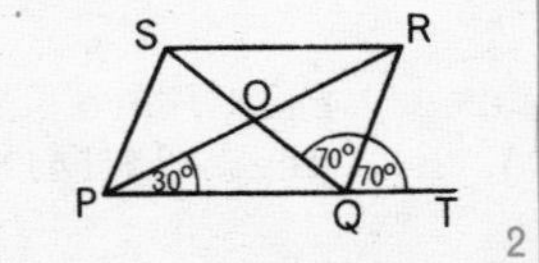

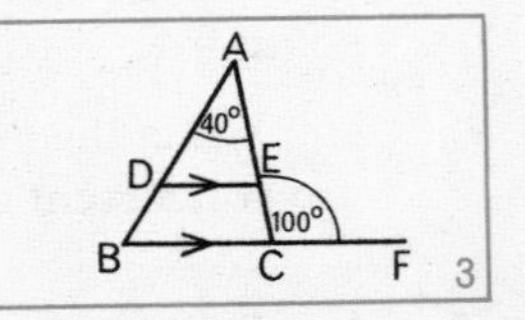

2 In Figure 1, $\angle PRS = 100°$, $\angle TPV = 80°$ and PV is parallel to QS. Explain why $\triangle PQR$ is isosceles.

3 In Figure 2, PQRS is a parallelogram. Mark in the sizes of all the angles in the diagram.

4 In Figure 3, $DE \parallel BC$. Mark in the sizes of all the angles.

5 O, P and Q are the points $(0,0)$, $(2,1)$ and $(2,4)$ respectively. If O, P and Q are vertices of a parallelogram, state all the possible positions of the fourth vertex.

Explain why in each case the area of the parallelogram is the same.

6 The vertices A and B of parallelogram ABCD are $(2,1)$ and $(6,-1)$. If its diagonals intersect at $(3,2)$, find the coordinates of C and D.

7 Which of the following properties are common to *both* parallelograms and rectangles?

a Diagonals bisect each other. *b* Diagonals are equal.
c All the angles are equal. *d* Opposite angles are equal.

8 ABCD is a parallelogram and E and F are the midpoints of BC and CD. Make a sketch of ABCD and its three images under separate half turns about E, C and F. What kind of figure is formed by ABCD and its three images? What is the ratio of the area of the complete figure to the area of ABCD?

9 If the universal set is the set of quadrilaterals (E), draw a Venn diagram to illustrate the relations between the set of parallelograms (P), rectangles (R) and squares (S).

10 If a parallelogram has one of its angles a right angle, explain why all of its angles must be right.

11 The diagonals of a parallelogram intersect at E. Use half turn symmetry to show that any line through E bisects the area of the parallelogram.

12 Under a transformation, $(x, y) \rightarrow (x+4, y-1)$. Find the images A_1 and B_1 of $A(3, -1)$ and $B(-1, 5)$ under the mapping. Verify that $A_1 B_1$ is equal and parallel to AB.

13 ABCD is a parallelogram with AB, BC, CD, DA produced to P, Q, R, S so that $BP = DR$ and $CQ = AS$. Prove that PQRS is a parallelogram.

Revision Topic 4
Reflection (also Rhombus and Kite)

Reminders

1 *Bilateral and central symmetry.* If a figure is mapped onto itself under *reflection in a line*, the line is an *axis of symmetry* and the figure has *bilateral symmetry* about the axis.

If a figure is mapped onto itself under *reflection in a point*, the point is called the *centre of symmetry*, and the figure has *half turn symmetry* about the point.

2 *Images of points under reflection, and associated matrices*

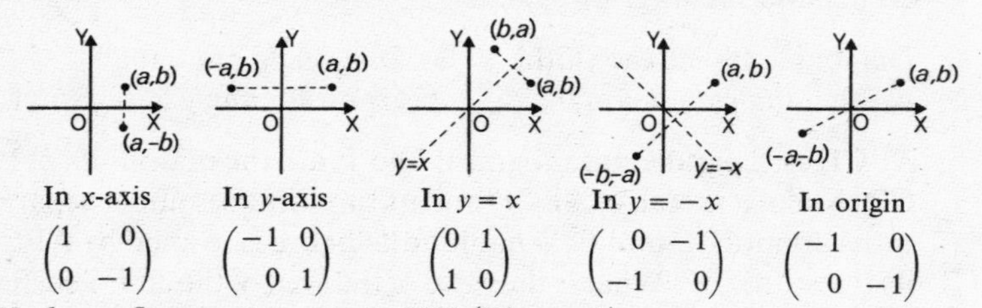

In x-axis $\begin{pmatrix}1 & 0\\0 & -1\end{pmatrix}$ In y-axis $\begin{pmatrix}-1 & 0\\0 & 1\end{pmatrix}$ In $y=x$ $\begin{pmatrix}0 & 1\\1 & 0\end{pmatrix}$ In $y=-x$ $\begin{pmatrix}0 & -1\\-1 & 0\end{pmatrix}$ In origin $\begin{pmatrix}-1 & 0\\0 & -1\end{pmatrix}$

3 *Under reflection in an axis XY* (Figure 1),
$A \leftrightarrow A_1$; $B \leftrightarrow B_1$; $AB \leftrightarrow A_1B_1$; $AB = A_1B_1$; $\angle ACY = \angle A_1CY$; AA_1 and BB_1 are bisected at right angles by XY; $AA_1 \parallel BB_1$; AB and A_1B_1 meet on XY, or are parellel to XY.

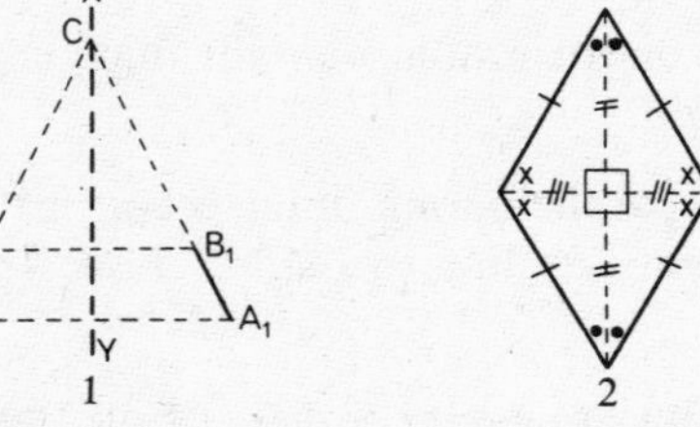

1 2

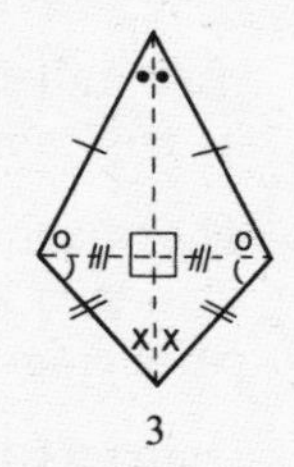

3

4 A *rhombus* (Figure 2) consists of two congruent isosceles triangles, base to base, and fits its outline in four ways. The diagonals are axes of symmetry.

5 A *kite* (Figure 3) consists of two isosceles triangles with equal bases and fits its outline in two ways. One diagonal is an axis of symmetry.

Exercise 4

1 Give the equation of the axis of reflection if the image of $(4, -3)$ is

a $(-4, -3)$ *b* $(4, 3)$ *c* $(-3, 4)$ *d* $(0, -3)$ *e* $(3, -4)$.

2 O is the origin, A is $(2, 0)$, B$(1, 1)$ and C$(0, 1)$.

a Make a sketch of the final figure if OABC is reflected in the x-axis, and OABC with its reflection is then reflected in the y-axis.
b What name would you give the final figure?
c How many pairs of parallel sides has it?
d Give the equations of its axes of symmetry.

3 ABCD is a parallelogram with AB = 2BC. E is the midpoint of AB, and F is the midpoint of DC. What shapes are AEFD and EBCF? Join AF and BF, and mark pairs of equal angles. Prove $\angle$AFB $= 90°$.

4 Triangle ABC has vertices A$(3, 3)$, B$(4, 3)$ and C$(4, 5)$. Show its image, triangle $A_1B_1C_1$, under reflection in the line $x = 5$ followed by reflection in the line $x = 10$. What single transformation would give the same result? State the coordinates of the image of triangle $A_1B_1C_1$ under reflection in the x-axis, then reflection in the y-axis.

5 a State the image of (x, y) under reflection in the y-axis.
b The equation of a curve is $y = (x-1)(x-3)$. Find the equation of the image of this curve under reflection in the y-axis.
c Sketch the two curves.

6 A is the point $(5, 5)$, B is $(3, 1)$ and C is $(1, 3)$.

a If ABDC is a kite, list the coordinates of four possible positions of D.
b State the equation of the locus of D.
c If ABDC is a rhombus, state the coordinates of D.
d Which point on the locus will make ABDC an isosceles triangle?

7 Show by means of: *a* a sketch *b* matrices, that reflection in the line $y = x$ followed by reflection in $y = -x$ is equivalent to a half turn about the origin. Hence find the image set of $\{(4, 3), (0, 0), (-1, -2)\}$ under the composite transformation.

8 $A = \begin{pmatrix} 1 & 0 \\ 0 & -1 \end{pmatrix}$ and $B = \begin{pmatrix} 0 & 1 \\ 1 & 0 \end{pmatrix}$. If $C = AB$ find the matrix C, and interpret the geometric transformations associated with A, B and C.

Revision Topic 5 Locus and Equations of Lines

Reminders

1 A *locus* can be regarded as a set of points defined in some way, or as the path traced out by a moving object.

2 The *gradient* of a line through O is given by the ratio

$$\frac{y\text{-coordinate of a point on the line}}{x\text{-coordinate of the same point}}.$$

3 The equation of a straight line:

a through the origin, with gradient m, is $y = mx$
b through the point $(0, c)$, with gradient m, is $y = mx + c$.

Exercise 5

1 A wheel with radius 1 cm runs round a rectangular frame 12 cm by 8 cm. Make sketches of the locus of the centre of the wheel if it rolls:

a round the outside *b* round the inside of the frame.

2 Make a sketch to show approximately the locus of a point on the rim of a wheel rolling along level ground. Is the locus periodic? If so, what is the period?

3 Make an accurate drawing to show the following:

a $R = \{P : AP \geqslant PB\}$ where A and B are fixed points 5 cm apart
b $S = \{Q : Q \text{ is equidistant from ED and EF}\}$ where $\angle DEF = 48°$.

4 a Draw a fixed straight line AB 10 cm long.
b Draw the locus of P if $\triangle PAB$ has an area of $15\,cm^2$.
c Find the possible positions of P if $AP = 4$ cm.

5 A is a point 10 cm from the straight line PQ. Make a drawing to show the set of points which are 6 cm or more from PQ and less than 6 cm from A.

6 Plot the points A(6, 0) and B(3, 6) and draw the medians of $\triangle AOB$. From your figure find the equation of the median from O.

7 The coordinates of the points $(p, 5)$ and $(-1, q)$ satisfy the equation $y = 2x - 1$. Calculate the values of p and q.

Plot these two points on squared paper and draw the straight line through them. Write down the coordinates of four other points on the line and verify that they also satisfy the equation.

8 Use set notation to define the locus of points in the shaded region of Figure 4.

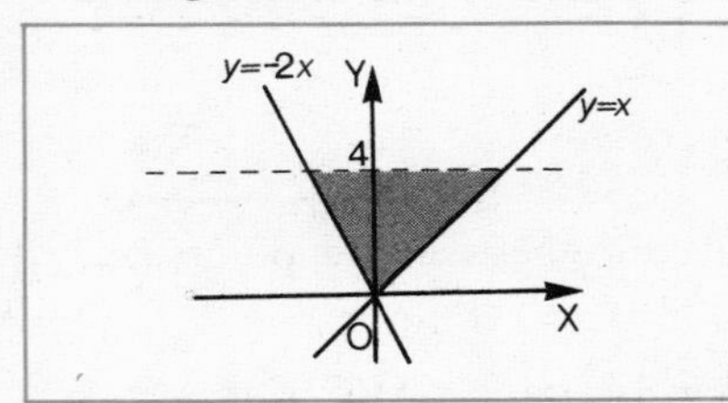

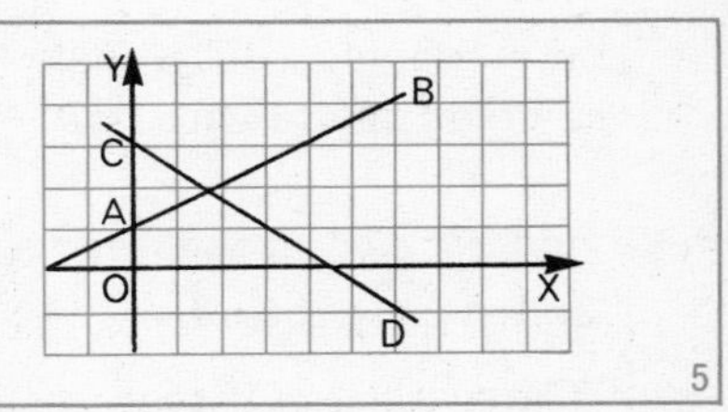

9 In Figure 5, find the equations of: *a* AB *b* CD.

10 Write down the equations of the lines through the following points, and having the following gradients:

a $(0, -3), 2$ *b* $(0, 4), -\frac{1}{2}$ *c* $(0, 1), -\frac{2}{3}$

Express your answers in the form $ax+by+c=0$, a, b and c integers.

11 State the gradient and the point of intersection with the y-axis of the straight lines with the following equations. Sketch the lines.

a $y=2x-3$ *b* $3x+y=4$ *c* $2x+3y=6$

12 A is the point $(4, 0)$ and B is $(0, 4)$. A point C is such that CA=CB.

a Write down the coordinates of four possible positions of C.
b What is the equation of the locus of C?
c When $\angle ACB = 40°$, state the sizes of the other two angles of $\triangle ACB$.
d What are the coordinates of C if A, C, B do not form a triangle?

13 Let $R = \{(x, y): x+y \geqslant 2\}$ and $S = \{(x, y): y-2x \geqslant 2\}$.

a Illustrate $R \cap S$ in a diagram and show that $(0, 2) \in R \cap S$.
b If $H(1, h) \in R \cap S$ find the possible values of h.
c Repeat for the points $K(2, k)$ and $L(3, l)$.

14 Sketch the locus defined by $A \cup B \cup C \cup D$ where

$A = \{(x, y): 0 \leqslant x \leqslant 1, y=1\}$,
$B = \{(x, y): 1 \leqslant x \leqslant 2, y=2\}$,
$C = \{(x, y): 2 \leqslant x \leqslant 3, y=3\}$
and $D = \{(x, y): 3 \leqslant x \leqslant 4, y=4\}$.

Revision Topic 6 Calculation of Distance

Reminders

1. *Pythagoras' theorem.* The square on the hypotenuse of a right-angled triangle is equal to the sum of the squares on the other two sides; *or*, in triangle ABC, if angle A is a right angle then $a^2 = b^2 + c^2$.

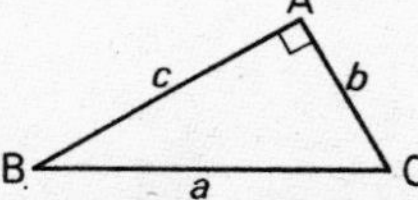

2. *The distance formula.* If d is the distance between the points (x_1, y_1) and (x_2, y_2), then $d = \sqrt{[(x_2 - x_1)^2 + (y_2 - y_1)^2]}$.

3. *The converse of Pythagoras' theorem.* In triangle ABC, if $a^2 = b^2 + c^2$, then angle A is a right angle.

Exercise 6

1. The greatest length of an extending ladder is 10 m. Calculate the greatest distance up a vertical wall the ladder can reach when the foot of the ladder is 6 m from the foot of the wall.

   When the ladder is adjusted to 8·5 m, it reaches a point 7·5 m above the ground. Calculate the distance of the foot of the ladder from the foot of the wall.

2. A man starting from A walks 5 km due east and then 4 km due north to B. Calculate, to one decimal place, the distance from A to B direct.

   From B he walks 5 km north and then 4 km west to C. Calculate, to one decimal place:

   a the distance from B to C   *b* the distance from A to C.

3. *a* Use squared paper to draw accurately a square with a diagonal of 10 cm.
   b Calculate the length of a side. (Let x cm be the required length.)
   c Verify your answer by measurement.

4. State the converse of Pythagoras' theorem. Prove that the triangle with sides PQ = 6 cm, PR = 2·5 cm and QR = 6·5 cm is right-angled. If S and T are the images of Q and R under the dilatation [P, 3], find the lengths of PT and QS.

5. The vertices of an isosceles triangle are O(0, 0), A(8, 6) and B (6, 8).

Find the area of $\triangle$OAB by calculating the lengths of AB and of the altitude from O to AB.

Verify your answer by using a circumscribing square.

6 A line of gradient $-\frac{4}{3}$ is drawn through the origin O. Write down the equation of this line.

A point A is chosen on the line with first coordinate 3. If B is the point $(5, 0)$, show that triangle OBA is isosceles and calculate its area.

If C is the point $(3, 4)$, calculate the area of OABC.

7 A and B are the points $(1, 2)$ and $(5, -2)$ respectively. $P(x, y)$ is a point such that $AP = PB$. Use the distance formula to find the equation of the locus of P. Describe the locus in relation to the line AB.

8 If the distance of the point (x, y) from the origin is always equal to the distance between $(3, -9)$ and $(-1, -6)$, find an equation connecting x and y. What locus does this equation represent?

9 By replacing x by $0, \frac{1}{2}, 1, 1\frac{1}{2}, 2, 2\frac{1}{2}$ make a sketch of the locus given by $A = \{(x, y): y^2 = 4x\}$.

Show that each of the points found above is the same distance from the point $(1, 0)$ as from the line $x = -1$.

10 S is the point $(0, 3)$ and DD_1 is the line with equation $y = -3$. Find the equation of the locus of $P(x, y)$ such that the distance from P to S is equal to the distance from P to DD_1.

Sketch the locus and give the equation of its axis of symmetry.

11 A is the point $(2, 0)$, $B(8, 0)$ and $P(x, y)$. PD is drawn perpendicular to the x-axis. Express the lengths of AD and DB in terms of x and use Pythagoras' theorem to calculate AP^2 and BP^2 in terms of x and y.

Hence show that the equation of the locus $\{(x, y): PB^2 = 4PA^2\}$ is $x^2 + y^2 = 16$.

12 A triangle has sides AB, BC and CA measuring 14, 48 and 50 units.

a Prove that the triangle is right-angled, and calculate its area.
b Calculate the length of the altitude from B to CA.

13 The points A, B and C have coordinates $(-3, 1)$, $(-2, -1)$ and $(4, 2)$ respectively. Show that the angle ABC is a right angle. Find the coordinates of D if:

a ABCD is a rectangle *b* ABDC is a parallelogram
c D is the centre of the circumcircle of $\triangle$ABC.

Revision Topic 7 Translation and Vectors

Reminders

1 A *translation* is a displacement of all points in the plane through the same distance in the same direction.

The translation may be represented by any one of a set of directed line segments with the same magnitude and direction.

The translation may be defined by components, e.g. $\begin{pmatrix} a \\ b \end{pmatrix}$.

2 *Translations may be combined* (or added) by:

a a head-to-tail arrangement of directed line segments ($\overrightarrow{PQ}+\overrightarrow{QR}=\overrightarrow{PR}$)

b adding components, $\begin{pmatrix} a \\ b \end{pmatrix}+\begin{pmatrix} c \\ d \end{pmatrix}=\begin{pmatrix} a+c \\ b+d \end{pmatrix}$.

3 *Vectors.* The set of all directed line segments with the same magnitude and direction is a *geometrical vector*.

The segments are *representatives* of the vector.

A vector may be written $\boldsymbol{u}$ or $\begin{pmatrix} a \\ b \end{pmatrix}$.

Vectors are added like translations; addition is commutative and associative.

4 *a* *Identity element.* $\boldsymbol{u}+\boldsymbol{0}=\boldsymbol{u}=\boldsymbol{0}+\boldsymbol{u}$

b *Additive inverse.* $\boldsymbol{u}+(-\boldsymbol{u})=\boldsymbol{0}=(-\boldsymbol{u})+\boldsymbol{u}$

c *Subtraction.* $\boldsymbol{u}-\boldsymbol{v}=\boldsymbol{u}+(-\boldsymbol{v})$

d *Multiplication by a number.* $k\boldsymbol{a}$ has $|k|$ times the magnitude of $\boldsymbol{u}$ and the same or opposite direction according as k is positive or negative.

e *Distributive laws.* $k\boldsymbol{u}+k\boldsymbol{v}=k(\boldsymbol{u}+\boldsymbol{v})$, $k\boldsymbol{u}+m\boldsymbol{u}=(k+m)\boldsymbol{u}$

f *Position vector.* If P is the point (h, k), $\overrightarrow{OP}$ represents the position vector $\boldsymbol{p}$ of P, and $\boldsymbol{p}=\begin{pmatrix} h \\ k \end{pmatrix}$.

5 *a* If $\boldsymbol{u}=\begin{pmatrix} a \\ b \end{pmatrix}$, then the *magnitude* of $\boldsymbol{u}$, $|\boldsymbol{u}|$, $=\sqrt{(a^2+b^2)}$.

b $\overrightarrow{PQ}$ represents $\boldsymbol{q}-\boldsymbol{p}$.

c If M is the *midpoint* of PQ, then $\boldsymbol{m}=\frac{1}{2}(\boldsymbol{p}+\boldsymbol{q})$. If P is the point (x_1, y_1) and Q is (x_2, y_2), then M is $[\frac{1}{2}(x_1+x_2), \frac{1}{2}(y_1+y_2)]$.

Exercise 7

1 In Figure 6, ABCDEF is a regular hexagon with centre O. $\vec{AB}$ represents $\boldsymbol{u}$ and $\vec{BC}$ represents $\boldsymbol{v}$. Write down in terms of $\boldsymbol{u}$ and $\boldsymbol{v}$ the vectors represented by:

a $\vec{OA}$ *b* $\vec{OB}$ *c* $\vec{EF}$ *d* $\vec{AD}$ *e* $\vec{AM}$, M the midpoint of AE.

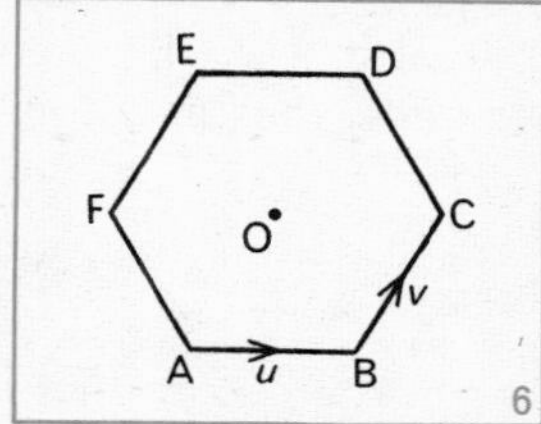

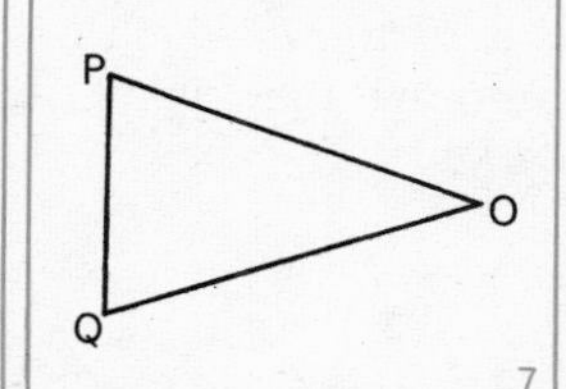

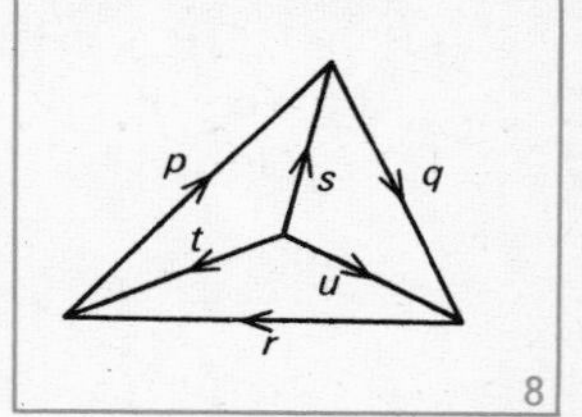

2 In Figure 7, △OPQ has OP = OQ. If $\boldsymbol{p}$ and $\boldsymbol{q}$ are position vectors of P and Q for origin O, which of these are true and which are false?

a $\boldsymbol{p}=\boldsymbol{q}$ *b* the direction of $\boldsymbol{p}$ is the same as the direction of $\boldsymbol{q}$
c $|\boldsymbol{p}|=|\boldsymbol{q}|$ *d* $\vec{PQ}$ represents the vector $\boldsymbol{q}-\boldsymbol{p}$.

3 From Figure 8 simplify:

a $\boldsymbol{s}+\boldsymbol{q}$ *b* $\boldsymbol{s}-\boldsymbol{t}$ *c* $\boldsymbol{p}+\boldsymbol{q}+\boldsymbol{r}$ *d* $\boldsymbol{r}-\boldsymbol{t}$ *e* $\boldsymbol{p}+\boldsymbol{q}-\boldsymbol{u}$ *f* $\boldsymbol{t}+\boldsymbol{p}+\boldsymbol{q}$

4 O is the origin and A and B are the points (4, 5) and (−1, 3).

a Express $2\boldsymbol{a}-3\boldsymbol{b}$ in component form.
b What are the components of $\vec{AB}$?
c What are the components of $\vec{OP}$ where P is the midpoint of AB?
d Calculate the length of AB, leaving your answer in surd form.

5 a Simplify $\frac{1}{3}\begin{pmatrix}3\\6\end{pmatrix}+\frac{1}{2}\begin{pmatrix}2\\-4\end{pmatrix}+\frac{1}{4}\begin{pmatrix}-8\\0\end{pmatrix}$.

b Find x and y from the vector equation $\begin{pmatrix}x\\2x\end{pmatrix}+\begin{pmatrix}2y\\-3y\end{pmatrix}=\begin{pmatrix}1\\9\end{pmatrix}$.

6 Relative to the origin O, A has position vector $\boldsymbol{a}=\begin{pmatrix}x\\4\end{pmatrix}$

and B has position vector $\boldsymbol{b}=\begin{pmatrix}x+2\\-2\end{pmatrix}$ *a* If OA = OB, find x.

b Express $\vec{AB}$ in component form. *c* If ∠AOB = 90°, find x.

7 A, B, C and D are the points (4, 1), (10, 3), (7, 7) and (1, 5) respectively.

a Find the components of the vectors represented by $\vec{AB}$ and $\vec{CD}$.
b What do these enable you to say about the figure ABCD?

Revision Topic 8 Dilatation and Similar Shapes

Reminders

1 For a given centre O and a scale factor k, a figure may be *enlarged* or *reduced* by mapping $P \rightarrow P_1$, where $\overrightarrow{OP_1} = k\overrightarrow{OP}$.

The transformation of points in a plane $(P \rightarrow P_1)$ by the relation $\overrightarrow{OP_1} = k\overrightarrow{OP}$ is called a *dilatation* which is denoted by $[O, k]$.

If the centre of dilatation is the origin O, then under the dilatation $[O, k]$, $P(a, b) \rightarrow P_1(ka, kb)$.

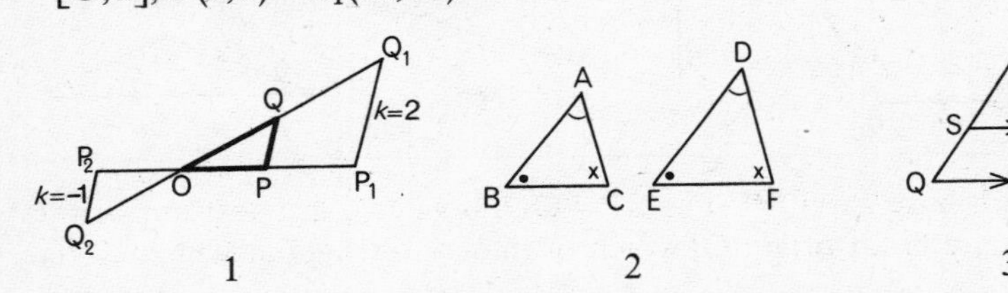

2 Two figures are *similar* if they are equiangular *and* have pairs of corresponding sides in proportion.

3 Triangles ABC and DEF are similar $\Leftrightarrow \angle A = \angle D$, $\angle B = \angle E$ and $\angle C = \angle F$, *or* $\frac{AB}{DE} = \frac{BC}{EF} = \frac{AC}{DF}$. In Figure 3, $\frac{PS}{SQ} = \frac{PT}{TR}$

4 The *gradient* of a line $AB = \frac{y\text{-component of AB}}{x\text{-component of AB}}$.

Exercise 8

1 A rectangular photograph, 10 cm wide and 15 cm high is similar to a rectangular mount. If the margins at the top and bottom are 3 cm and 6 cm wide respectively, calculate the width of the mount and the breadth of the equal margins at the sides.

2 a Why are two regular pentagons similar?

b Explain why an isosceles triangle with an angle of 52° can be similar to another isosceles triangle with an angle of 76°.

c Explain why a square need not be similar to: (*1*) a rhombus (*2*) a rectangle.

3 In Figure 9, parallel lines are shown by arrows. Find x in each part.

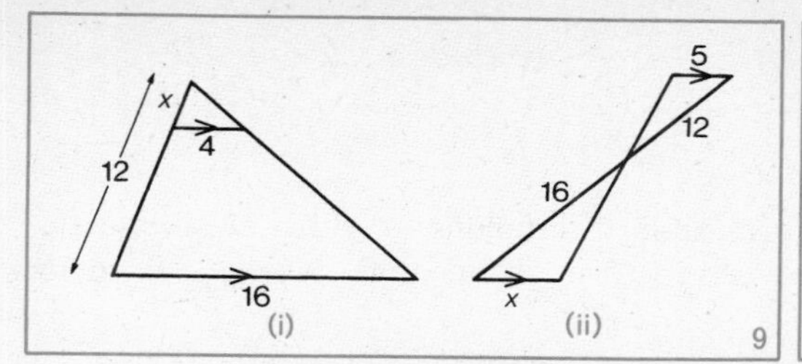

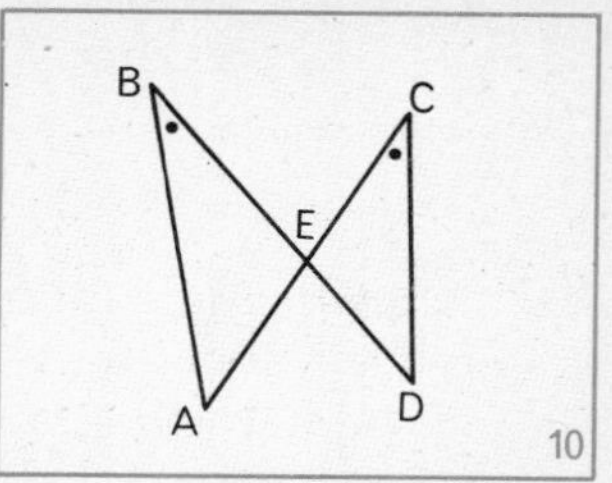

4 In Figure 10, ∠ABE = ∠DCE. Explain why triangles ABE and DCE are similar.

If AB = 10 cm, DC = 6 cm, EC = 4·5 cm and BD = 12 cm, calculate the lengths of BE and ED.

5 A triangle is transformed by a dilatation with scale factor $\frac{1}{2}$. Which of the following are true and which false?

a Its sides are halved. *b* Its angles are halved. *c* Its area is halved.
d The image triangle is congruent to the original triangle.
e The image is similar to and has its sides parallel to the original.

6 Write down the coordinates of the images of the point (3, 2) under dilatations with the following centres and scale factors.

a (2, 1), 2 *b* (−1, −1), 3 *c* (3, 2), −5 *d* (9, 8), $-\frac{1}{2}$

7 In Figure 11, ABCD is a parallelogram. P is the midpoint of DB, and Q is the midpoint of PB. The parallelograms ARPS and STQV are drawn. Find the centre and scale factor of the dilatation which maps:

a ASPR to ABCD *b* SVQT to ASPR *c* SVQT to ABCD.

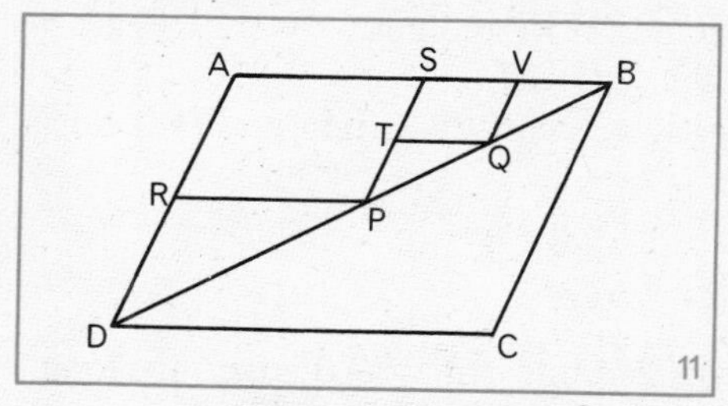

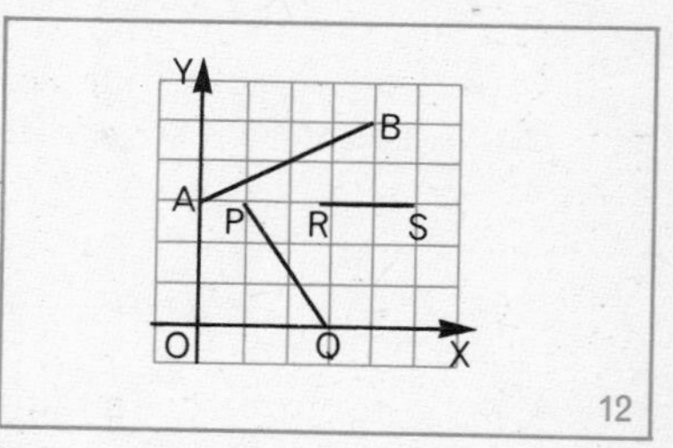

8 In Figure 12, find the gradients of: *a* AB *b* BA *c* RS *d* PQ. Calculate also the gradient of the line joining (−1, 4) to (6, −2).

9 OP is a piece of elastic with one end O fixed at the origin. Q is the midpoint of OP. If the elastic remains stretched state the locus of Q as P traces out a straight line with equation $x+2y=32$.

Revision Topic 9 Rotation and Circles

Reminders

1 A *rotation* is a transformation of the plane in which all points are rotated about a given centre through an angle which is fixed in magnitude and sense.

Every line in the plane is rotated through the same angle.

The centre of rotation is an *invariant* point.

2 *Rotational properties of the circle.* In equal circles or in the same circle,

equal angles at the centre
⇔ equal arcs subtending the angles
⇔ equal chords cutting off the arcs
⇔ equal sectors

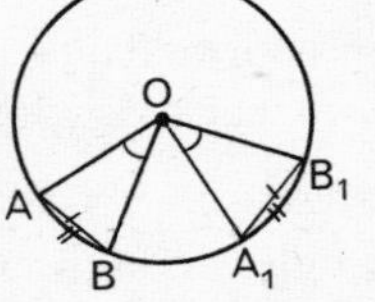

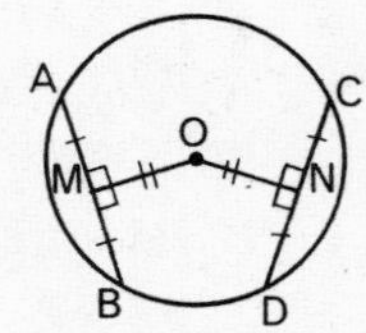

3 *Equal chords* of a circle are equidistant from the centre. Chords of a circle which are equidistant from the centre are equal.

4 A *regular polygon* of n sides can be inscribed in a circle by drawing n equally spaced radii and joining their ends.

5 *Properties of the circle by bilateral symmetry.*

a The perpendicular from O to AB bisects AB.

b The join of O to the midpoint of AB is perpendicular to AB.

c The perpendicular bisector of AB passes through O.

We may use Pythagoras' theorem or trigonometrical calculations in triangles like △OAM.

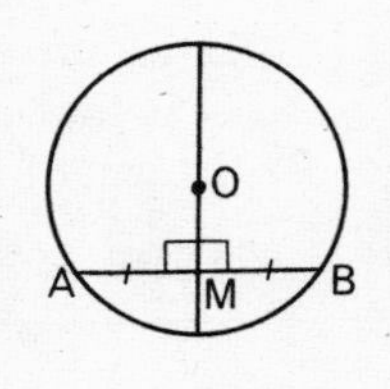

6 The *equation of a circle* with its centre at the origin and radius r is $x^2+y^2=r^2$.

Exercise 9

1 A circle, centre P, has a radius of 24 units. An arc AB is 6π units long. Calculate the size of ∠APB.

2 In Figure 13, ∠OBC = 36° and OA = 10 cm. Calculate:

a ∠AOB *b* ∠COB *c* the length of arc BC
d the area of sector AOB.

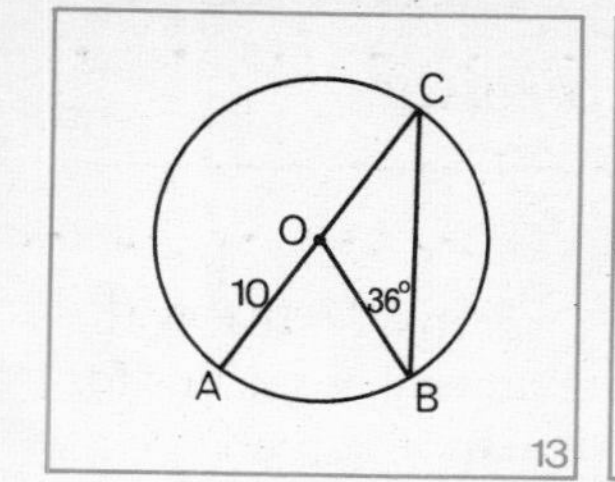

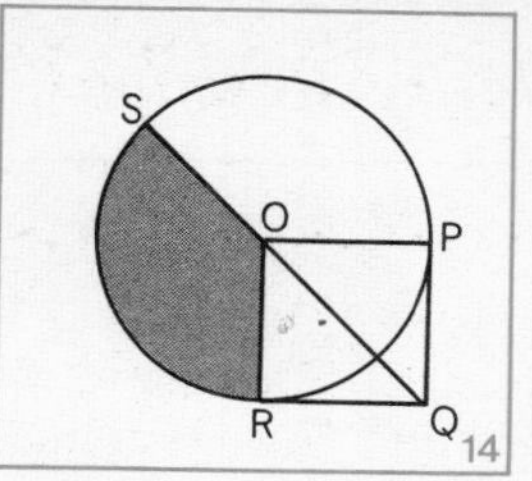

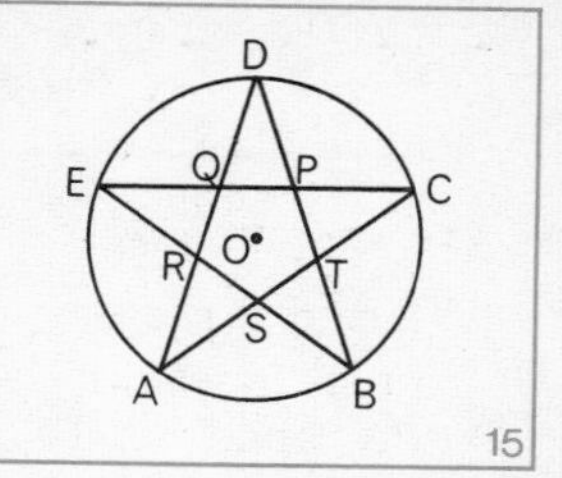

3 In Figure 14, the radius is 4 cm, and OPQR is a square. Show that the shaded area is $6\pi\,\text{cm}^2$, and that its perimeter is $(3\pi+8)$ cm.

4 In Figure 15, A, B, C, D, E are vertices of a regular pentagon.

a By means of a rotation, show that PQRST is a regular pentagon.
b Calculate the angle at each vertex of the five-pointed star shown.

5 Calculate the distance of the origin O from the point A$(-1, 2)$. OA is rotated through $-90°$ to a position OA_1. Find the coordinates of A_1, and the area of $\triangle OAA_1$.

6 $\triangle$ABC is acute-angled at A. ABDE and ACFG are squares drawn outside the triangle. Describe a rotation which will map $\triangle$EAC onto $\triangle$BAG, and state the size of the angle between EC and BG.

7 A circle centre O has radius 40 mm and P is a point inside the circle 10 mm from O. Calculate the lengths of the longest and shortest chords through P.

8 T is the centre of a circle with radius r units. AB is a chord of the circle and S is a point on AB such that AS $= x$ units and SB $= 3x$ units. If TS $= p$ units, show that $p^2 = r^2 - 3x^2$.

If Q is a point on the circle such that SQ $= x\sqrt{3}$ units, prove that $\angle QST = 90°$.

9 On the same diagram on squared paper show the sets:

a $\{(x, y): x^2+y^2 \leqslant 16\} \cap \{(x, y): -3 < x < -2\}$
b $\{(x, y): x^2+y^2 \leqslant 16\} \cap \{(x, y): 2 < x < 3\}$.

Does the resulting figure possess rotational symmetry? If so, state the centre and order.
Does the figure possess bilateral symmetry? State axes of symmetry.

10 Find the equations of the circles produced from the circle $x^2+y^2=9$ under the dilatations *a* $[O, 3]$ *b* $[O, \frac{1}{2}]$ *c* $[O, \frac{1}{4}]$ *d* $[O, 4]$.

Cumulative Revision Exercises

Exercise 1A

1 A and B are the points $(-1, 3)$ and $(5, -3)$ respectively. If P is the point (x, y), use the distance formula to write down expressions for PA^2 and PB^2. If $PA^2 = PB^2$, find a simplified equation in x and y.

2 PQRS is a square on a side of 12 cm. A and D are points on PQ and SP respectively such that PA = 8 cm and DP = 4 cm. The images of A and D under reflection in SQ are A_1 and D_1 respectively.

a Calculate the area of AA_1D_1D.
b Explain why DA, D_1A_1, SQ are concurrent.

3 A triangle ABC is such that $\overrightarrow{BA}$ represents vector $\boldsymbol{u}$ and $\overrightarrow{BC}$ represents vector $\boldsymbol{v}$. BC is produced its own length to P, and Q is taken on AC so that AQ:QC = 1:2. Express in terms of $\boldsymbol{u}$ and $\boldsymbol{v}$: *a* $\overrightarrow{BP}$ *b* $\overrightarrow{AP}$ *c* $\overrightarrow{BQ}$.

4 $\boldsymbol{u} = \begin{pmatrix} 1 \\ 2 \end{pmatrix}$, $\boldsymbol{v} = \begin{pmatrix} 3 \\ -1 \end{pmatrix}$, $\boldsymbol{w} = \begin{pmatrix} 4 \\ -6 \end{pmatrix}$. Find p, q such that $p\boldsymbol{u} + q\boldsymbol{v} = \boldsymbol{w}$.

5 Under a certain transformation, $P(x, y) \rightarrow P_1(x+y, x-y)$. Describe the image of the set of points on the line $y = x$ under the transformation. What combined rotation and dilatation would be equivalent to the transformation?

6 In Figure 16, AB∥PQ, RA = 6 cm, AP = 2 cm, AB = 5 cm and RB = 4 cm. Calculate the lengths of BQ and PQ.

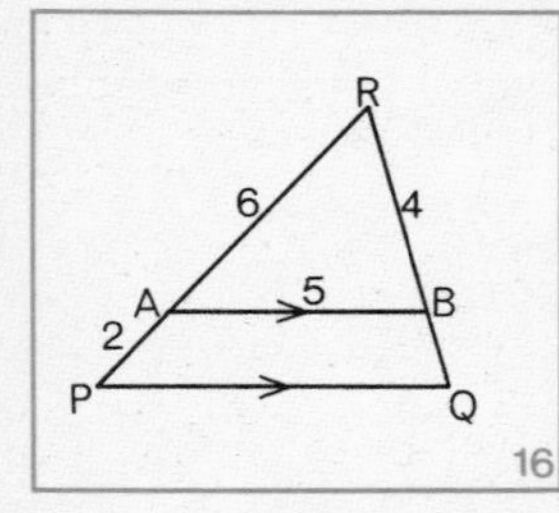

16

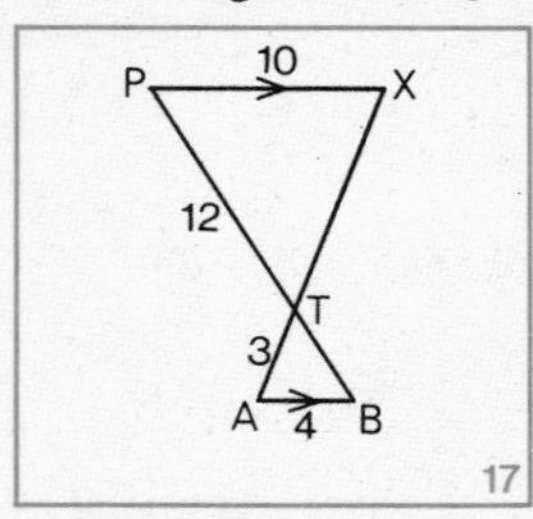

17

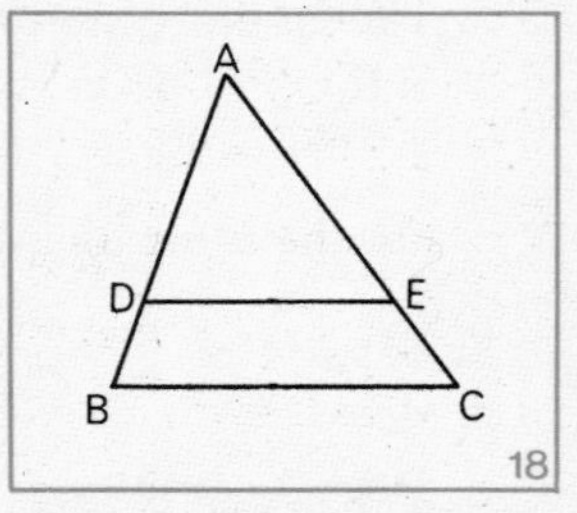

18

7 In Figure 17, PX∥AB. PX = 10 mm, PT = 12 mm, AB = 4 mm and AT = 3 mm. Calculate the lengths of XT and TB.

8 In Figure 18, D and E divide AB and AC in the ratio $m:n$. If $\overrightarrow{BC} = k\overrightarrow{DE}$ express k in terms of m and n. What can you say about BC and DE?

Exercise 1B

1 A and B are the centres of two circles of radii 5 and 3 units respectively. AB has length 20 units. L and M are points on the circles centres A and B respectively, such that LM touches the circle centre B at M and LM is parallel to AB. Calculate the length of LM. (There are two answers.)

2 $\vec{OP}$ represents the vector $\begin{pmatrix} a \\ b \end{pmatrix}$ and $\vec{OQ}$ represents $\begin{pmatrix} r \\ s \end{pmatrix}$ relative to rectangular axes.

a Given that O, P and Q are collinear, show that $as = br$.
b Given that OP is perpendicular to OQ, show by using the distance formula that $ar + bs = 0$.

3 In Figure 19, RS is parallel to BC. AR = 12 units and RB = 8 units. Use similar triangles to calculate the ratio in which SB divides RC.

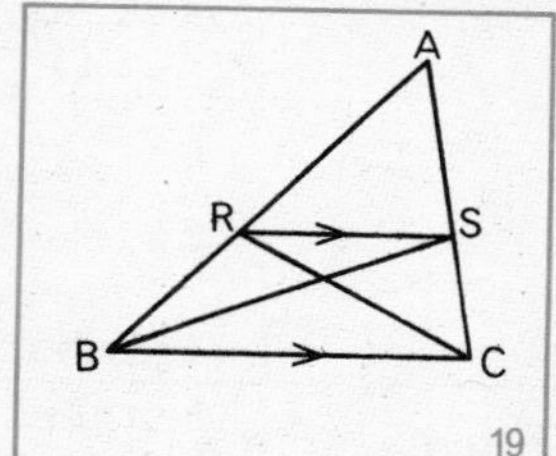

19

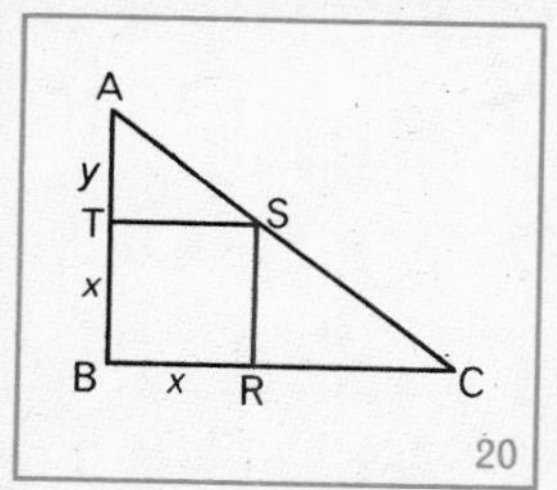

20

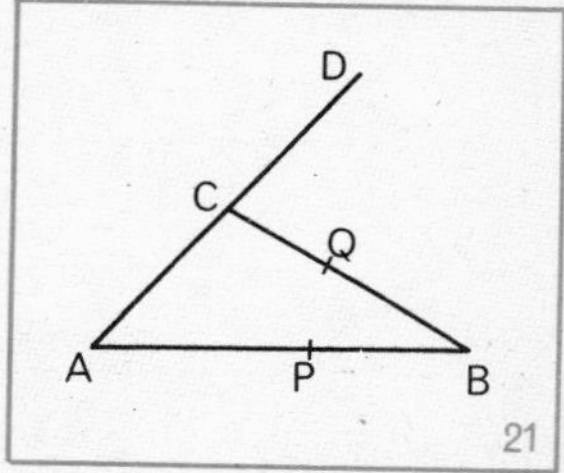

21

4 In Figure 20, BRST is a square of side x units. RC = 2 units and AT = y units. By considering similar triangles, prove that the area of $\triangle$ATS is $\frac{1}{4}x^3$ units^2.

5 In Figure 21, $\vec{AB}$ represents the vector $7\boldsymbol{u}$ and $\vec{BC}$ the vector $5\boldsymbol{v}$. If also $\frac{AP}{PB} = \frac{4}{3}$ and $\frac{BQ}{QC} = \frac{3}{2}$ express the vector represented by $\vec{PQ}$ in terms of $\boldsymbol{u}$ and $\boldsymbol{v}$.

D is a point on AC produced such that CD = AC. Express the vectors represented by $\vec{AD}$ and $\vec{PD}$ in terms of $\boldsymbol{u}$ and $\boldsymbol{v}$. What can you deduce about P, Q and D?

6 Under the dilatation [O, 4], the curve $8y = 3x^2$ is mapped to the curve $ay = bx^2$. Find a and b, giving the smallest positive integers.

7 On a diagram shade, and find the area of, the region

$$\{(x, y): x^2 + y^2 \geqslant 100\} \cap \{(x, y): -10 \leqslant y \leqslant 10\} \cap \{(x, y): 0 \leqslant x \leqslant 10\}.$$

Exercise 2A

1 A and B are fixed points 4 cm apart. T is the set of points nearer B than A and S is the set of points less than 3 cm from A. Show in a diagram $S \cap T$. Show also that the greatest distance apart of any two members of $S \cap T$ is $2\sqrt{5}$ cm.

2 O is the origin, A is $(3, -1)$, B$(5, 5)$. Given $\overrightarrow{OC} = \overrightarrow{AB}$, find the coordinates of C. Find also the midpoint of AC and the length of AC.

3 In Figure 22, AC, BQ and RP are parallel; QP and BR are parallel. $\overrightarrow{BQ} = 2\overrightarrow{AC}$ and $\overrightarrow{BP} = 2\overrightarrow{AB}$. $\overrightarrow{AB}$ and $\overrightarrow{AC}$ represent vectors $\boldsymbol{a}$ and $\boldsymbol{b}$ respectively. Express in terms of $\boldsymbol{a}$ and $\boldsymbol{b}$ the vectors represented by:

a $\overrightarrow{BC}$ *b* $\overrightarrow{BQ}$ *c* $\overrightarrow{BP}$ *d* $\overrightarrow{BR}$. Why are C, B, R collinear?

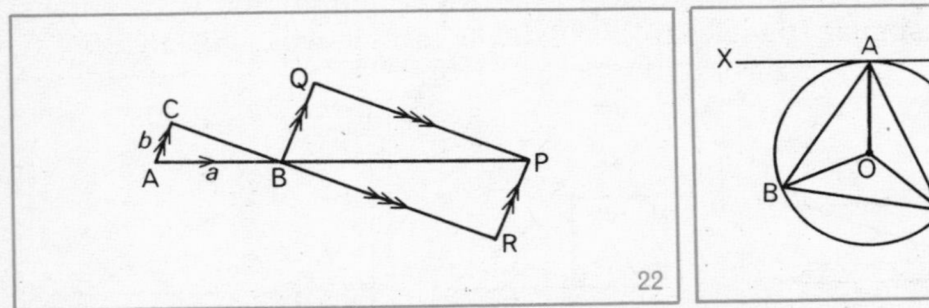

22

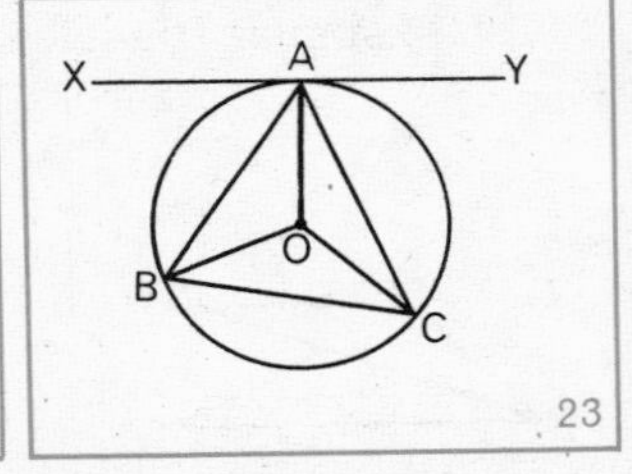

23

4 In Figure 23, XAY is a tangent. $\angle AOC = 130°$ and $\angle AOB = 110°$.

a Calculate the size of each angle in the figure.
b Name an angle equal to $\angle XAB$ and an angle equal to $\angle YAC$.

5 Under a dilatation centre (2, 2) and scale-factor 3, the circle with equation $x^2 + y^2 = 4$ becomes a circle with centre A and radius c. Find the coordinates of A and the value of c.

6 One rectangle measures 4 cm by 3 cm and another measures 6 cm by 8 cm. Are they similar? What is the ratio of lengths of corresponding sides? Calculate the ratios of: *a* their perimeters *b* diagonals *c* areas.

7 Show the locus $A = \{(x, y): x^2 + y^2 \leqslant 25, 3 \leqslant x \leqslant 5\}$ on squared paper. Define in set notation the locus obtained by reflecting A in the origin.

8 In quadrilateral ABCD, $\overrightarrow{AB}$ represents vector $\boldsymbol{u}$, $\overrightarrow{BC}$ vector $\boldsymbol{v}$, $\overrightarrow{DC}$ vector $k\boldsymbol{u}$. What can you say about the quadrilateral? E and F are the midpoints of AD and BC. What vectors are represented by:

a $\overrightarrow{AD}$ *b* $\overrightarrow{AE}$ *c* $\overrightarrow{AF}$ *d* $\overrightarrow{EF}$? What can you deduce about EF?

Exercise 2B

1. Show the set $A = \{(x, y): x^2 + y^2 = 25\}$ on squared paper. Draw also the image of A under reflection in the line $x = -3$, and state the coordinates of the centre of the image.

2. In Figure 24, the equation of the circle is $x^2 + y^2 = 64$, and the line PQ cuts the circle at A and B on the axes as shown.

   a Find the equation of PQ.
   b State in set notation the locus indicated by the shaded area.
   c Calculate in terms of π, the area of the shaded part.

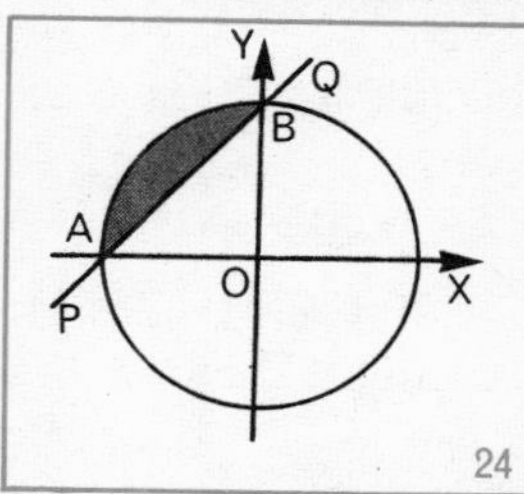

24

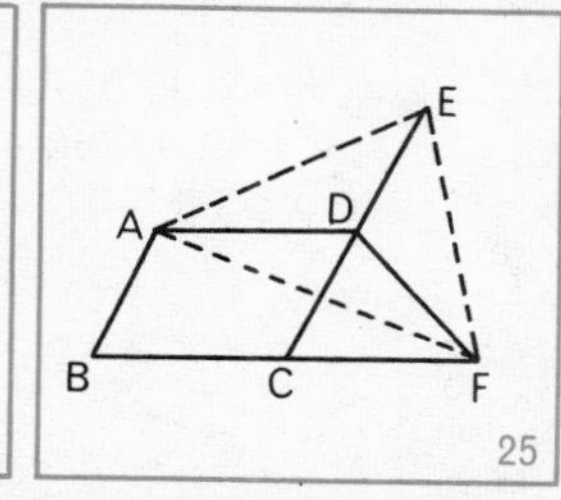

25

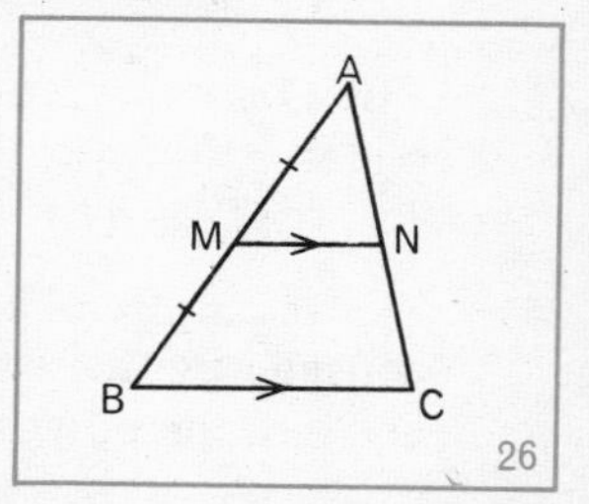

26

3. In Figure 25, ABCD is a parallelogram. BC = CF; CD = DE. If $\overrightarrow{BA}$ represents the vector $\boldsymbol{u}$ and $\overrightarrow{BC}$ represents the vector $\boldsymbol{v}$, express in terms of $\boldsymbol{u}$ and $\boldsymbol{v}$ the vectors represented by $\overrightarrow{AE}$, $\overrightarrow{DF}$, $\overrightarrow{EF}$ and $\overrightarrow{FA}$. Check that $\overrightarrow{AE} + \overrightarrow{EF} + \overrightarrow{FA} = \boldsymbol{O}$.

4. In Figure 26, M is the midpoint of AB, and MN$\parallel$BC. $\overrightarrow{AB}$ represents vector $\boldsymbol{c}$, and $\overrightarrow{AC}$ represents $\boldsymbol{b}$. Show that $\overrightarrow{MN}$ represents $\frac{1}{2}(\boldsymbol{b} - \boldsymbol{c})$. What can you say about N?

5. *a* P is a point outside a circle, centre O. PS and PT are tangents touching the circle at S and T. Prove that quadrilateral OSPT is cyclic.

   b If P is a point outside a circle, centre O, describe how you could construct two tangents from P to the circle.

6. O is the centre of two concentric circles of radii 1 and 2 units. A diameter PQORS of the larger circle cuts the smaller at Q and R. A chord PTV of the larger touches the smaller at T.

   a Prove that triangles PTO and PVS are similar.
   b Calculate the length of SV.
   c Show that $\angle OPT = 30°$ and calculate $\angle TOS$.
   d Show that $\triangle OVS$ is equilateral.

Exercise 3A

1 The graph of sin $x°$ is drawn and continued without limit in both positive and negative directions of the x-axis.

a Describe: (*1*) axes of symmetry (*2*) centres of symmetry.

b Answer the same questions for the section of the graph where: (*1*) $0 \leqslant x \leqslant 180$ (*2*) $0 \leqslant x \leqslant 360$ (*3*) $-180 \leqslant x \leqslant 180$.

c State some translations that will fit the curve onto itself.

2 $A = \{(x, y) : x^2 + y^2 \leqslant 9\}$, $B = \{(x, y) : x \geqslant 1\}$, $C = \{(x, y) : x \leqslant -1\}$. Draw a diagram to show $A \cap (B \cup C)$. If R is the centre of set A, and $T \in A \cap (B \cup C)$, find the greatest and least lengths of TR.

3 ABCD is a quadrilateral, and T is a point inside it. P, Q, R and S are midpoints of AB, BC, CD and DA. Explain why:

a $\vec{TP} = \frac{1}{2}(\vec{TA} + \vec{TB})$ *b* $\vec{TP} + \vec{TQ} + \vec{TR} + \vec{TS} = \vec{TA} + \vec{TB} + \vec{TC} + \vec{TD}$

4 Z is the centre of a regular polygon ABC...T with twenty sides. By considering the sizes of the angles of $\triangle$ABZ, calculate the size of an angle of the polygon.

5 a On squared paper draw the rhombus HKLM where H is (2, 10), K(5, 11), L(8, 10) and M(5, 9).

b If $H_1K_1L_1M_1$ is the image of HKLM under a translation such that M_1 is (13, 9), draw $H_1K_1L_1M_1$ and state the coordinates of L_1.

c Draw $H_2K_2L_2M_2$ the image of $H_1K_1L_1M_1$ under a half turn about (10, 8).

d Find the point about which HKLM can be rotated to $H_2K_2L_2M_2$.

6 ABCD is a square of side 1 unit. $\vec{AB}$ represents vector $\boldsymbol{u}$, $\vec{BC}$ represents $\boldsymbol{v}$. Name line segments representing $-\boldsymbol{u}$, $-\boldsymbol{v}$, $\boldsymbol{u}+\boldsymbol{v}$ and $\boldsymbol{u}-\boldsymbol{v}$. Calculate $|\boldsymbol{u}|+|\boldsymbol{v}|$ and $|\boldsymbol{u}+\boldsymbol{v}|$.

7 Draw a triangle on squared paper with A the point (2, 2), B(6, 3) and C(3, 5). $\triangle A_1B_1C_1$ is the image of $\triangle$ABC under a dilatation [O, 3]. $\triangle A_2B_2C_2$ is the image of $\triangle A_1B_1C_1$ under reflection in the x-axis. Write down the coordinates of A_1, B_1, C_1, A_2, B_2 and C_2. Is the composition of these two operations commutative?

8 P is a point 17 cm from O, the centre of a circle of radius 8 cm. Calculate the length of the tangents PX and PY drawn from P.

9 Two chords AB and DC in a circle are produced to meet outside the circle at E. If EB = 5 cm, BD = 7·5 cm, EA = 12 cm, EC = 6 cm, calculate AC and ED.

Exercise 3B

1 ABCD is a parallelogram in which $\vec{AB}$ represents $\boldsymbol{u}$ and $\vec{AD}$ represents $\boldsymbol{v}$. AC and BD intersect in H. P is a point on AB such that AP:PB = 1:3 and PH produced meets DC in Q. Find in terms of $\boldsymbol{u}$ and $\boldsymbol{v}$ the vectors represented by $\vec{DQ}$ and $\vec{PH}$.

2 A and B are the points $(-1, 1)$ and $(3, 4)$ respectively.
a Write down the components of the position vectors $\boldsymbol{a}$ and $\boldsymbol{b}$.
b Write down the components of the vector represented by $\vec{AB}$.
c Calculate the magnitude of $\boldsymbol{b}-\boldsymbol{a}$.

3 A is the point $(4, -1)$, B is $(11, 5)$ and C is $(6, 2)$. Find the coordinates of D such that: *a* $\vec{CD} = 2\vec{AB}$ *b* $\vec{CD} = 2\vec{BA}$.

4 a Write down the equation of the line L joining the origin to the point A(4, 2). What is the gradient of this line?

b Find: (*1*) A_1, the image of A under reflection in the x-axis
(*2*) the equation of L_1, the image of L under reflection in the x-axis.

c If L_2 is the image of L_1 under reflection in the line $y = 2$, give a transformation which maps L to L_2, and find the equation of L_2.

5 Find the 2×2 matrix which maps $P(x, y)$ to $P_1(x+y, x-y)$. Find the image of the set of points on: *a* the x-axis *b* the y-axis.

6 Two tangents AB and AC are drawn to a circle centre O and radius 10 cm. AO = 25 cm. Use your tables to calculate $\angle$BAC.

7 In Figure 27, $\angle AOB = 138°$. Calculate the size of $\angle$BDC.

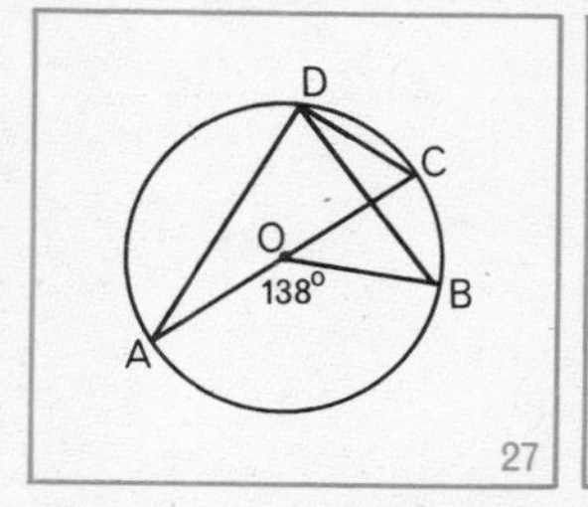

27

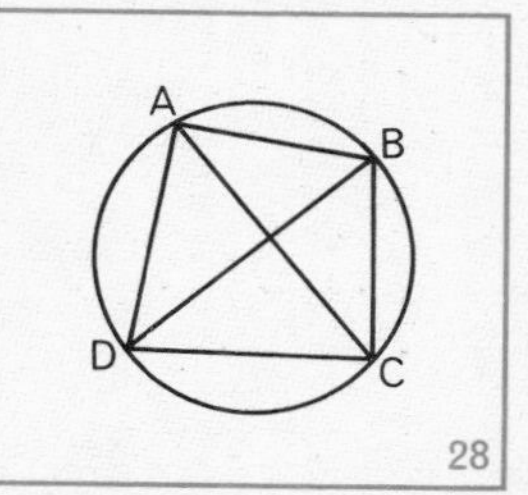

28

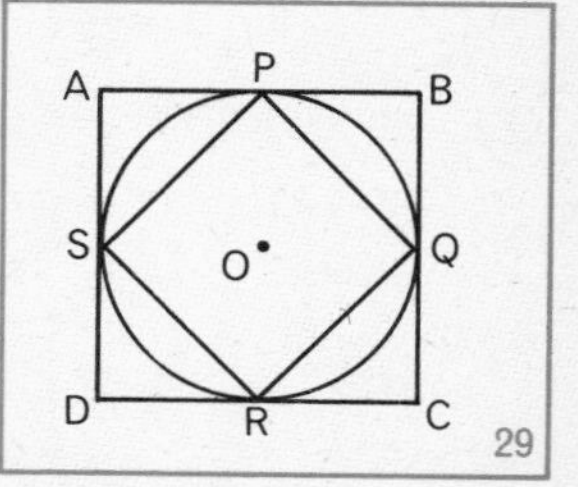

29

8 Copy Figure 28; mark four pairs of equal angles at the circumference. Prove the opposite angles of a cyclic quadrilateral supplementary.

9 a Show that area of square ABCD: area of circle: area of square PQRS$=4:\pi:2$ in Figure 29.

b If $T \circ V$ maps PQRS to ABCD, and V is a rotation of $+45°$ about O, what transformation does T represent?

Arithmetic

1 Estimation of Error

1 Approximation

As we have seen, all measurement is approximate, and measures of length, mass, time, area, etc., should always be given to a reasonable degree of approximation. There are three main ways in which this is done; by *rounding off* to:

(i) the nearest appropriate unit
(ii) an appropriate number of decimal places
(iii) an appropriate number of significant figures.

(i) *To the nearest unit*

The rules for rounding off a number are as follows: If the following figure is greater than 5, increase the round-off figure by 1; if the following figure is 5, round off to the nearest even number; otherwise leave the round-off figure as it is.

Examples

a 14·7 kg = 15 kg, rounded off to the nearest kilogramme.
b 10·13 s = 10·1 s, rounded off to the nearest tenth of a second.
c 128·5 m = 128 m, rounded off to the nearest metre.
d 128·51 m = 129 m, rounded off to the nearest metre.

(ii) *Decimal places*

Approximations are not only applied to measurements; sometimes it is convenient to round off a decimal by expressing it to a given number of *decimal places.*

For example, 5·20735 = 5·2074, rounded off to 4 decimal places
= 5·207, rounded off to 3 decimal places
= 5·21, rounded off to 2 decimal places
= 5·2, rounded off to 1 decimal place.

(iii) *Significant figures*

A convenient way to indicate the degree of approximation is by means of the number of figures used. We say that 67·3 cm has 3 *significant figures*, and that 67 has 2 *significant figures.*

A zero is a significant figure except when it is used simply to indicate the position of the decimal point.

a **2·40 m. The 0 indicates that the length has been measured to the nearest hundredth of a metre, and is significant. 3 significant figures.**

b **0·0810 km. The first two 0s show the position of the point, and are not significant. The third 0 shows that the length has been measured to one tenth of a metre, and is significant. 3 significant figures.**

Exercise 1

1 Round off 673 843 to the nearest:

a 10 *b* 100 *c* 1000 *d* 10 000

2 Round off to 1 decimal place:

a 8·72 *b* 11·29 *c* 507·01 *d* 39·08 *e* 0·45 *f* 0·09 *g* 4·98

3 Round off each of the following to 2 decimal places and to 2 significant figures:

a 8·123 *b* 16·091 *c* 2·468 *d* 0·375 *e* 1·001

4 Round off these to the number of significant figures shown in brackets:

a 6·135 (2) *b* 5·007 (3) *c* 18 918 (2) *d* 18 918 (3)

e 0·005 18 (2) *f* 4·821 (1) *g* 10·001 (4) *h* 3·1416 (3).

5 Write down the number of significant figures in each of these:

a 564 *b* 5064 *c* 3·9 *d* 0·9 *e* 2·70

6 Express $1\frac{3}{7}$ as a decimal, rounded off to:

a 2 decimal places *b* 3 decimal places

c 2 significant figures *d* 3 significant figures.

2 *Counting and measuring; absolute error*

In each of the following the number involved is exact. There is one and only one correct answer to each and this is obtained by *counting*.

The number of eggs in a dozen.
The number of pence in exchange for a £1 note.
The number of goals scored by the 1st Hockey XI last Saturday.
The number of new houses completed in Scotland in 1973.
The number of pupils in your school.

Contrast this with the situation when the numbers obtained are the result of *measurement*, as illustrated below.

The height of a person is 176 centimetres.
The mass of a packet of cereal is 345 grammes.
The volume of liquid in a bottle is 1 litre.

No matter how careful we are in carrying out a measurement, we can never know what the correct measure is; nevertheless it is convenient to imagine that such a measure exists. The difference between this true measure and the one obtained by measurement is called the *error*, even although we have made no *mistake* in the actual measurement. The size of this error can be reduced by using more accurate instruments, but measurements can never be exact, and therefore errors can never be completely eliminated.

Since this is so, it is most important that we should know in any situation the extent to which we can rely on our measurements, i.e. we should know the maximum possible error involved.

Exercise 2

Which of the following measures are exact (found by *counting*) and which are approximate (found by *measurement*)?

1 The score in a football game.

2 The annual rainfall in a certain city.

3 The speed of a space-craft between the earth and the moon.

4 The sum of money collected by a class for charity.

5 The distance from Aberdeen to London.

6 The number of pupils in your class.

7 The mass of a load of coal.

8 The price of a packet of potato crisps.

9 The winner's time in a race at the school sports.

10 The number of degrees in one right angle.

11 The number of days in one calendar year.

12 The length of a day as measured by the time of rotation of the earth about its axis.

* * * *

All measurements are necessarily inexact. Consider the measurement of a straight line. If we use a ruler graduated in centimetres we may give the length of the line as 5 cm. This does not mean that the length is exactly 5 cm. We adopt the convention that (unless otherwise stated) this measurement is *correct to 1 significant figure*, or *correct to the nearest centimetre*, and we say that the *least unit of measurement* is 1 cm. Thus the true length is nearer to 5 cm than to 4 cm or 6 cm, i.e. it may lie anywhere between 4·5 cm and 5·5 cm, and the *error* may be as much as 0·5 cm.

We say that the *absolute error* is 0·5 cm. Notice that the absolute error is half the least unit of measurement.

Also, the upper limit of this length of the line is 5·5 cm, and the lower limit is 4·5 cm.

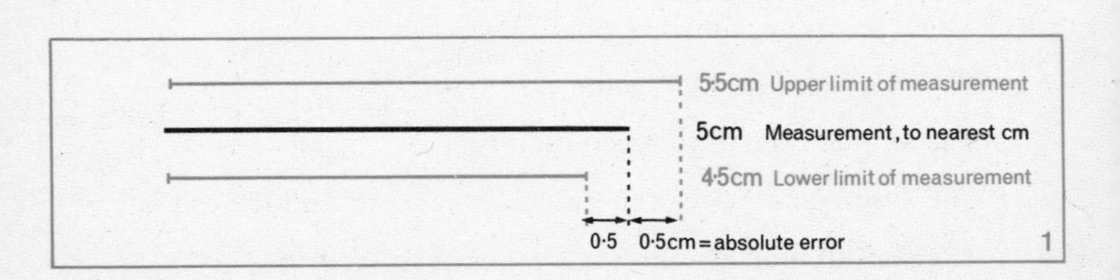

Example 1. For a mass of 15·8 kg,

least unit of measurement = 0·1 kg

So absolute error = $\frac{1}{2}$ of 0·1 kg = 0·05 kg

Upper limit of mass = 15·85 kg

Lower limit of mass = 15·75 kg

Example 2. For a volume of 2·24 litres,

least unit of measurement = 0·01 litre

So absolute error = 0·005 litre

Upper limit of volume = 2·245 litres

Lower limit of volume = 2·235 litres

Exercise 3

Make a table under the headings shown for the measurements in questions ***1–12***.

Measurement	*Least unit of measurement*	*Absolute error*	*Upper limit of measurement*	*Lower limit of measurement*
15 seconds	1 s	0·5 s	15·5 s	14·5 s

1 8 cm *2* 124 m *3* 234 km *4* 13 kg

5 7·5 cm *6* 17·8 kg *7* 18·2 cm *8* 1·6 cm

9 3·1 litres *10* 1·03 cm *11* 51·2 h *12* 10·24 s

3 Relative error; percentage error

We have seen that no measurement is exact, and that errors of different amounts may be expected with different measuring instruments. However, the *same* error may be of greater importance in some cases than in others. Consider, for example, a groundsman marking out the side lines of a football pitch. An error of 1 cm—or even

1 m—is relatively unimportant, but an error of 1 cm by a carpenter would completely ruin his work. Then again, an engineer engaged in high-precision work may be required to work to thousandths of a centimetre. Thus we often need to consider error in relation to the measurement itself. So we calculate the *relative error*, given by

$$\text{relative error} = \frac{\text{absolute error}}{\text{measurement}}$$

For example, if the length of a line is measured as 2·5 cm (to the nearest 0·1 cm) the absolute error is 0·05 cm so that

$$\textit{relative error} = \frac{0{\cdot}05}{2{\cdot}5} = \frac{5}{250} = \frac{1}{50}.$$

If we wish to express this as a percentage, we can multiply by 100% (i.e. by 1), giving

$$\textit{percentage error} = \frac{1}{50} \times 100\% = 2\%$$

Example. What is the percentage error in giving a mass as 1·50 kg?

It is to be noted here that the zero in the second decimal place is significant. This zero implies that the measurement is given to the nearest 0·01 kg. If the mass were given as 1·5 kg, the least unit of measurement would be 0·1 kg.

$$\text{Absolute error} = 0{\cdot}005\,\text{kg}$$

$$\text{Relative error} = \frac{0{\cdot}005}{1{\cdot}50} = \frac{5}{1500} = \frac{1}{300}$$

$$\text{Percentage error} = \frac{1}{300} \times 100\% \doteqdot 0{\cdot}33\%$$

Note

(i) We have of course been calculating the *maximum* absolute, relative, and percentage errors; in practice, the error in a measurement is likely to be less than this.

Throughout this chapter we take absolute, relative, and percentage errors to be the maximum error in each case.

(ii) The absolute error is a quantity of the same kind as the measurement itself, whereas the relative and the percentage errors are numbers.

Exercise 4

Find the absolute error and the relative error for:

1 125 m *2* 25 kg *3* 15 km *4* 2·5 m

Find the absolute error and the percentage error, rounded off to 2 significant figures, for:

5 6 cm *6* 12 kg *7* 3·6 litres *8* 4·4 m

Find the percentage error, to 2 significant figures, for:

9 3 cm *10* 3·0 cm *11* 3·00 cm *12* 25 kg

13 25 g *14* 25 tonnes *15* 8·5 m *16* 15·2 cm

17 1·2 kg *18* 9·8 s *19* 3960 km (to the nearest 10 km)

20 168 000 000 km (to the nearest million km)

4 *Tolerance*

In modern industry using mass-production methods, components are often manufactured in separate factories and sent to a central point for assembly. It is essential therefore to ensure that those components are made sufficiently accurately to 'fit' when assembled. In order to do this it is usual to specify the maximum error which will be allowed in making the parts.

For example, if bolts are required with a machined diameter of 8 mm, the specification might allow for diameters between 7·8 mm and 8·2 mm. The difference between these limits, 0·4 mm, is called the *tolerance* in the measurement, and in this case could be given by the expression $(8 \pm 0{\cdot}2)$ mm.

The tolerance in a measurement is the difference between the greatest and least acceptable measurements.

Example. For a mass of $(15 \pm 0{\cdot}5)$ g the greatest and least acceptable masses are 15·5 g and 14·5 g, and the tolerance is 1 g.

Exercise 5

1 Give the greatest and least acceptable measurements for the following:

a (12 ± 1) g *b* (76 ± 2) m *c* $(4{\cdot}3 \pm 0{\cdot}1)$ cm

d $(6{\cdot}3 \pm 0{\cdot}1)$ s *e* $(4{\cdot}8 \pm 0{\cdot}5)$ kg *f* $(2 \pm 0{\cdot}2)$ cm^2

g $(1{\cdot}4 \pm 0{\cdot}05)$ s *h* $(15 \pm 0{\cdot}25)$ m

2 Find the tolerance, given that acceptable measurements lie between:

a 6 cm and 8 cm *b* 27 g and 28 g

c 4·2 cm and 4·4 cm *d* 8·7 kg and 8·4 kg

e 4·86 cm^2 and 5·00 cm^2 *f* 6·9 cm^3 and 7·0 cm^3

3 The range of measurements between 9·8 m and 10·1 m can be written $(9{\cdot}95 \pm 0{\cdot}15)$ m. Express the following ranges in the same way:

a 5 mm to 9 mm *b* 79 m to 83 m *c* 11 kg to 14 kg

d 5·4 kg to 5·8 kg *e* 4·6 kg to 4·7 kg *f* 1·26 cm to 1·28 cm

g 0·85 s to 0·95 s *h* $1{\cdot}3\text{ cm} \leqslant x\text{ cm} \leqslant 1{\cdot}4\text{ cm}$

4 The masses of certain packages must be within the range (500 ± 20) g. Which of the following are acceptable?

a 487 g *b* 519 g *c* 478 g *d* 480 g *e* 500 g

5 Pieces of tubing are required with lengths given by $(6 \pm 0{\cdot}2)$ cm. Which of the following will be accepted, and which rejected?

a 6·3 cm *b* 5·6 cm *c* 6·09 cm *d* 5·82 cm *e* 5·98 cm

6 Figure 2 shows sketches of three metal frames; the lengths are given to the nearest mm, the tolerance in each case being specified by $\pm 0{\cdot}5$ mm. Calculate the greatest and least acceptable perimeters.

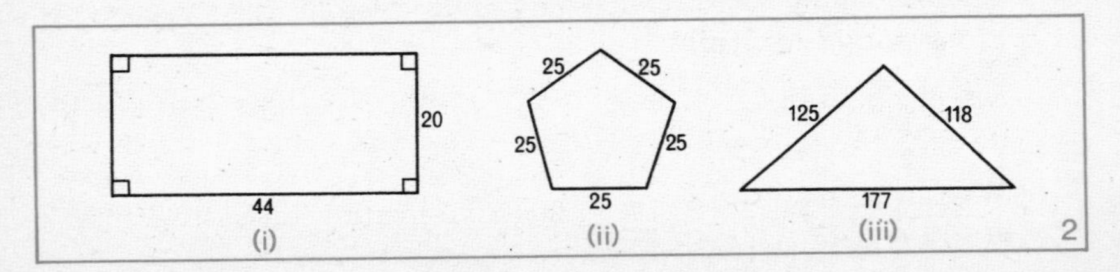

5 *The sum and difference of measurements*

(i) *Addition of measurements*

Example 1. What is the sum of the measurements 5 cm and 3 cm, each being given to the nearest centimetre?

The first length lies within the range $(5 \pm 0{\cdot}5)$ cm, i.e. 4·5 cm to 5·5 cm.

The second length lies within the range $(3 \pm 0{\cdot}5)$ cm, i.e. 2·5 cm to 3·5 cm.

Hence the maximum sum $= 5{\cdot}5 + 3{\cdot}5 = 9$ cm, and

the minimum sum $= 4{\cdot}5 + 2{\cdot}5 = 7$ cm.

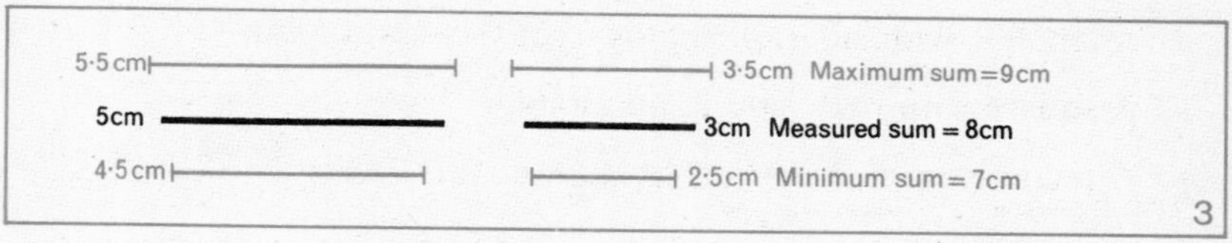

5·5 cm 3·5 cm Maximum sum = 9 cm
5 cm 3 cm Measured sum = 8 cm
4·5 cm 2·5 cm Minimum sum = 7 cm
3

Note that the apparent sum of 8 cm has an absolute error of 1 cm, which is equal to the sum of the absolute errors in the original measurements.

Example 2. Two rods are 3·2 cm and 1·6 cm long, to the nearest 0·1 cm. Between what lengths are they when put end to end?

The first length is in the range $(3{\cdot}2 \pm 0{\cdot}05)$ cm, i.e. 3·15 cm to 3·25 cm.

The second length is in the range $(1{\cdot}6 \pm 0{\cdot}05)$ cm, i.e. 1·55 cm to 1·65 cm.

The maximum length $= 3{\cdot}25 + 1{\cdot}65 = 4{\cdot}90$ cm, and

the minimum length $= 3{\cdot}15 + 1{\cdot}55 = 4{\cdot}70$ cm.

Notice that the apparent sum of 4·8 cm has an absolute error of 0·10 cm.

When measurements are added, the absolute error is the *sum* of the errors in the original measurements.

Exercise 6

1 Find the maximum and minimum sums of the following measurements:

a 6 cm and 8 cm *b* 12 g and 17 g *c* 4·3 m and 4·7 m

d 4·6 cm and 11·8 cm *e* 1·42 kg and 0·90 kg *f* 2·7 g and 1·4 g

2 What is the absolute error in the sum of these measurements?

a 5 cm and 8 cm *b* 24 g and 19 g *c* 4·8 mm and 5·9 mm

3 Find the limits between which the perimeters of the following metal shapes must lie:

a A triangle with sides of lengths 3 cm, 4 cm and 5 cm.

b A square with each side 12 mm long.

c A rectangle with length 12 m and breadth 8 m.

4 Ten lengths of rail, each 25 m long, are laid end to end. What are the limits of their total length?

5 A rectangle has length $(7 \pm 0{\cdot}5)$ cm and breadth $(4 \pm 0{\cdot}5)$ cm. Between what limits does its perimeter lie?

6 The average number of hours of sunshine per day in a certain week was 5·8 to the nearest 0·1 hour. What are the maximum and minimum possible number of hours of sunshine for the week?

7 In a month of 30 days the average rainfall per day was found to be 3 mm to the nearest 0·5 mm. Find the limits between which the rainfall for that month must lie.

8 What is the minimum length of wire that must be purchased in order that you can be sure that you will have sufficient to make the following skeleton models?

a a regular hexagon of side 12 cm
b a cube of edge 4 cm
c a circle of radius 3 cm.

(ii) *Subtraction of measurements*

Example. What is the difference between the measurements 5 cm and 3 cm, each being given to the nearest centimetre?

5 cm lies within the range $(5 \pm 0{\cdot}5)$ cm, i.e. 4·5 cm to 5·5 cm.
3 cm lies within the range $(3 \pm 0{\cdot}5)$ cm, i.e. 2·5 cm to 3·5 cm.

The maximum difference between the lengths occurs when the smallest value of the second is subtracted from the largest value of the first, as shown in Figure 4.

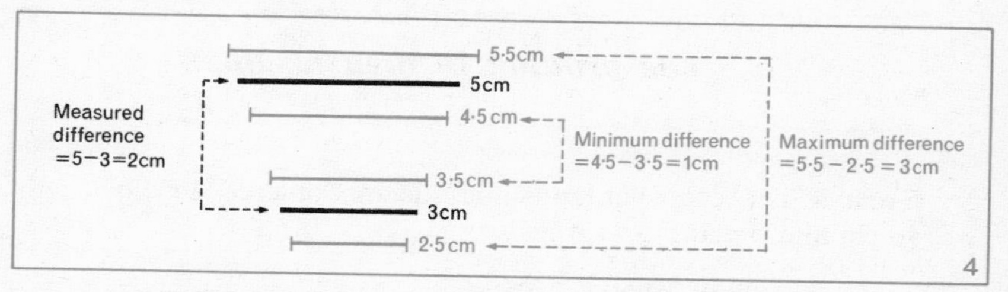

Study Figure 4, and you will see that the maximum difference = $5{\cdot}5 - 2{\cdot}5 = 3$ cm, and the minimum difference = $4{\cdot}5 - 3{\cdot}5 = 1$ cm.

Note that the apparent difference of 2 cm has an absolute error of 1 cm, which is equal to the sum of the absolute errors in each of the original measurements.

When measurements are subtracted, the absolute error is the *sum* of the errors in the original measurements.

Exercise 7

1 Find the limits of the difference between the following measurements:

a 4 cm and 8 cm *b* 5 g and 8 g *c* 3 s and 9 s

d 9·8 cm and 4·6 cm *e* 2·7 kg and 1·4 kg *f* 1·42 m and 0·90 m

2 What is the absolute error in the difference of these measurements?

a 4 km and 2 km *b* 22 cm and 17 cm *c* 3·2 g and 1·7 g

3 A length of 20 cm is cut from a metal rod 28 cm long. What are the limits of the remaining length?

4 12 kg of flour are removed from a container holding 50 kg. What are the limits of the mass of the remaining flour?

5 A metal strip 11 cm long is cut from a piece 50 cm long. What are the limits of the length left?

6 If in question **5** four such strips are cut off, what are the limits of the remaining length?

7 A bottle of acid is stated to contain 1 litre ± 5 ml. If ten volumes of 25 ml each, to the nearest 0·1 ml, are removed, what are the limits of the volume of acid left in the bottle?

6 *The product of measurements*

Example. Between what limits does the area of a rectangle of length 4·1 cm and breadth 2·9 cm lie?

The maximum possible area

$= 4{\cdot}15 \times 2{\cdot}95 \text{ cm}^2$
$= 12{\cdot}2425 \text{ cm}^2$

The minimum possible area

$= 4{\cdot}05 \times 2{\cdot}85 \text{ cm}^2$
$= 11{\cdot}5425 \text{ cm}^2$

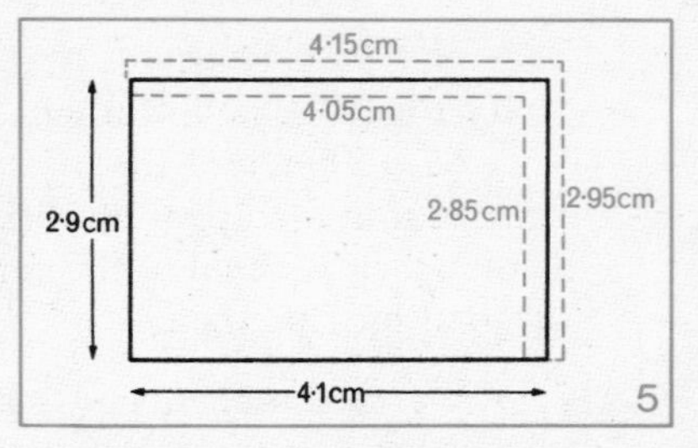

Thus the 'true' area lies between $12{\cdot}2425 \text{ cm}^2$ and $11{\cdot}5425 \text{ cm}^2$. Also, the apparent area $= 4{\cdot}1 \times 2{\cdot}9 \text{ cm}^2 = 11{\cdot}89 \text{ cm}^2$.

Exercise 8

Find the limits between which the areas of the shapes in questions ***1–5*** lie:

1 A rectangle with length 5 m and breadth 4 m.

2 A rectangle with length 9 cm and breadth 2 cm.

3 A square with side of length $(6 \pm 0{\cdot}2)$ mm.

4 A right-angled triangle with shorter sides 3 cm and 2 cm long.

5 A triangular field with 'base' 4 km and 'altitude' 3 km.

6 Calculate the limits of the perimeter and area of a square with side of length $(5 \pm 0{\cdot}1)$ m.

7 Calculate the area of a square of side 1·0 metre. What is the relative error in the given measurement? What is the relative error in the area?

This example suggests that the relative error of a product is approximately the sum of the relative errors of the factors; this is in fact true. Compare this result with the result in the case of a sum or difference.

Summary

1 *Approximation*

(i) 0·05037 = 0·050 to 3 decimal places.
(ii) 0·05037 = 0·0504 to 3 significant figures.

2 *The measurement* given as 2·3 g is correct to the nearest 0·1 g.
The *least unit of measurement* is 0·1 g, and
the *absolute error* = $\frac{1}{2}$ (least unit of measurement) = 0·05 g.
The upper limit of measurement = (2·3 + 0·05) g = 2·35 g,
the lower limit of measurement = (2·3 − 0·05) g = 2·25 g.

The *relative error* $= \frac{\text{absolute error}}{\text{measurement}} = \frac{0{\cdot}05}{2{\cdot}3} = \frac{1}{46}$

The *percentage error* = relative error × 100%

$$= \frac{1}{46} \times 100 \doteqdot 2{\cdot}2\%.$$

3 *Tolerance* specified by (2·3 ± 0·01) m gives
an upper acceptable limit of 2·31 m, and
a lower acceptable limit of 2·29 m.

Tolerance = upper limit − lower limit = 0·02 m.

4 *Measurement* given as 3·2 m and 1·6 m:

(i) Upper limit of sum = (3·25 + 1·65) m
= 4·90 m
Lower limit of sum = (3·15 + 1·55) m
= 4·70 m

(ii) Upper limit of difference = (3·25 − 1·55) m
= 1·70 m
Lower limit of difference = (3·15 − 1·65) m
= 1·50 m

(iii) Upper limit of product = 3·25 × 1·65 m^2
= 5·3625 m^2
Lower limit of product = 3·15 × 1·55 m^2
= 4·8825 m^2

The *absolute error of a sum or difference of measurements* is equal to the sum of the absolute errors of the individual measurements.

2 Counting Systems

1 *Counting on our fingers*

How did primitive man start to count? Possibly by matching objects against small stones laid out on the ground, one stone for each object (and our word 'calculate' comes from the Latin word *calculus* which means a stone); or by cutting a notch or a 'tally' on a piece of stick to correspond with each object; or by using his fingers, counting one finger, two fingers, and so on. In fact he set up a 'one-to-one correspondence' between objects and stones, tally marks or fingers.

When numbers are small it is easy to recognize the patterns formed by stones or fingers: when you play 'Dominoes' you do not require to count the dots—you instantly recognize the pattern as indicating three, five, etc., and you studied some of these patterns in Book 5.

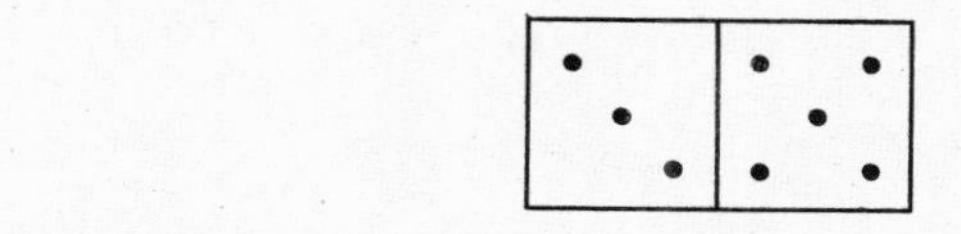

1

For numbers greater than ten, the patterns were not so easily recognized, and all the fingers were used up, so man probably started to count in 'double-handfuls', that is in tens. Thus he would regard thirty-five as three double-handfuls and five fingers, and so was committed to a *decimal* system. (However, in a few primitive civilizations, in warm countries, man appears to have used his toes as well, so for example the Maya of Central America seem to have had a system based, not on ten, but on twenty.)

When man found it necessary to deal with still larger numbers, he devised the *abacus* as a calculating aid.

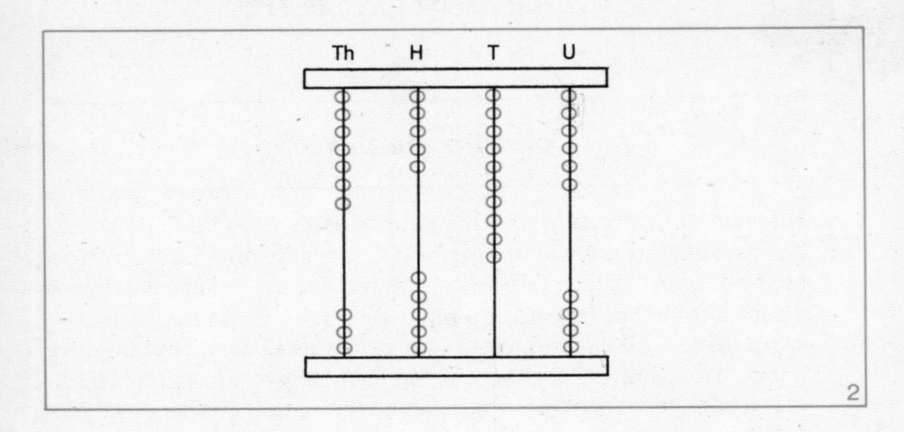

The abacus illustrated in Figure 2 has ten beads on each wire. The right-hand wire indicates units, the next tens, and so on; Figure 2 shows three thousands, five hundreds, and four units. It may surprise you to know that the use of zero as a placeholder was not introduced into Europe until the fifteenth or sixteenth century.

In our modern notation the number represented on the abacus is:

$$\begin{aligned} &3 \text{ thousands} + 5 \text{ hundreds} + 0 \text{ tens} + 4 \text{ units} \\ &= (3 \times 10^3) + (5 \times 10^2) + (0 \times 10) + 4 \\ &= 3504 \end{aligned}$$

Notice that if we add 6 units to this number we have to bring down 6 beads on the units wire, making ten down on this wire; these ten can now be pushed up and the equivalent tens bead brought down on the tens wire.

When our present set of numerals $\{0, 1, 2, 3, \ldots, 9\}$ came into common use about five hundred years ago, it became possible to do simple calculations without the abacus, using the positional notation instead.

Sketch an abacus showing 2345. What happens when you use the abacus to add 2345 to 2345? And to subtract 467 from 683?

In fact this is how calculation was done for hundreds of years, and abaci rather like this are still in common use in certain parts of the world.

Things to make and do

1 In this chapter you will be finding out about counting systems with base ten, base two, base three, base five, base seven, base eight, and base twelve. Calculations in all of them can be done by means of an abacus, having ten, two, three, five, seven, eight or twelve beads respectively on each wire.

Make a model of one of the above kinds of abacus. Figure 3 gives you some suggestions.

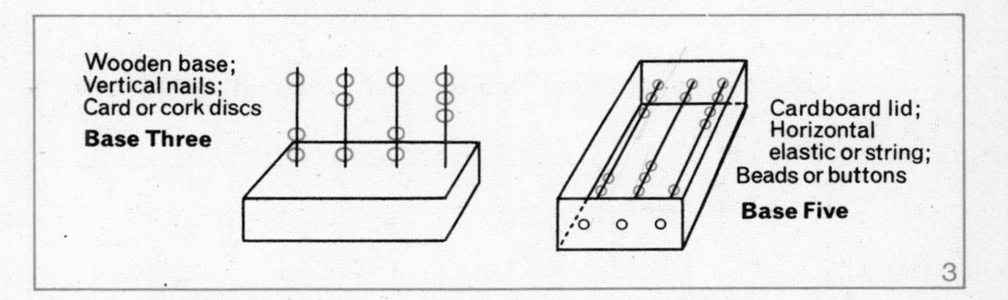

2 Find out from books in the school or local library how numerals were introduced to represent various numbers, e.g. to write six as VI, 6, etc.; and how zero was introduced as a cipher or placeholder.

3 The abacus used by a Chinese shopkeeper does not have ten beads on each wire. Find out what his abacus looks like and how it is operated.

4 Finger multiplication is a device which was used until recently in central Europe to avoid having to learn the multiplication tables beyond the 'five times' table. As an example, to multiply 7 by 9, subtract 5 from each, giving 2 and 4; on one hand hold 2 fingers up and the rest down; on the other hand hold four fingers up and the rest down. *Add* the *up* fingers and call the answer tens, that is 60; *multiply* the *down* fingers, that is 3. Add these answers to give 63!

Try this with other products such as 6×8. The question is, why does it work?

2 *Base two*

A binary abacus has two beads on each wire, those on the right-hand wire representing units, as in Figure 4 (i). We can bring down one bead to indicate 1 as shown. Now when we bring down another bead to indicate 2, all the beads on this wire are down, so we push them up and bring down one bead on the second wire: thus the second wire shows twos, and 2 in our *decimal* scale is represented by 10 on the *binary* scale. We have $2_{ten} = 10_{two}$ ('one zero to the base two').

Proceeding in this way we see that the third wire shows fours, the next eights and so on. In other words, we are counting in powers of two.

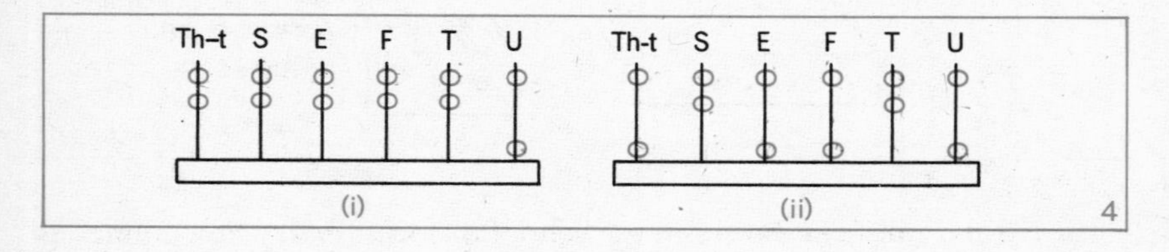

The abacus of Figure 4 (ii) shows $101\,101_{two}$, which means

$$\begin{aligned}&(1\times 2^5)+(0\times 2^4)+(1\times 2^3)+(1\times 2^2)+(0\times 2)+1\\ &= 1 \text{ thirty-two} + 0 \text{ sixteens} + 1 \text{ eight} + 1 \text{ four} + 0 \text{ twos} + 1 \text{ unit}\\ &= 32+0+8+4+0+1 \text{ in the decimal scale}\\ &= 45_{ten}\end{aligned}$$

Computers and the binary scale

It is clear that in the binary scale we need only two symbols, 1 and 0, to represent any number, a fact that is of importance in 'communicating' with a computer, as we have seen. A computer can accept instructions only if they consist of numbers in binary form. The reason is that the logical circuits used in the construction of computers are of a two-state type, that is they are based on true–false, yes–no, on–off type responses. Therefore the information required to operate them can be rapidly transmitted between the various parts of the computer as electrical impulses whose presence or absence can be represented by two symbols. If 1 and 0 are chosen as these symbols, messages in binary form can be readily processed.

Numbers and numerals

The same number can be represented by many different symbols. These symbols are called *numerals*. For example, the number of dots in Figure 5 can be represented by the numerals 12_{ten}, 1100_{two}, XII and many others.

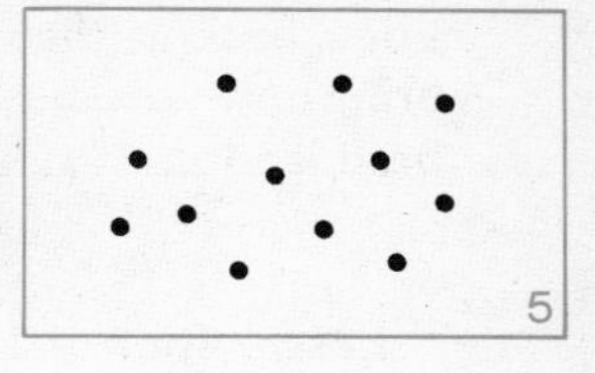

Numbers represented by base-two and base-ten numerals

Example 1. Convert 10101 from binary to decimal form.

SEFTU
1 0 1 0 1 means $16+4+1=21$ in the decimal system.

$$\text{i.e. } 10101_{two} = 21_{ten}$$

(This reads 'one, zero, one, zero, one to base two is equal to twenty-one to base ten', or 'one, nought, one, nought, one to base two is equal to twenty-one to base ten'.)

Example 2. Convert 26 from decimal to binary form.

We do this by arranging 26_{ten} as powers of 2, in order. Two methods are given below.

a We find the largest power of 2 less than 26, then the next largest, etc. (It is worth memorizing the powers: 1, 2, 4, 8, 16, 32, 64.)

$26 = 16+10$
$\quad = 16+8+2$

so TU SEFTU
2 6 = 1 1 0 1 0

i.e. $26_{ten} = 11010_{two}$

b

2	26	ones
2	13	twos + 0 units
2	6	fours + 1 two
2	3	eights + 0 fours
2	1	sixteen + 1 eight
	0	thirty-twos + 1 sixteen

So $26_{ten} = 11010_{two}$, from the remainders.

Addition and multiplication tables in the binary scale

Look back at the binary abacus in Figure 4. To add $1+1$ we first bring down one bead on the units wire. We then bring down a second bead, and must now 'carry' one to the second wire, giving

$$1_{\text{two}} + 1_{\text{two}} = 10_{\text{two}}$$

For addition and multiplication in the binary system we need only remember the following simple tables:

+	0	1
0	0	1
1	1	10

×	0	1
0	0	0
1	0	1

Example 3

(i) $\begin{array}{r} 10001 \\ +1011 \\ \hline 11100 \\ \hline \end{array}$ (ii) $\begin{array}{r} 10101 \\ -1110 \\ \hline 111 \\ \hline \end{array}$ (iii) $\begin{array}{r} 101 \\ \times 101 \\ \hline 101 \\ 10100 \\ \hline 11001 \\ \hline \end{array}$ (iv) $\begin{array}{r} 110 \\ 101\overline{)100000} \\ 101 \\ \hline 110 \\ 101 \\ \hline 10 \text{ (remainder)} \end{array}$

Exercise 1

(*All numbers are in binary form unless otherwise stated.*)

1 Write down the first nine natural numbers as binary numerals.

2 Convert from base two to base ten:

a 101 *b* 1101 *c* 100110 *d* 111000110

3 Convert from base ten to base two:

a 23 *b* 37 *c* 48 *d* 65 *e* 127

Calculate:

4 $101+110$ *5* $1101+111$ *6* $11010+10110$

7 $110-101$ *8* $1101-111$ *9* $11011-10110$

10 101×11 *11* 1101×110 *12* 10101×1001

13 $1011 \div 11$ *14* $11011 \div 101$ *15* $11011 \div 111$

16 Check your answers to questions ***5***, ***8***, ***11***, and ***14*** by converting the numerals and answers to base ten.

Calculate:

17 $1+10+101$ *18* $101+110-111$

19 $(11011+1101)\times 11$ *20* $(11011-1101)\div 111$

21 Find the perimeters of squares with sides of length:

a 101 cm *b* 11011 cm *c* 1111 cm

22 Find the areas of rectangles with lengths and breadths:

a 110 cm and 101 cm *b* 1101 cm and 111 cm

23 Calculate:

a $(1001+110)\times 101$ *b* $1001+(110\times 101)$

24 Which of the following are true and which are false?

a $10101>11010$ *b* $10^{10}=100$

c $100^{10}=1000$ *d* $(110\times 1010)\div 100=1111$

e 1010101 represents an odd number

25 Calculate:

a 1010×111 *b* $1010101-111111$ *c* $1001101\div 1101$

26a Subtract 123_{ten} as often as possible from 615_{ten}. Hence write down the value of $615_{ten}\div 123_{ten}$.

b By repeated subtraction find $10010_{two}\div 110_{two}$. Verify by division.

27 How can you tell when a binary numeral represents:

a an even number *b* a number divisible by 4?

Advantages of the binary scale

1 Only two symbols, 0 and 1, are necessary to represent any number.
2 Tables of addition and multiplication are easy to learn and use.
3 It is very useful in connection with computers.

Disadvantage of the binary scale.

Numbers require many digits to represent them, e.g. $734_{ten}=1011011110_{two}$.

Topics to explore

1 Here is an amusing and interesting method of multiplication. Multiply 179 by 346.

179	346	*check*:	346
89	692		179
44	(1384)		3114
22	(2768)		24220
11	5536		34600
5	11072		61934
2	(22144)		
1	44288		
	61934		

Method. Write the two numbers side by side. Now in succession divide by 2 in the first column (discarding all remainders) and multiply by 2 in the second column. Stroke out every number in the second column which is opposite an *even* number in the first column. Add the remaining numbers in the second column to give the answer.

Try this with a few multiplications to check that it always gives the right answer. The question is, why? The curious fact is that it is the remainders which you discarded that you are actually multiplying by! The process used for the left-hand column is the process you would use to turn 179_{ten} into the binary number, 10110011_{two}: and the numbers remaining in the right-hand column are the results of multiplying 346 by 1, by 2, by 16, by 32 and by 128 (and $1+2+16+32+128=179$).

2 *The Nim Game*

One version of this consists in placing any number of matches in three piles. Two players take it in turn to remove at least one match from one pile. The player who removes the last match wins.

Suppose there are 11_{ten}, 13_{ten} and 6_{ten} matches in the piles, i.e.

$$\begin{array}{r} 1011_{two} \\ 1101_{two} \\ 110_{two} \end{array}$$

It can be shown that a player will win if he can arrange things so that the columns in the binary numerals add up to even totals (in the above case 2, 2, 2, 2) when his opponent has to play. See *The Gentle Art of Mathematics*, Pedoe (E.U.P. and Penguin) for more details.

3 *Other bases less than ten*

It is now clear that there is no special reason why ten should always be chosen as a base for numeration, and that in fact it may not be the best choice.

We have seen that in the

denary system (base *ten*) we count in ones, tens, hundreds, etc.;
binary system (base *two*) we count in ones, twos, fours, etc.

In the same way, in the

ternary system (base *three*) we count in ones, threes, nines, etc.;
octal system (base *eight*) we count in ones, eights, sixty-fours, etc.

Examples.

1 $2101_{\text{three}} = (2 \times 3^3) + (1 \times 3^2) + (0 \times 3) + 1$

$= 54 + 9 + 0 + 1 = 64_{\text{ten}}$

2 $1024_{\text{five}} = (1 \times 5^3) + (0 \times 5^2) + (2 \times 5) + 4$

$= 125 + 0 + 10 + 4 = 139_{\text{ten}}$

3 Convert 259_{ten} to an octal number.

a $259_{\text{ten}} = (4 \times 64) + 3$

$= (4 \times 64) + (0 \times 8) + 3$

$= (4 \times 8^2) + (0 \times 8) + 3$

$= 403_{\text{eight}}$

or

b

8	259
8	32, 3
8	4, 0
	0, 4

So $259_{\text{ten}} = 403_{\text{eight}}$

4

$$\begin{array}{r} 356_{\text{seven}} \\ +644_{\text{seven}} \\ \hline 1333_{\text{seven}} \end{array}$$

('Four and six, ten; i.e. one, three to base seven, etc.')

5

$$\begin{array}{r} 815_{\text{nine}} \\ \times 5_{\text{nine}} \\ \hline 4477_{\text{nine}} \end{array}$$

('Five fives, twenty-five; i.e. two, seven to base nine, etc.')

Exercise 2

1 a Sketch an abacus which can be used for calculations with base three.

b How many different symbols are necessary to represent any whole number in base three?

c Make up tables for addition and multiplication of ternary (base-three) numerals.

2 Write out the notation which shows the meaning of the following, and state the equivalent of each in the denary (base-ten) system:

a 210_{three} *b* 2120_{three} *c* 2202_{three}

3 a Convert to base ten:

(*1*) 2122_{three} (*2*) 11200_{three} (*3*) 22112_{three}

b Convert from base ten to base three:

(*1*) 59 (*2*) 60 (*3*) 243

c What multiplication in the ternary scale is equivalent to multiplication by 3_{ten}, 9_{ten}, 27_{ten}?

4 Calculate the following, the numerals being to base three:

a $102+212$ *b* $2102+21$ *c* $2102-1021$

d $1212-1121$ *e* 221×21 *f* $1000\div 121$

* * * *

5 a What symbols are necessary for numeration with base five?

b Write out the notation which shows the meaning of the following, and state the equivalent of each in the decimal scale:

(*1*) 23_{five} (*2*) 412_{five} (*3*) 2310_{five}

6 a Convert these base-five numerals to base ten:

(*1*) 234 (*2*) 130 (*3*) 1400 (*4*) 2434

b Convert these base-ten numerals to base five:

(*1*) 123 (*2*) 270 (*3*) 3300 (*4*) 4125

c If a decimal numeral ends in 0 or 00, must the equivalent base-five numeral end in 0 or 00 respectively? Why? Is the converse true? Why?

7 Calculate the following, the numerals being to base five:

a $341+234$ *b* $4203+1332$ *c* $212-121$

d $300-143$ *e* 231×41 *f* $2134\div 3$

* * * *

8 Which numerals in base ten have exactly the same meaning in base eight?

9 Convert the following from base eight to base ten:

a 10 *b* 43 *c* 126 *d* 700 *e* 1031

10 Convert the following from base ten to base eight:

a 10 *b* 27 *c* 193 *d* 426 *e* 1000 *f* 4096

11a If an octal numeral ends in 0, what can you say about the equivalent decimal numeral?

b A decimal numeral ends in 000; how do you know that its octal equivalent must end in 0?

12 Perform the following calculations in the octal scale:

a $123+25$ *b* $256+127$ *c* $235-172$ *d* $1000-777$

e 32×6 *f* 346×5 *g* $150\div 3$ *h* $1000\div 7$

* * * *

13 Express 100_{ten} in the scales of two, three, five, eight, nine, eleven, twelve, twenty, one hundred.

14a What numbers have exactly the same representation in all scales?

b Is it possible to have a base-one system?

15 In which bases have the following calculations been done?

a $12+3=21$ *b* $12-3=6$ *c* $12\times 3=41$

d $12\div 3=2$ *e* $231+132=413$ *f* $432-234=165$

16 If $29_{\text{ten}} = x_{\text{eight}} = y_{\text{six}} = z_{\text{five}} = w_{\text{three}}$, find x, y, z, w.

Topics to explore

1 *Constructing a nomogram*

A nomogram is a device which is used to give results quickly when a large number of similar calculations have to be done. Let us make a nomogram for addition in the scale of eight. Draw three parallel lines equal distances apart, as shown. The outer two are marked with numerals on the scale of eight. The middle one is marked with these same numerals but with the unit length half of that on the others, as shown in Figure 6.

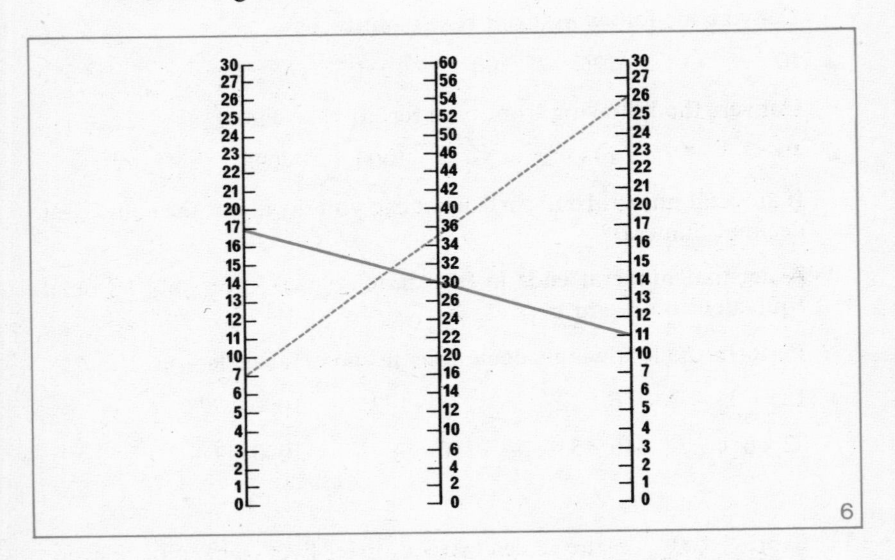

6

Use the nomogram to add 17+11 in the scale of eight as follows.

Place a ruler from 17 on the left-hand line to 11 on the right-hand line. The answer 30 is shown where the ruler crosses the middle line as shown on the diagram. The broken line in the diagram demonstrates $7+26=35$ in the scale of eight.

Use your nomogram for some other additions. Can you use the nomogram to subtract numerals?

Your knowledge of similar triangles in geometry should show you why this nomogram works.

2 Numbers represented by base-eight and base-two numerals

Check that $1101_{two} = 15_{eight}$; $110101_{two} = 65_{eight}$; and $10110111_{two} = 267_{eight}$.

Can you see that the units digit in the octal numeral can be obtained from the three digits on the right in the binary numeral? This process can be extended.

Thus $$10110111_{two} = 10\ 110\ 111_{two}$$
$$= 2\quad 6\quad 7_{eight}$$

Find out why this should be so, and work out some examples showing the conversion both ways. Note that an octal numeral requires about one-third the number of digits of the equivalent binary numeral.

4 *Base twelve*

Dantzig in his book *Number, the Language of Science* (Allen and Unwin) writes 'Everywhere the ten fingers of man have left their permanent imprint'. But in fact a few other imprints suggest that man has had ideas of other numeration systems. The word 'score' for twenty suggests a system like the Maya one; again why do we so commonly use 'a dozen'? Why do we count hours in twelves or twenty-fours? It has been suggested that when counting out eggs, we can quickly grasp 3 eggs in each hand and so two double-handfuls give us a dozen, and a very rapid means of counting them (in fact this was the method used, certainly before the days of automation, to count out the reels of thread by the employees of a famous Scottish thread-making firm). We also have the word 'gross' for $12 \times 12 = 12^2$. So there are indications of the use of the base twelve for calculation. The Latin word for twelve is *duodecim*, so we talk about the *duodecimal* scale.

For the abacus in duodecimals we require twelve beads on each wire, and we must have *twelve* symbols to represent any number. We have only ten at present, so we must invent two more: let us call them t and e so that we now count

0, 1, 2, 3, 4, 5, 6, 7, 8, 9, t, e,
10, 11, 12, 13, 14, 15, 16, 17, 18, 19, 1t, 1e,
20, 21, 22, 23, 24, 25, 26, 27, 28, 29, 2t, 2e,
etc.

The numeral $5t2e_{\text{twelve}}$ means $(5\times 12^3)+(10\times 12^2)+(2\times 12)+11$ in the decimal scale (and is read 'five, tee, two, ee, base twelve').

Numbers represented by base-twelve and base-ten numerals

Example 1. Convert $3t4_{\text{twelve}}$ to base ten.

$$3t4_{\text{twelve}} = (3\times 12^2)+(10\times 12)+4$$
$$= 556_{\text{ten}}$$

Example 2. Convert 659_{ten} to base twelve.

a 659_{ten}

$= (4\times 12^2)+(6\times 12)+11$

$= 46e_{\text{twelve}}$

or

b

12	659	ones
12	54	twelves + 11(e)
12	4	144s + 6×12
	0	1728s + 4×12^2

So $659_{\text{ten}} = 46e_{\text{twelve}}$

Exercise 3

1 Convert from base twelve to base ten:

a 53 *b* 90 *c* 8t *d* ett *e* $2t9e$

2 Convert from base ten to base twelve:

a 27 *b* 100 *c* 180 *d* 1000 *e* 3587

3 What can you say about the decimal equivalent of a duodecimal numeral which ends in 0? What about a duodecimal which ends in 00?

4 Calculate the following in the duodecimal scale:

a $42e+9tt$ *b* $t894+e97e$ *c* $357-319$

d 896×3 *e* $tet\times 7$ *f* $5tt1\times e$

Advantage of the duodecimal scale

Fewer digits than in the decimal scale are needed to represent a given number.

Disadvantages of the duodecimal scale

1. More symbols are necessary than in the denary scale.
2. Tables are larger than in the denary scale.

Summary

1 In the *decimal* system we count in ones, tens, hundreds, etc.

Thus $4567_{\text{ten}} = 4 \text{ thousands} + 5 \text{ hundreds} + 6 \text{ tens} + 7$

$$= (4 \times 10^3) + (5 \times 10^2) + (6 \times 10) + 7$$

2 In the *binary* system we need only two symbols, 0 and 1, and we count in ones, twos, fours, eights, etc.

Thus $1011_{\text{two}} = (1 \times 2^3) + (0 \times 2^2) + (1 \times 2) + 1$

Numbers expressed in binary form enable instructions to be conveyed to computers.

3 Any number greater than 1 may be taken as the *base* of a counting system. If, for example, eight is chosen, we require eight symbols 0, 1, 2, 3, 4, 5, 6, 7, and

$$3047_{\text{eight}} = (3 \times 8^3) + (0 \times 8^2) + (4 \times 8) + 7$$

4 Twelve is the base of the *duodecimal* scale. We require twelve symbols and so must invent two. If we call these t(for ten) and e(for eleven) we count

$0, 1, 2, 3, 4, 5, 6, 7, 8, 9, t, e, 10, 11, 12, \ldots$

$$5t2e_{\text{twelve}} = (5 \times 12^3) + (10 \times 12^2) + (2 \times 12) + 11$$

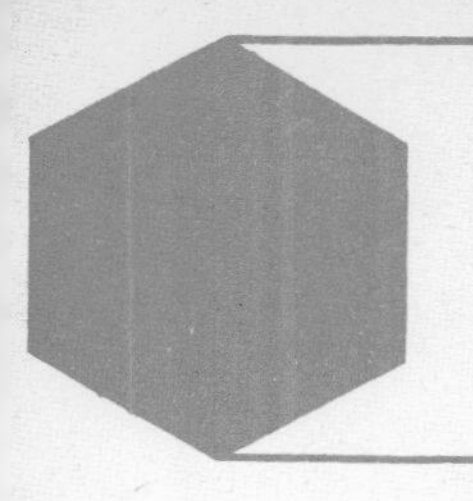

Revision Exercises

Revision Exercise on Chapter 1

Estimation of Error

1 Complete the following table:

Measurement	*Least unit of measurement*	*Absolute error*	*Upper limit of measurement*	*Lower limit of measurement*
a 5 cm				
b 4·8 m				
c 5·3 g				
d 28·2 cm				
e 8·72 m²				

2 A boy's height is 148 cm, correct to the nearest centimetre. Between what limits may his height actually lie?

3 Find the relative error and the percentage error in measurements given as:

a 11 hours *b* 0·8 g *c* 6·2 kg *d* 0·000 023 cm

4 State the tolerance, given the acceptable measures must lie between:

a 10·30 cm and 10·45 cm *b* 2·425 m and 2·430 m

c 3·1420 m² and 3·1415 m² *d* 99 ml and 100 ml

5 State the greatest and least acceptable measures for the following:

a (1056 ± 3) cm³ *b* $(1{\cdot}59 \pm 0{\cdot}004)$ cm

6 Find the upper and lower limits of the true sum and of the true difference of the following measurements:

a 7·6 g and 2·9 g *b* 5·2 m and 2·6 m

c 1276 km and 291 km *d* 2·6 mm and 0·9 mm

7 The length and breadth of an envelope are measured to the nearest centimetre and found to be 12 cm and 10 cm. Find the possible range of the perimeter of the envelope.

8 The hours of sunshine on a certain day at four towns were 10, 2, 1 and 3. If each of these measurements has an absolute error of 0·1 hour, find the maximum and minimum values of the mean number of hours of sunshine for the four towns.

9 5 lengths of wire each 3·2 cm long are to be cut from a length of 25·0 cm. If each of these six lengths has a tolerance of 0·4 cm, what are the limits of the remaining piece?

10 From a map I reckon that the distance between two towns is 7 km, to the nearest kilometre. If my speed of walking is 4 km/h, to the nearest 0·5 km/h, find the range of times I could take for the journey.

11 The length of a coil of wire is given as (250 ± 10) m. I want to cut pieces 6 m long from the coil but the measurement of each piece has an absolute error of 0·1 m. Within what limits will the number of pieces obtained lie?

12 Find the limits between which the areas of the following must lie:

a a rectangle of length 7 cm and breadth 3 cm

b a square of side 1·1 cm.

Revision Exercise on Chapter 2

Counting Systems

1 Express the following denary numerals to base two:

a 7 *b* 22 *c* 32 *d* 65 *e* 83

2 Express to the base ten:

a 1011_{two} *b* 1201_{three} *c* 256_{eight} *d* $2te_{twelve}$

3 Express 123_{ten} to base:

a three *b* eight *c* twelve.

4 Perform the following calculations to base two:

a $1101+111$ *b* 1101×101 *c* $1101\div101$

d $1101-1001$ *e* 101×11 *f* $10101\div11$

5 Copy and complete the following table. Each row represents the same number expressed in different ways.

Base	twelve	ten	eight	five	three	two
		27				
	2*e*					
						11101

6 The length and breadth of a rectangle are 1101_{two} cm and 11_{two} cm. Find its perimeter and area as numbers in the base two.

7 A square has a side of length 120_{three} mm. Find its perimeter and area.

8 Express the following to the base eight:

a 1011_{two} *b* 111000_{two} *c* 10101101_{two}

9 Perform these calculations in base three:

a $121+212$ *b* $210-21$ *c* 202×21

d $2022+212$ *e* $2211-1202$ *f* 222×102

10a Write out whole numbers to base five as far as 22_{five}.

b Convert 1123_{five} to base ten, and 376_{ten} to base five.

11 Perform these calculations in base five:

a $134+222$ *b* $234-143$ *c* 23×32

d $1000-432$ *e* 134×22 *f* $3443\div 31$

12 Perform these calculations in base eight:

a $725+123$ *b* $504-471$ *c* 47×21

d 123×46 *e* $761-73$ *f* $(22)^2$

13a 'Multiplying by two in a base-two system is like multiplying by ten in a base-ten system.' Why?

b A base-eight numeral ends in zero. What can you conclude?

c When will a number, expressed in octal form, be divisible by four?

14 Perform the following calculations, all numerals being to base twelve:

a $416+72t$ *b* $351-1te$ *c* $e47\times t$

15 $\begin{array}{r} 231 \\ +\underline{152} \\ \underline{423} \end{array}$ The addition shown is correct; what base is used? Calculate the difference between the numbers to the same base.

16 $\begin{array}{r} 342 \\ -\underline{163} \\ \underline{157} \end{array}$ Find the base of the numeration system in which the subtraction shown is correct. Calculate the sum of the numbers to the same base.

17 If $28_{\text{nine}}=35_x=101_y=122_z$, find x, y, and z.

18 If $27_x=32_y$, find the smallest possible replacements for x and y.

19 Can you show that no whole-number replacements exist for x and y such that $43_x=26_y$?

20 Show that 121_a, where a is a natural number greater than 2, must be a square number.

21 Show that for numbers to base two, $100^{10}=10^{100}$. Is $1000^{10}=10^{1000}$?

22 Show that $43-34$ is a multiple of 7 if the numbers are in the scale of eight, and is a multiple of 5 if the numbers are in the scale of six; check that these are also true for $52-25$. Can you make a general statement?

Cumulative Revision Section

Cumulative Revision Section (Books 1–7)

Revision Topic 1 Calculations by slide rule, logarithms and calculating machine

Exercise 1

Calculate:

1 $3{\cdot}62 \times 52{\cdot}7$

2 $35{\cdot}4 \times 0{\cdot}734$

3 $0{\cdot}374 \times 0{\cdot}0954$

4 $\dfrac{82{\cdot}4}{2{\cdot}57}$

5 $\dfrac{3{\cdot}58}{11{\cdot}5}$

6 $\dfrac{0{\cdot}654}{4{\cdot}27}$

7 $(5{\cdot}75)^2$

8 $(0{\cdot}394)^3$

9 $\sqrt{257}$

10 $\sqrt{0{\cdot}005\,94}$

11 $\dfrac{6{\cdot}84 \times 3{\cdot}75}{52{\cdot}1}$

12 $\dfrac{0{\cdot}012 \times 0{\cdot}464}{0{\cdot}87}$

13 $\dfrac{78{\cdot}6 \times (0{\cdot}44)^2}{24{\cdot}6}$

14 $\dfrac{(5{\cdot}02)^2 \times \sqrt{1{\cdot}04}}{91{\cdot}3}$

15 $\left(\dfrac{5{\cdot}24 \times 6{\cdot}84}{1{\cdot}12}\right)^2$

16 $\sqrt{\left(\dfrac{74{\cdot}9 \times 0{\cdot}0101}{2{\cdot}57}\right)}$

17 $\dfrac{11{\cdot}6 \times 2{\cdot}93}{4{\cdot}76 \times 8{\cdot}51}$

18 $\sqrt{\left(\dfrac{0{\cdot}574 \times 0{\cdot}024}{0{\cdot}101 \times 0{\cdot}964}\right)}$

19 $\dfrac{2{\cdot}42 \times 15{\cdot}6}{0{\cdot}777 \times 0{\cdot}84}$

20 $\sqrt[3]{\left(\dfrac{78{\cdot}4 \times 11{\cdot}5}{934 \times 2{\cdot}74}\right)}$

Revision Topic 2 Whole numbers, fractions and decimals

Reminders

1 The *commutative law*: $a+b=b+a$; $ab=ba$.

2 The *associative law*: $(a+b)+c=a+(b+c)$; $(ab)c=a(bc)$.

3 The *distributive law*: $a(b+c)=ab+ac$

4 *Identity elements*: *addition*, 0; $a+0=0+a=a$.

multiplication, 1; $a\times 1=1\times a=a$.

5 *Addition and subtraction of fractions.*

$$5\tfrac{3}{4}-1\tfrac{5}{6}=4\tfrac{9}{12}-\tfrac{10}{12}=3\tfrac{21}{12}-\tfrac{10}{12}=3\tfrac{11}{12}$$

6 *Multiplication and division of fractions*

$$\frac{3}{4}\div\frac{5}{6}\qquad or \qquad \frac{3}{4}\div\frac{5}{6}$$

$$=\frac{3}{{}_{2}\not{4}}\times\frac{\not{6}^{3}}{5}\qquad\qquad =\frac{\frac{3}{4}\times 12}{\frac{5}{6}\times 12}$$

$$=\frac{9}{10}\qquad\qquad =\frac{9}{10}$$

7 *Approximation*

a All measurements are approximate. A length given as 3·3 cm means that the length is 3·3 cm to the nearest 0·1 cm.

b *Rounding off*

36·728 = 36·7 to 1 decimal place, or 3 significant figures

36·75 = 36·8 to 1 decimal place, or 3 significant figures.

8 *Standard form.* This is the form $a\times 10^n$, where $1\leqslant a<10$.
n is given by the number of places the point is displaced from the standard position (after the first figure), e.g.

Distance to moon = 382 000 km = $3{\cdot}82\times 10^5$ km

A certain wavelength = 0·000 75 mm = $7{\cdot}5\times 10^{-4}$ mm

Exercise 2

1 Add all the prime numbers between 40 and 60.

2 If * means 'Square the first number and divide by the second', find the values of: *a* $9 * 6$ *b* $1\frac{1}{3} * \frac{2}{3}$ *c* $1{\cdot}2 * 0{\cdot}018$

3 Calculate: *a* $2\frac{1}{2} \times$ £7·38 *b* $\frac{3}{4}$ of £0·64 *c* $8\frac{1}{4} \times$ £1·34

4 Find the values of:

a $3\frac{3}{4} + 7\frac{5}{8}$ *b* $\frac{3}{4} - \frac{1}{6}$ *c* $4\frac{1}{4} - 1\frac{5}{6}$ *d* $2\frac{1}{2} \times \frac{3}{5}$ *e* $3\frac{3}{4} \div \frac{5}{8}$ *f* $2\frac{5}{8} \div 1\frac{7}{8}$

5 *Write down* the answers to the following:

a $1{\cdot}8 \times 100$; $0{\cdot}057 \times 20$; $876{\cdot}5 \times 1000$; $0{\cdot}05 \times 0{\cdot}2$

b $14{\cdot}7 \div 10$; $917 \div 1000$; $1{\cdot}232 \div 0{\cdot}08$; $0{\cdot}96 \div 0{\cdot}4$

6 *a* Express in terms of decimals: $\frac{3}{4}$, $1\frac{3}{8}$, $4\frac{1}{20}$, $\frac{7}{5}$

b Express as common fractions: 0·8, 0·08, 3·25

7 Round off 478·2649 to:

a 4 significant figures *b* the nearest 10

c 2 decimal places *d* the nearest thousandth

8 By taking each number to only one significant figure, calculate the following approximately, giving answers to one significant figure:

a 57×23 *b* $28{\cdot}3 \times 0{\cdot}72$ *c* $97{\cdot}6 \times 18{\cdot}3$

d $3{\cdot}14 \times 492 \div 61{\cdot}7$ *e* $(3{\cdot}14 \times 2{\cdot}71^2) - (3{\cdot}14 \times 1{\cdot}06^2)$

9 Find the exact values of:

a $2{\cdot}9 \times 0{\cdot}53$ *b* $91 \div 5{\cdot}2$ *c* $0{\cdot}119 \div 2{\cdot}8$

d $\dfrac{(0{\cdot}4)^2 \times 1{\cdot}2}{48}$ *e* $\dfrac{(0{\cdot}2)^3 \times (0{\cdot}3)^2}{0{\cdot}012}$

10 How many wavelengths each 0·0034 mm long will fit into a length of 10 mm? Give the answer rounded off to the nearest 100.

11 Express in standard form to the number of significant figures shown:

a 1230; 3 *b* 280 000; 2 *c* 5 862 000; 3

d 0·0056; 2 *e* 0·000 082 0; 1 *f* 0·000 000 082 6; 2

Revision Topic 3 Metric system; rectangular areas and volumes

Reminders

1 *Length* 1 cm = 10 mm

1 m = 100 cm = 1000 mm

1 km = 1000 m

2 *Area* 1 hectare (ha) = 10 000 m^2

3 *Volume* 1 litre = 1000 ml = 1000 cm^3

4 *Mass* 1 g = 1000 mg

1 kg = 1000 g

1 tonne = 1000 kg

5 *Area of rectangle*: $A = lb$

6 *Volume of cuboid*: $V = Ah = lbh$

7 *Area of triangle*: $A = \frac{1}{2}bh$

Exercise 3

1 *a* Express in m: 25 cm, 300 cm, 148 cm, 345 mm

b Express in m: 2·5 km, 1·76 km, 0·54 km, 0·0075 km

c Express in kg: 2500 g, 450 g, 31 g, 7 g

d Express in litres: 5000 cm^3, 4800 cm^3, 125 cm^3

e Express in cm^3: 4 litres, 3·7 litres, 0·875 litre, 0·007 litre

f Express in m^2: 30 ha, 2·7 ha, 0·5 ha

2 A floor is 8·7 m long and 6·9 m broad, and is to be tiled with square tiles of side 15 cm. How many rows of tiles will be required, and how many tiles will there be in each row? How many tiles would be required?

3 A cuboid has length 10 cm, breadth 3·5 cm and height 2·2 cm. Find:

a the sum of all its edges *b* its surface area *c* its volume.

4 A development site in a new town is in the shape of a rectangle 480 m long and 350 m broad. Find its area in hectares.

5 A tank in the shape of a cuboid is 2·5 m long, 45 cm broad and 24 cm deep. How many litres of water can it hold?

6 Find the area of the shaded part of each diagram in Figure 1. (All angles are right angles.)

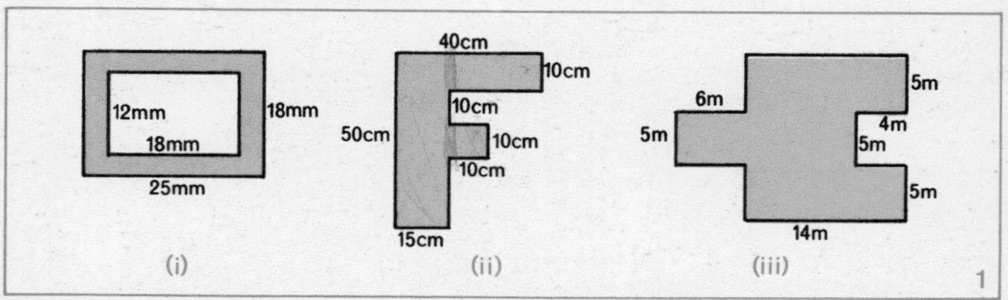

7 A cuboid is 24 cm long, 8 cm broad, and 9 cm high. How many cuboids each 3 cm by 2 cm by 6 cm can be packed into it?

8 Calculate with as little working as possible the area of each shape in Figure 2. (Assume each shape to be symmetrical.)

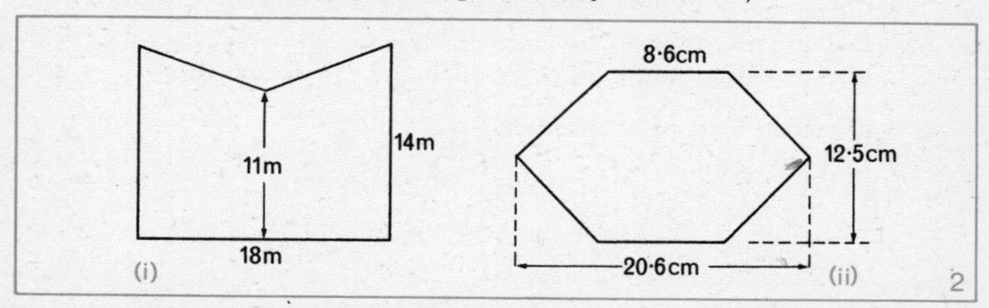

9 Calculate the altitude of a triangle with area $18{\cdot}4\,m^2$ and base 6·4 m.

10 Calculate the altitude and the area of an equilateral triangle of side 8 cm.

11 Figure 3 shows a concrete block with vertical sides, a horizontal rectangular base, and a sloping top. Calculate its volume.

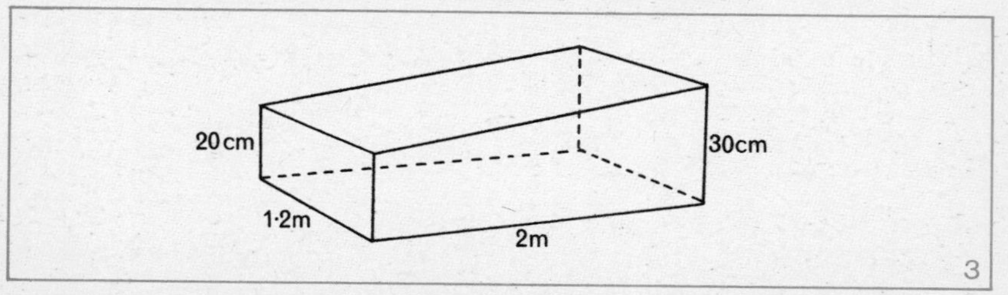

Revision Topic 4 Squares and square roots

Reminders

1 *Area of a square*: $A = l^2$

Squares of numbers can be found from tables:

e.g. $315^2 = (3{\cdot}15 \times 100)^2 = 3{\cdot}15^2 \times 100^2 = 99\,200$

$$0{\cdot}75^2 = \left(\frac{7{\cdot}5}{10}\right)^2 = \frac{56{\cdot}25}{100} = 0{\cdot}5625$$

2 *Length of side of a square*: $l = \sqrt{A}$

Square roots can be found by:

a Estimation: 75 lies between 64 and 81,

so $\sqrt{75} \doteqdot 8{\cdot}6$ or $8{\cdot}7$

b Graph

c Tables For numbers between 1 and 100 found direct.

d Slide rule For other numbers:

e.g. $\sqrt{1570} = \sqrt{(15{\cdot}7 \times 100)} = 3{\cdot}96 \times 10 = 39{\cdot}6$

$$\sqrt{0{\cdot}0157} = \sqrt{\frac{1{\cdot}57}{100}} = \frac{\sqrt{1{\cdot}57}}{10} = \frac{1{\cdot}25}{10} = 0{\cdot}125$$

e An iterative method

Example Calculate $\sqrt{8{\cdot}25}$ to 3 significant figures.

First estimate $= 2{\cdot}8$

$$\frac{8{\cdot}25}{2{\cdot}8} = 2{\cdot}946$$

2 2·8 3

4 8·25 9

$$\begin{array}{r} 2{\cdot}946 \\ 28\overline{)82{\cdot}500} \\ \underline{56} \\ 265 \\ \underline{252} \\ 130 \\ \underline{112} \\ 180 \end{array}$$

$$\textit{Second estimate} = \frac{2{\cdot}8 + 2{\cdot}946}{2}$$

$$= 2{\cdot}873$$

$\sqrt{8{\cdot}25} = 2{\cdot}87$ to 3 significant figures.

Exercise 4

1 *a* Find the *exact* squares of 4·3, 123, 0·45, 3000.

b Use tables to find the squares of:

3·79, 8·96, 25·9, 780, 0·563, 0·098.

2 Find the total area of two squares whose sides are 7·9 m and 12·8 m, rounding off the answer to 3 significant figures.

3 *a* Give the exact square roots of 14 400, 2·25, 0·01 and 1·96.

b Between which integers do the square roots of 29, 181 and 58·6 lie?

c Estimate to 1 significant figure $\sqrt{0{\cdot}59}$, $\sqrt{0{\cdot}07}$, $\sqrt{77{\cdot}7}$.

4 Find from your tables or slide rule the square roots of all the prime numbers less than 100 whose units digit is 7.

5 *a* Find the lengths of the sides of the squares whose areas are 74·0 cm^2 and 3·56 cm^2.

b A rectangle is three times as long as it is broad, and its area is 576 cm^2. Find its dimensions.

6 Use tables to find the square roots of:

a 23·4 *b* 234 *c* 2340 *d* 23 400

e 0·567 *f* 0·0345 *g* 0·007 16 *h* 0·000 003 59

7 *a* Estimate $\sqrt{14}$ to 2 figures and find by the method of iteration a better approximation (to 3 significant figures).

b Repeat ***a*** for $\sqrt{590}$ and $\sqrt{0{\cdot}08}$.

8 Find from your square-root tables $\sqrt{3{\cdot}47}$. Use this approximation as a first estimate in an iterative process to find $\sqrt{3{\cdot}47}$ to 5 significant figures. (One division only is necessary; a calculating machine may be used.)

9 Find the values of:

a $\sqrt{(1^2+4^2+8^2)}$ *b* $\sqrt{(2{\cdot}7^2+4{\cdot}3^2)}$

Revision Topic 5 Ratio and proportion

Reminders

1 *Direct proportion*:

Example. What is the value of 42 francs in British money if the rate of exchange is 11·2 francs to the £?

Number of francs	*Number of £*
11·2	1
42	$1 \times \frac{42}{11 \cdot 2} = £3 \cdot 75$

In general

First variable	*Second variable*
a	x
b	y

If

$$\frac{a}{b} = \frac{x}{y},$$

the variables are in direct proportion.

When one *increases*, the other *increases* in the same ratio.

2 *Inverse proportion*

If

$$\frac{a}{b} = \frac{y}{x},$$

i.e. $ax = by$, the variables are in inverse proportion.

When one *increases* the other *decreases* in the same ratio.

3 The *graph* of a *direct*-proportion relationship between two sets of variables is a set of points which may be joined by a straight line through the origin.

4 The scale of a map is given by the *Representative Fraction*

$$\text{R.F.} = \frac{\text{distance on map}}{\text{distance on ground}}$$

Exercise 5

1 Find in simplest form the ratio of:

a 1 cm to 1 km *b* 5 mm to 5 cm *c* £2·80 to £1·28

2 Say which of the following pairs of quantities are in direct proportion (*d*), inverse proportion (*i*), or neither (*n*).

a The time taken for a given journey and the speed.

b A boy's age in years and his height in centimetres.

c The number of similar articles bought and their total cost.

d The number of cattle and the number of days they can be fed from a given quantity of feeding stuff.

3 On a demolition site a mechanical shovel takes $7\frac{1}{2}$ hours to remove 350 tonnes of rubble. At the same rate how long will it take to remove a further 140 tonnes?

4 Sixty-six French pupils plan a bus excursion which is to cost them 12·60 francs each. If the total expenses remain the same but an additional 18 pupils take part, find the new cost per head.

5 The first edition of a book contains 184 pages, with an average of 540 words to the page. The second edition, in smaller print, has 100 more words to the page. How many pages are there in the second edition?

6 Given that £1 = 2·40 dollars, express £3·50 in dollars, and 3·50 dollars in £ to the nearest p.

7 Given that £1 = 2·40 dollars and £1 = 145 pesetas, find the value of 1 dollar in pesetas, rounded off to 1 decimal place. Hence find the value of 14·50 dollars in pesetas to the nearest peseta.

8 A model caravan is made to the scale of 1:50. If the dimensions of the caravan are: length 4200 mm, breadth 2100 mm and height 2400 mm, find the dimensions of the model.

9 The R.F. of a map is 1:100 000. If two villages are 18·5 km apart, how far apart are they on the map? If two towns are 255 mm apart on the map, how far apart are they on the ground?

10 A plan has a scale of 1:5000. Find the equivalents on the ground of:

a lengths of 8 cm and 3·4 cm on the map

b an area of 4 cm^2 on the map.

11 Two cubes have edges of lengths 2 cm and 3 cm. Find the ratios of:

a the total lengths of all their edges *b* their areas *c* their volumes.

Revision Topic 6 Time, distance, speed

Reminders

1 A *time* given as 09 53 means

9.53 am, or 7 minutes to 10 in the morning.

A *time* given as 16 30 means

4.30 pm, or half past four in the afternoon.

2 $$\textit{Average Speed} = \frac{\text{Distance}}{\text{Time}}$$

3 If D represents distance, S represents speed and T represents time,

$$D = ST$$

Exercise 6

1 Find the time taken for each of the following journeys. Note that in ***d*** and ***e*** departure and arrival are on successive days.

	a	*b*	*c*	*d*	*e*
Time of departure	09 45	12 35	03 05	21 20	22 45
Time of arrival	11 15	20 40	21 25	03 35	09 20

2 Copy and complete the following table:

	a	*b*	*c*	*d*
Time	$2\frac{1}{2}$ h	1 h 12 min		3·39s
Distance	150 km		24 km	12 mm
Speed		65 km/h	60 km/h	

3 A train leaves Glasgow at 18 30 and arrives in Edinburgh 68 km away at 19 14. Calculate its average speed, to the nearest km/h.

4 I have to reach a car ferry 525 km away at 18 00 hours. When must I leave if I estimate that I can achieve an average speed of 70 km/h while on the road, and also allow myself a stop of 50 minutes for a meal?

5 A space probe takes 39 hours to reach the moon, which is 384 000 km from the earth. Find its average speed in km/s to 2 significant figures.

6 A missile is fired at a target 178 500 km from the earth. If its average speed is 17 000 km/h, and it is fired at 06 30 hours, when will it reach its target?

7 I can maintain an average speed of 84 km/h on an autobahn. If I leave at 08 30 hours, have a stop lasting 1 hour 40 minutes, and wish to finish my journey by 17 30 hours, how many kilometres can I cover?

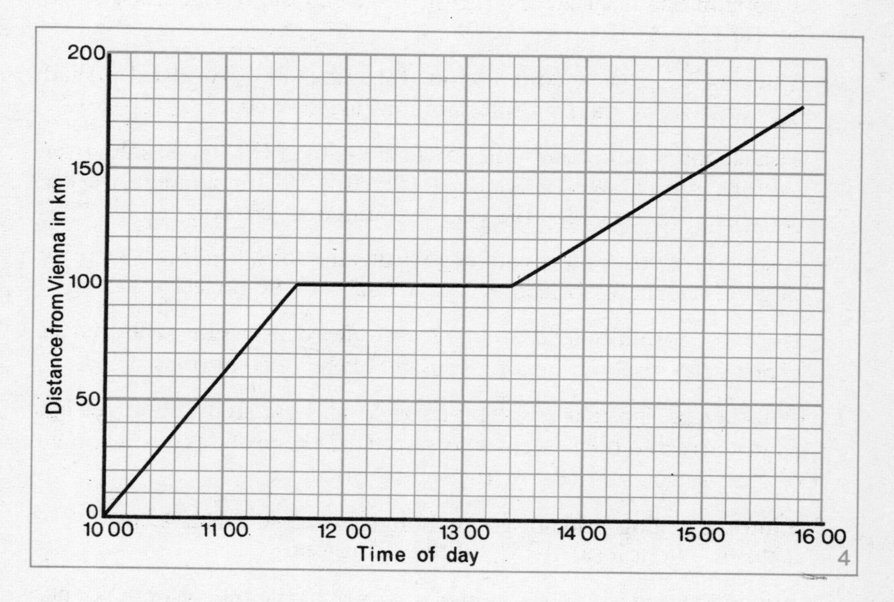

4

8 A motorist leaves Vienna at 10 00 hours travelling towards Graz 180 km away. On the way he has a breakdown, and after stopping for repairs he continues at a reduced speed. The graph in Figure 4 illustrates his progress.

a When, and how far from Vienna, did he have the breakdown?

b How long did he stop for repairs?

c What were his average speeds before and after the breakdown?

Copy the graph on 5-mm squared paper.

A second motorist leaves Graz at 13 48 and travels towards Vienna at 100 km/h. By drawing a second graph on the diagram, find when and where they meet.

Revision Topic 7 Social arithmetic — 1

Exercise 7

1 Convert to fractions: *a* 30% *b* $112\frac{1}{2}$% *c* 0·5%

2 Convert to percentages: *a* $\frac{7}{8}$ *b* $1\frac{1}{3}$ *c* 0·04

3 A hotel makes a service charge of 10%. Find the service charge on a bill for £21·73.

4 A salesman is given a commission at the rate of $1\frac{1}{2}$% on all sales. Find the commission paid on sales amounting to £750.

5 A salesman is paid a commission at the rate of $1\frac{1}{4}$% on all sales over £10 000, plus a weekly wage of £22·50. Find his average weekly earnings if his sales for the year amounted to £88 000.

6 A shop is offering a discount of 15% on all goods. Find the sale price of goods normally priced at: *a* £1·65 *b* 96 pence.

7 If goods normally sold at £1·12 are offered at a sale at 98 pence, express the discount as a percentage of the normal price.

8 Potatoes are bought at 105p per 50 kg and are sold at $3\frac{1}{2}$p per kg. Find, correct to the nearest whole number, the profit as a percentage of: *a* the cost price *b* the selling price.

9 Find the selling price of goods bought at £2·80 and sold at:
a a profit *b* a loss, of 15% of the cost price.

10 A man insured his house, valued at £8500, at the rate of 13 pence per £100, and its contents, valued at £2800, at 25 pence per £100. Find the annual premium he had to pay.

11 A man aged 25 finds that his insurance company can offer him an endowment policy (with profits) at £3·38 per £100 (25 annual payments). Find the annual premium on a policy for £3500, and also the total amount he will pay the company over the 25 years.

12 Find the interest when:
a £78 is invested for 10 months at 5% per annum
b £350 is invested for 4 months at $2\frac{1}{2}$% per annum.

13 Find the compound interest on:
a £2000 for 3 years at 5% p.a. *b* £345 for 2 years at 8% p.a.

Revision Topic 8 Social arithmetic—2

Exercise 8

1 Find the total rates due on a house with rateable value £145, given that the local rates are £0·84 in the £.

2 A town has to raise £2 200 000 and the rateable value of the property in the town is £2 800 000. What rate in the £ must be levied?

3 A man owns a house whose rateable value is £165. He hears that the rates are to be increased by 12p in the £. Find the amount of the increase he would have to pay for the year.

4 An insurance company quotes a premium of £85 to insure a car. From this premium is deducted a discount of 10% on account of the owner's occupation. On the remainder he is allowed a no-claims bonus of 60%. What does he actually pay?

5 A man's total income is £1700. His allowances free of tax are £795. On the remainder he has to pay tax at the rate of 30p in the £. How much tax does he pay?

6 A man has a salary (paid monthly) of £3150 per annum. His total allowances free of tax are £1150. On the remainder he has to pay tax at the rate of 30p in the £. What is his net monthly salary after tax deduction?

7 Shares are quoted at 18 pence each. These shares are sold in multiples of 10 only. Find the number of shares a man could purchase with £100.

8 Find the cost of 1000 shares if they are quoted at 18 pence each.

9 A man owns 900 shares. The company announced a dividend of 1·2 pence per share. Find the gross income from his holding. If this income is all taxed at the rate of 30p in the £, find his net return.

10 If you had £1000 available for investment, which of the following options would be the more profitable as regards annual income?

a A building society which gives a dividend of $6\frac{1}{4}$%, free of tax.
b Shares which are quoted at 18 pence and are expected to give an annual dividend of 1·8 pence per share—subject to tax at 30 pence in the £. (Shares can only be bought in multiples of 10.)

Revision Topic 9 Areas and volumes associated with the circle

Reminders

1 *Circle* Circumference $C = 2\pi r = \pi d$ Area $A = \pi r^2$

2 *Cylinder* Curved surface area $A = 2\pi rh$ Volume $V = \pi r^2 h$

3 *Cone* Curved surface area $A = \pi rs$ Volume $V = \frac{1}{3}\pi r^2 h$

4 *Sphere* Area $A = 4\pi r^2$ Volume $V = \frac{4}{3}\pi r^3$

5 Approximations for π are $\frac{22}{7}$ and 3·14 (to 3 significant figures)

Exercise 9

1 Calculate the perimeter and area of each of the shapes in Figure 5. In *a* the curves are semicircles, in *b* quarter-circles, and in *c* O is the centre of the circle whose arc is shown.

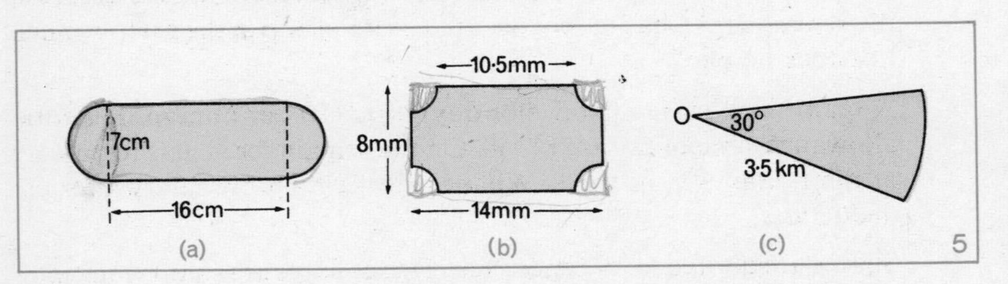

What is the radius of a circle of area 5 cm^2?

Calculate the volume of a cylindrical can of diameter 9 cm and height 9·7 cm. Calculate also the area of a label which will fit round the can.

Find the radius of a cylinder of volume $6{\cdot}28 \times 10^6$ cm^3, height 200 cm.

A cylindrical cardboard carton, closed at both ends, is to be 22 cm high and is to have a radius of 12 cm. What area of cardboard will be used, assuming that there will be no overlaps at the joins?

A cone has a circular base of radius 5 cm and a vertical height of 12 cm. Calculate its volume, and the area of the curved surface.

Calculate the area and the volume of a sphere of radius 10·4 cm.

Revision Topic 10 Sequences

Exercise 10

1 a Give three more terms for each of the following sequences:

(*1*) 2, 5, 8, 11,... (*2*) 1, 4, 9, 16,... (*3*) 8, 4, 2, 1,...

(*4*) 1, 1, 2, 3, 5, 8,... (*5*) 0, 3, 8, 15,... (*6*) 2, 3, 5, 7, 11,...

b Write down your rule for finding the next term in each of the above.

2 Write down the first five 'square' numbers and the first five 'triangular' numbers. Show by means of a sketch that every square number can be expressed as the sum of two triangular numbers. Which two triangular numbers have as their sum the twentieth square number?

3 Find the first four terms of the sequence of:
a 'triangular pyramid' numbers *b* 'square pyramid' numbers.

4 Write down the first three terms of the sequences with *n*th terms:
a $2n+1$ *b* n^2-1 *c* $\frac{1}{2}n(n+1)$ *d* $\dfrac{1}{n+1}$ *e* $\frac{1}{6}n(n+1)(2n+1)$

Which of these appear to be the sequences in questions ***2*** and ***3***?

5 Give a formula for the *n*th term of each of the following sequences:

a the natural numbers *b* the whole numbers

c the odd numbers *d* the even numbers

e $\frac{1}{3}, \frac{1}{4}, \frac{1}{5}, \frac{1}{6}, \ldots$ *f* 2, 5, 10, 17,...

6 Write down two possible numbers for the fourth term in each of the following: *a* 1, 2, 4,... *b* 1, 3, 9,...

7 In a Fibonacci sequence, every term after the second is the sum of the previous two terms. Which sequence in question ***1*** is of this type?

Use a slide rule to calculate as a decimal the ratio of each term to the following term in this sequence. (Use the folded scales if your slide rule has them.) When you reach the stage where no further adjustment of the slide is necessary *a* check that each of the next few terms appears on the C scale opposite the preceding term on the D scale *b* notice that the sequence of ratios thus appears to have a *limit* *c* estimate this limit. This limit is called the *Golden Mean*.

Revision Topic 11 Probability

Reminders

1 The *probability* of an outcome in a random experiment is the limit of the relative frequencies of the outcome in a large number of trials.

When an experiment has several *equally likely* outcomes,

$$\text{probability of a 'favourable' outcome} = \frac{\text{number of favourable outcomes}}{\text{number of possible outcomes}}$$

2 If the probability of an outcome of an experiment is P, then the probability the outcome will *not* happen is $1-P$.

3 $0 \leqslant P \leqslant 1$

4 *Expected frequency* = probability of outcome × number of trials.

5 For *mutually exclusive* outcomes A and B of an event,

$$P(A \text{ or } B) = P(A) + P(B)$$

6 For *independent* outcomes A and B of an event,

$$P(A \text{ and } B) = P(A) \times P(B)$$

Exercise 11

1 A bag contains a number of coloured balls. A ball is picked out, the colour noted and the ball replaced. The table shows the results of various numbers of trials of the experiment.

Number of trials	10	50	100	500	2000
Number of times white is picked	3	10	18	84	339

a Calculate the relative frequency of choosing a white ball for each set of trials, to 2 decimal places.

b Does the relative frequency appear to approach a limit?

c Estimate the probability of picking a white ball.

2 A bag of mixed toffees contains 20 cream, 15 treacle, 10 nut caramels.

a If I select one toffee at random, what is the probability that it is a nut caramel?

b If the first toffee is not a nut caramel, and I eat it, what is the probability that the next one I pick will be a nut caramel?

3 One of the letters of the word MISSISSIPPI is chosen at random. What is the probability that it is:

a M *b* P *c* I *d* not P *e* not I?

4 Show as an array of ordered pairs all possible outcomes in throwing a pair of dice. Hence find the probability of getting:

a two sixes *b* at least one six

c a total score of six *d* a total score of *not* six

e a total score of one *f* a total score of more than one

g a total score which is a multiple of three.

What is the most likely total score? How many times would you expect to get this score in 144 throws?

5 a What is the probability of an event which is certain to happen?

b What is the probability of an event which is impossible?

c Given that the probability that an event will happen is a $(0 \leqslant a \leqslant 1)$, what is the probability that the event will *not* happen?

6 An airfield has a beacon which flashes a code of three characteristics, either dots or dashes, e.g. ·——. Assuming that all possible codes are equally likely, what is the probability that a code selected at random consists of:

a 3 dashes *b* 2 dots followed by a dash *c* 2 dots and a dash?

7 All possible two-digit numbers are formed from the digits 1, 2, 3, 4, 5, 6. Find the probability that one of these numbers chosen at random will be divisible: *a* by 5 *b* by 6.

8 In an arithmetic test the lower-quartile, median, and upper-quartile marks are 41, 55, 67 respectively. What is the probability that a pupil chosen at random scored: *a* less than 41 *b* between 41 and 67?

Revision Topic 12 Statistics

Exercise 12

1 During a period of 75 days a Fire Brigade attended incidents as below:

Number of incidents in one day	0	1	2	3	4	5	6	7	8
Frequency	10	21	15	13	8	5	2	0	1

a Find, to the nearest whole number, the mean number of incidents attended per day during the period. Find also the median of the distribution.

b Find the probability that on any day chosen at random from the sample the brigade attended no incident.

c If the sample is assumed to be typical of the whole year, estimate the number of days in the year when the brigade has no calls.

2 A sample of 200 instruments is tested for dial tension. Here 6–8 means greater than or equal to 6 but less than 8, and so on.

Tension	6–8	8–10	10–12	12–14	14–16	16–18	18–20
Frequency	10	26	41	53	36	29	5

a Draw a histogram to illustrate the data. Calculate the mean tension.

b If all instruments showing a dial tension of 16 and more are to be rejected, what percentage of the sample is acceptable?

3 Draw a cumulative-frequency diagram for the data in question **2**, and estimate the semi-interquartile range of the distribution of tensions.

4 The number of words in each of the first 50 sentences of a particular book were as follows:

44	30	12	9	10	28	47	26	13	39
32	9	37	10	39	12	28	31	9	5
11	10	8	17	11	7	11	8	17	25
7	9	7	38	27	48	39	7	23	36
43	40	22	8	31	13	10	6	5	23

a Illustrate the above data more concisely as a grouped-frequency distribution starting 1–5, 6–10, . . . and from this distribution draw an appropriate histogram.

b State the modal class and calculate the mean length of sentence used, using your grouped data.

c Calculate the probability that if a sentence is chosen from the book at random it will contain more than 40 words, assuming this to be a typical sample.

d Make a similar investigation by choosing a sample: (*1*) from a favourite book (*2*) from a magazine or a newspaper article. (You may find it preferable to use class intervals different from those above.) Can you draw any conclusions from your results?

5 A class of 27 pupils sat three examinations of different types in one subject with results as tabulated:

Mark	*Frequencies* *Type A*	*Type B*	*Type C*	*Mark*	*Frequencies* *Type A*	*Type B*	*Type C*
85–89	0	0	2	55–59	2	5	3
80–84	0	1	1	50–54	7	3	4
75–79	1	2	2	45–49	2	0	5
70–74	3	2	1	40–44	8	0	5
65–69	1	7	2	35–39	2	0	0
60–64	1	7	1	30–34	0	0	0
				25–29	0	0	1

(*continued on the right*)

a Find the mean mark for each examination.

b Make cumulative-frequency tables, draw the cumulative-frequency curves, and estimate the interquartile ranges of the distributions of marks.

c Which examination gave the highest mean mark, and which gave the greatest spread of marks?

d Use the cumulative-frequency curves to estimate the pass mark in each examination that would allow 20 pupils to pass.

Cumulative Revision Exercises

Exercise 1A

1 Express $\frac{5}{16}-(0{\cdot}55)^2$ as a decimal.

2 Express $600\times3{\cdot}25\times20$, and $0{\cdot}08\times0{\cdot}03\times0{\cdot}15$, in the form $a\times10^n$.

3 Find to the nearest necessary metre the side of the largest square field whose area does not exceed $2000\,\text{m}^2$.

4 Find the simple interest on £525 invested for 10 months at $4\frac{1}{2}\%$ per annum.

5 Express 76_{eight} in the scales of ten and of two.

6 A rectangle of length 5·7 cm has an area of $18\,\text{cm}^2$. Calculate its breadth to 2 significant figures.

7 Calculate the representative fraction for a map on which a length of 1·72 cm represents a distance of 0·86 km.

8 During the five working days in a certain week a building contractor employed casual labourers as follows: Monday, 35 for 8 hours each; Tuesday and Wednesday, 28 for $6\frac{1}{4}$ hours each; Thursday, 30 for 6 hours each; Friday, 24 for $5\frac{1}{2}$ hours each. Calculate the total number of man-hours worked in the week.

If labourers are paid 80 pence per hour, and in addition the contractor has to pay £25 per hour for the hire of a bulldozer during the working hours, find (to the nearest £) the contractor's total outlay for the week.

9 The total rateable value of a city was £3 800 000 and the rate levied was 105 pence in the £. The corresponding figures for a neighbouring small town were £196 000 and 75 pence in the £. Calculate the amount raised by the rates in the two places.

If the small town were incorporated within the city, and the total rateable value and the total sum raised were to remain the same as before, what common rate in the £ would have to be imposed? (Answer to the nearest penny.)

Exercise 1B

1 Express $\frac{0{\cdot}008}{400}$ and $\frac{2{\cdot}4 \times 10^5}{600}$ in standard form.

2 A sheet of lead 4·5 mm thick weighs 150 kg. If lead weighs 13·6 g/cm^3, find the area of the sheet in cm^2.

3 After 14 completed innings a batsman's average was 17·7. In his next innings he scored 43. What was his average for the 15 innings?

4 The outside measurements of a picture frame are 25 cm and 20 cm and it is made of wood 1·5 cm broad. Find the inside measurements of the frame and the picture area.

5 A householder pays rates at 95 pence in the £ on a rateable value of £176. How much does he pay in rates?

6 I throw two dice, one after the other. Make an array of possible outcomes, and find the probability that I score:

a exactly 8 *b* at least 8.

7 Multiply the binary numbers 101 and 1011, and then express your answer in the scale of ten.

8 A merchant spent £2250 in buying 1000 articles. He fixed the selling price to allow himself a profit of 20% on his cost, and sold four-fifths of the articles at this price. He then reduced his selling price by one-third and sold the remainder of his stock at this new price. Calculate his profit as a percentage of his outlay.

9 A man receives a legacy of £1000. He considers two ways of investing it, as shown below. Calculate his net income in each case.

a In a Savings Bank he can invest £50 in the Ordinary branch, bearing 4% per annum interest, and the rest in the Special branch, bearing 7% p.a. interest. Interest received in the Ordinary branch is tax free; interest received in the Special branch is subject to tax at the standard rate of 30 pence in the £.

b He can buy oil shares costing £2·50 each which pay a dividend annually of 12·5 pence per share, all of which is liable to income tax at the standard rate.

Exercise 2A

1 Simplify $\dfrac{(0{\cdot}4)^2 \times 2{\cdot}5}{32 \times 0{\cdot}02}$,

expressing your answer in decimal form.

2 At what rate per cent per annum must I invest £4000 to bring in a quarterly income of £52·50? What will be the net annual income after deduction of income tax at 30 pence in the £?

3 Calculate $\sqrt{31{\cdot}6}$ correct to 5 significant figures by finding a first estimate from slide rule or tables and then performing one division.

4 A confectioner sold two orange ice lollies for every lemon, and three strawberry for every orange. What is the probability of a sale, taken at random, being that of a strawberry?

5 A metal square of side 5 cm is filed down to make a square of side 4 cm. What percentage of the metal is waste?

6 Repeat question **5** for the case of a cube of side 5 cm reduced to a cube of side 4 cm.

7 A wheel of diameter 12 cm makes 3000 revolutions per minute. Find in metres per second the speed of a point on the circumference, to 2 significant figures.

8 Every week a salesman is paid a basic salary of £22. In addition he receives commission at the rate of $\frac{4}{5}\%$ on the first £1000 of his weekly sales and at the rate of 2% on weekly sales over £1000. Calculate:

a his earnings in a week in which his sales totalled £1600

b his sales in a week when his total earnings amounted to £35.

9 A sequence expressed in the scale of two is 1, 10, 11, 101, 1000, 1101,...,

a What particular kind of sequence is this?

b Find the next two terms of the sequence.

c Express the sequence in the scale of ten, then three.

Exercise 2B

1 A television set cost a retailer £50. He offers to sell it either for a cash price of £65, or for a down payment of 25% of the cash price followed by 24 monthly instalments of £2·50. Calculate his percentage profit calculated on his cost price and his selling price for each method of selling.

2 The following shows the scores for each pair of shots by 8 friends in a game of Tenpin Bowling:

3	0	8	8	9	4	0	6	8	8	8	10	10	0	10	5
7	6	10	5	10	2	9	9	6	10	1	0	6	7	7	7
7	10	10	6	9	9	9	8	3	1	9	7	1	3	7	7
8	5	0	7	10	0	8	5	10	7	8	9	7	8	3	9
10	7	10	5	6	3	5	9	9	7	3	6	4	10	2	8

a Construct a frequency table, and draw a histogram and frequency polygon.

b Calculate the mean score for each pair of shots, to 1 decimal place.

c Draw a cumulative-frequency curve, and hence find the median, quartiles and semi-interquartile range for the distribution.

3 A cylindrical jar of internal diameter 3 cm is 12 cm in height. Six spherical metal balls, each of radius 1 cm, are placed in the jar.

a Find the volume of water that can now be poured into the jar.

b Find the total mass of the contents, given that 1 cm^3 of water has mass 1 g and 1 cm^3 of the metal has mass 4·6 g.

4 a Calculate, in the scale of three, $1201 + 222$ and $100 - 21$.

b In what scales are the following calculations performed?

(*1*) $\begin{array}{r} 137 \\ +\underline{266} \\ \underline{414} \end{array}$ (*2*) $\begin{array}{r} 266 \\ -\underline{137} \\ \underline{127} \end{array}$

(*3*) $3 \times 3 = 10$
(*4*) $3 \times 3 = 11$
(*5*) $3 \times 3 = 12$

5 Wire is made of a material which has mass 4·6 g/cm^3; it is circular in cross-section and of diameter 0·30 cm. Calculate to two significant figures the length of wire in a coil which has mass 40 kg.

6 Two pipes of diameters 6 cm and 8 cm have to be replaced by a single pipe with the same cross-sectional area. Find the diameter of this pipe.

Exercise 3A

1 Calculate:

a $\dfrac{23{\cdot}7 \times 1{\cdot}09}{102{\cdot}8}$ *b* $\sqrt{[(0{\cdot}315)^2 + (0{\cdot}426)^2]}$

2 I motored 22 km from Kendal to Carnforth in 22 minutes, then 187 km on the M6 motorway for 1 hour 50 minutes. Find my average speed for each part of the journey, and for the whole journey.

3 An area of ground on a map is in the shape of a trapezium (i.e. a quadrilateral with one pair of sides parallel). The parallel sides are 1 cm and 3 cm long, and are 2 cm apart. If the scale of the map is 1:200 000, find the actual lengths in km and the actual area in km^2.

4 a Write down the first four terms and the tenth term of the sequences whose nth terms are defined by:

(*1*) $3n-1$ (*2*) n^2+1 (*3*) 2^n+1

b Find three terms to continue each of the following sequences, and state your rule for finding the next term:

(*1*) 3, 5, 7, 9, ... (*2*) 3, 6, 12, 24 ... (*3*) 0, 3, 8, 15, 24 ...

5 a Taking the radius of the earth to be 6400 km, calculate the distance round the equator.

b How long will it take an aircraft to circle the equator at 2700 km/h? (Neglect its height above the earth.)

c Find the speed of a satellite which circles the equator at a height of 240 km in $1\frac{1}{4}$ hours. (Give your answer to 2 significant figures.)

6 A town has a total rateable value of £1 million, and it is necessary to raise £874 000 to cover expenditure. What is the rate the council must impose, correct to the necessary penny, at least to cover this expenditure, and what surplus may be expected if this rate is imposed?

How much will a man have to pay in this case if his house has a rateable value of £200 and he is allowed $2\frac{1}{2}\%$ discount if he pays before December 1?

7 Two metal ingots in the shape of cuboids each 4 cm by 3 cm by 2 cm are melted down. One is recast as a circular disc 0·5 cm thick, the other as a sphere. Calculate the diameters of the disc and the sphere.

Exercise 3B

1 Calculate:

a $\dfrac{23{\cdot}4 \times 0{\cdot}165}{0{\cdot}003\,67}$ *b* $\sqrt{(5{\cdot}36 \times 10^{3})}$

c $(5{\cdot}36 \times 10^{-2})^{2}$ *d* $\sqrt{\left(\dfrac{8{\cdot}35}{485}\right)}$

2 The cost of setting up the type for a school magazine is £135. The cost of running the machines to print it is £10·50 per 100 copies. The cost of paper, ink, etc., is 12 pence per copy. The magazines are sold at 20 pence each. If 700 copies are printed and 596 are sold, what is the minimum sum to be obtained from advertisements to prevent a loss on the venture?

3 One cubic centimetre of gold weighs 19·2 g. If a gold ingot weighing 100 g is rolled out into a square of gold leaf 0·02 cm thick, what is the length of one side of the square?

4 Two bells start tolling at the same instant. One strikes at intervals of 4 seconds, the other at intervals of 5 seconds. One second after the first two start, a third bell starts tolling at intervals of 3 seconds. By writing out sequences, find when the three bells first strike simultaneously.

5 A boiler is in the shape of a cylinder surmounted by a hemisphere. The diameter of the cylinder is 0·5 m and the total height of the boiler is 1·2 m. Find the volume of water it holds. Given that the empty boiler weighs 26 kg, and that water weighs 1000 kg per m^3, find the total weight of the boiler when it is full of water.

6 A man borrows £4000 and at the end of each of the three following years he pays the lender £400; part of this is in payment of interest at 5% per annum on the amount of his debt during the year, and the rest reduces the debt. How much does he still owe at the end of the third year?

7 At 09 00 hours a car passes a certain point, travelling at an average speed of 60 km/h. At 10 30 hours a second car starts off from that point to travel at an average speed of 100 km/h in pursuit of the other. Find when the second car overtakes the first, and at what distance from the starting point of the second car.

Trigonometry

1 Three Dimensions

1 *Intersections of lines and planes in space*

Exercise 1 (*for class discussion*)

1 Using two pencils to represent lines, demonstrate the following:

a Two lines in space which intersect.
b Two lines in space which are parallel.
c Two lines in space which do not intersect and are not parallel.
Are there any other possibilities?

2 Using a pencil to represent a line and a sheet of paper to represent a plane, demonstrate the following:

a A line intersecting a plane.
b A line parallel to a plane.
Are there any other possibilities?

3 Using two notebooks or sheets of paper to represent planes, demonstrate the following:

a Two planes meeting in a line.
b Two planes which do not meet in a line. How would you describe the relationship between the planes?

4 Give examples from the classroom of walls, wires, etc., which could represent the following:

a Two parallel planes.
b Two planes which intersect in a line.
c Three planes which intersect in a point.
d A line meeting a plane.
e A line parallel to a plane.

5 Figure 1 shows a storage box. Name:
a Two lines that intersect.
b Two parallel lines.
c Two lines which are not parallel and do not intersect.
d A line and a plane that intersect.
e A line and a plane that are parallel.
f Two planes that intersect in a line.
g Two planes that are parallel.
h Three planes that intersect in a point.
i A plane that meets two parallel planes.

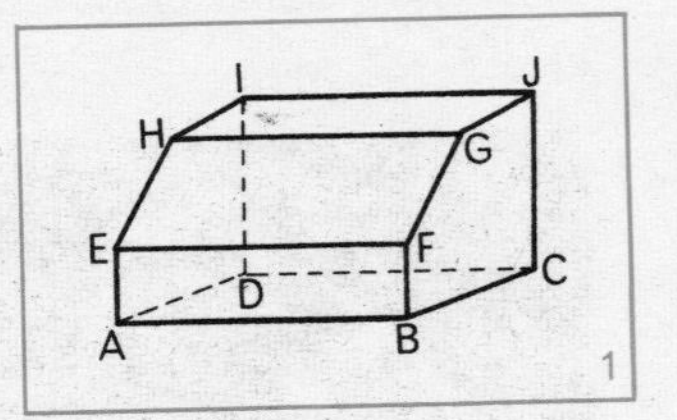

1

Some hints on drawing lines and planes in space

(i) *Draw vertical lines (if any) up and down the page.*
(ii) *Represent lines which are parallel by parallel lines.*
(iii) *Indicate lines which are not 'seen' by broken lines.*
Notice that a rectangle seen at an angle appears as a parallelogram.
(iv) *Mark right angles on your diagrams by completing a small parallelogram with the vertex as one corner.*
(v) *If your figure is not helpful, scrap it and start again. A preliminary rough sketch may help to avoid this state of affairs.*

6 Using these hints, draw cubes, cuboids and pyramids on plain paper.

2 *One, two and three dimensions*

(i) To fix the position of a point on a line, we require an origin and a measure of the distance of the point from the origin, relative to a chosen positive direction on the line. This distance can be specified by one coordinate, so we say that a line is *one-dimensional.*

P is 3 units distant from O and has coordinate 3. Q has coordinate -1.

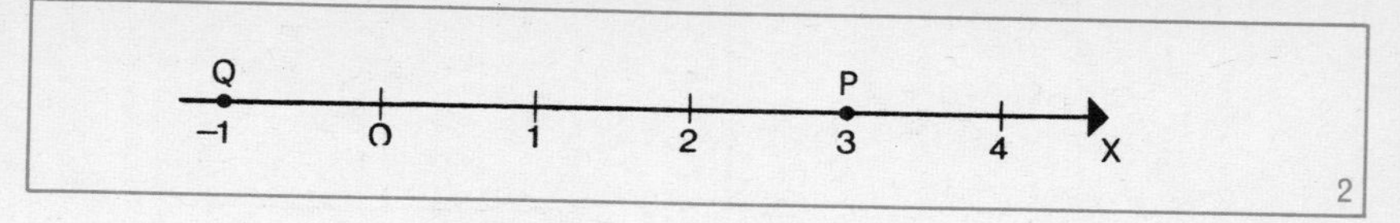

2

(ii) To determine the position of a point on a plane, two coordinates are required, and for this reason we say that a plane is *two-dimensional.*

In Figure 3, P is given by its Cartesian coordinates (3, 1), and in Figure 4, Q is given by its polar coordinates (5, 37°).

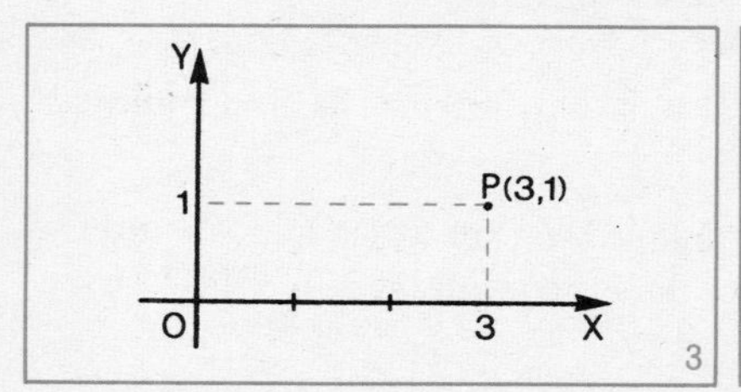

3

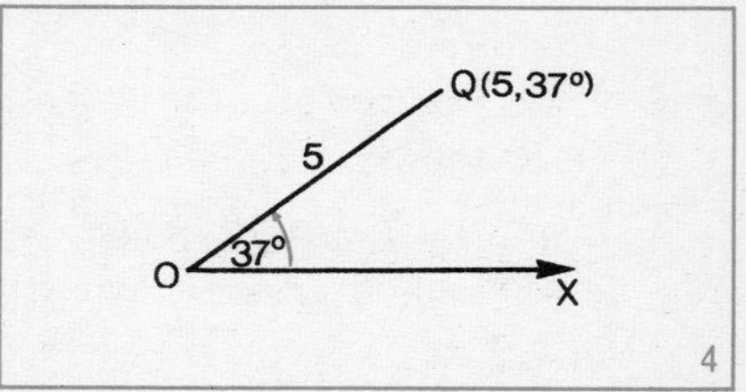

4

(iii) To give the position of a point in space, three coördinates are necessary, and we say that space is *three-dimensional.*

We may choose three axes, as shown in Figure 5, so that P is the point (3, 1, 4); note that the x-coordinate is measured parallel to the x-axis, the y-coordinate parallel to the y-axis, and the z-coordinate parallel to the z-axis. In this chapter we use axes which are perpendicular to one another (rectangular axes) since this makes the calculation of distance easier.

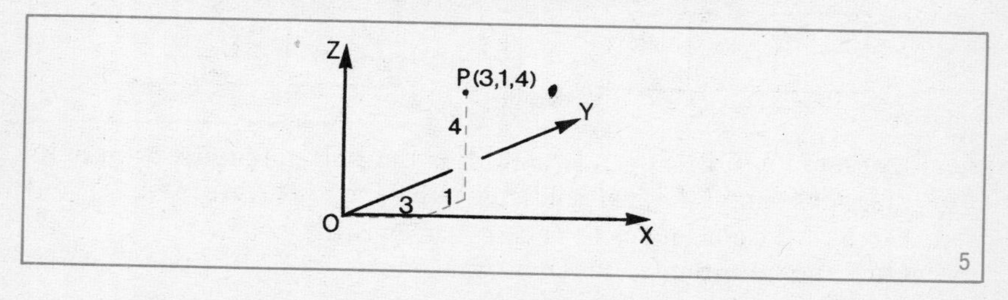

5

Figure 6 shows the roof and attic of a doll's house. If we take A as origin and axes along AB and AD, then with the dimensions given and with 1 cm as unit distance, M is the point (12, 13) in the x, y plane. If AZ is taken to be a vertical axis, then P is the point (12, 13, 10), and in three dimensions M is (12, 13, 0).

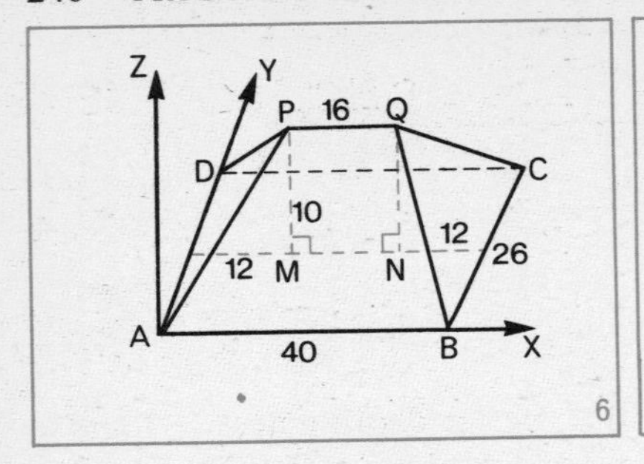

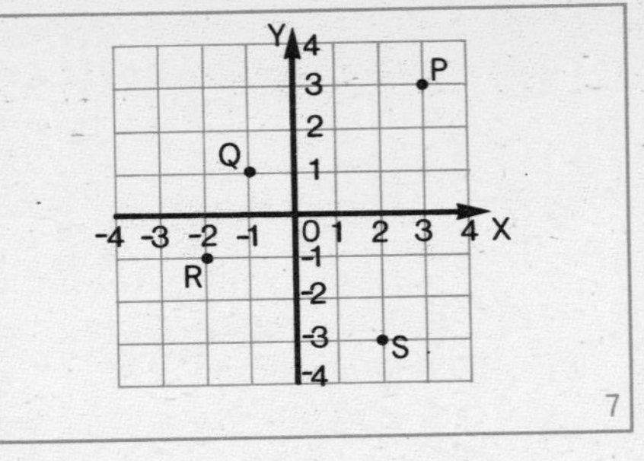

Exercise 2

1 From Figure 7 give the coordinates in two dimensions of the points P, Q, R and S.

2 In Figure 8, OPQR, STUV is a cuboid so that axes OX, OY and OZ are at right angles to each other. P is the point (4, 0, 0), Q(4, 3, 0) and U(4, 3, 2).

a Give the coordinates in three dimensions of the points T, S and V.

b Give the coordinates of the point of intersection of:

(*1*) the diagonals of rectangle PQUT

(*2*) the space diagonals of the cuboid.

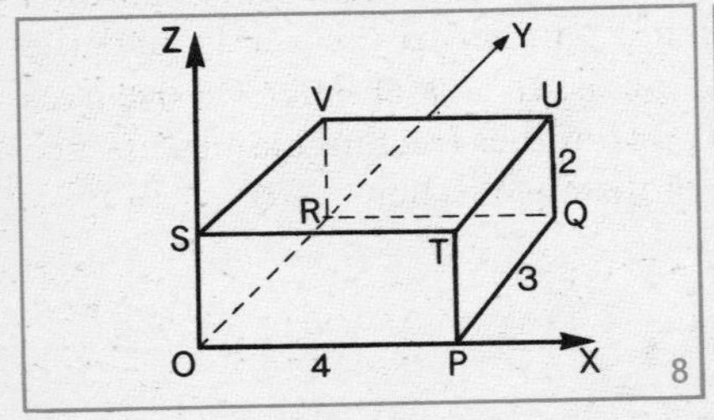

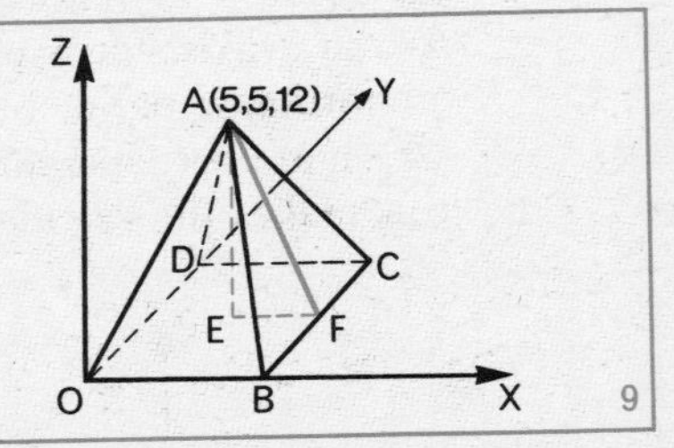

3 In Figure 9, OBCD is a square, E is the point of intersection of its diagonals, and EA is perpendicular to plane XOY. Axes OX, OY and OZ are at right angles to each other. A is the point (5, 5, 12).

a Give the coordinates in three dimensions of E, B, C and of the midpoint F of BC.

b Using Pythagoras' theorem in triangle AEF, calculate the length of AF.

4 Make a copy of Figure 6. Start by drawing the rectangle ABCD as a parallelogram, then mark the centre line parallel to AB and the points

M and N on it, and draw the vertical lines MP and NQ. Then complete the diagram. State the coordinates of B, C, N and Q.

5 A pyramid V,OPQR has a 2-unit square base OPQR in a horizontal plane. V is 3 units vertically above the centre-point B of the base. Sketch the pyramid, and taking O as origin, the x-axis along OP, the y-axis along OR and the z-axis as the vertical through O, state the coordinates of B and V.

6 An aircraft is 50 km north and 20 km east of its aircraft-carrier base, and is flying at a height of 8 km. Make a diagram, and indicate several ways in which its position relative to its base might be given.

3 *The distance from a point to a plane*

If the plane ABCD is horizontal, as in Figure 10, the perpendicular from P to the plane is the vertical line through P, and the easiest way in practice to obtain the direction of PQ is to suspend a plumb-line from P.

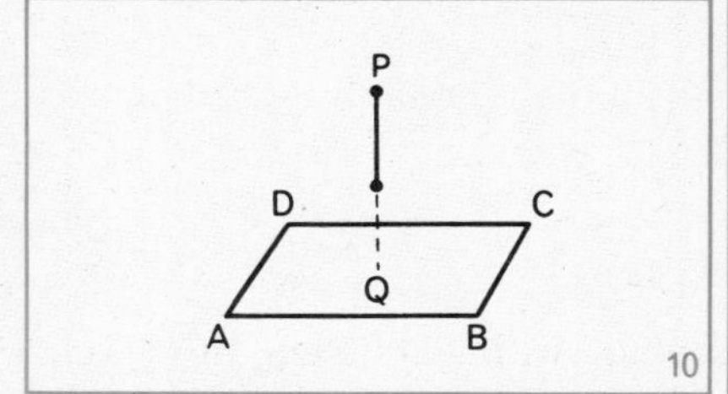

10

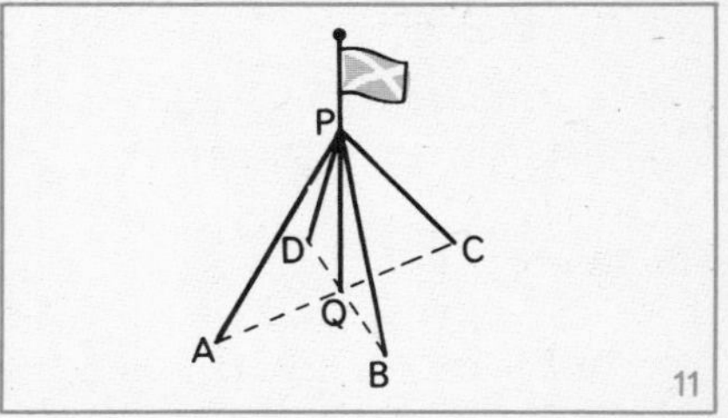

11

The flagpole PQ in Figure 11 is held vertical on a horizontal platform by four wires. The direction of the flagpole could be checked by means of a plumb-line. What size would you expect each of the angles PQA, PQB, PQC and PQD to be?

Notice that *vertical* and *horizontal* are terms related to the direction of gravity: they really belong to the physical world. Perpendicular, however, is a geometrical term. We know what it means to say two lines are perpendicular. What is meant by a line being *perpendicular to a plane*? When the plane is not horizontal, a plumb-line does not help at all.

In Figure 12, two pieces of cardboard PQR and PQS, each having two edges at right angles to each other, rest on the plane and are in

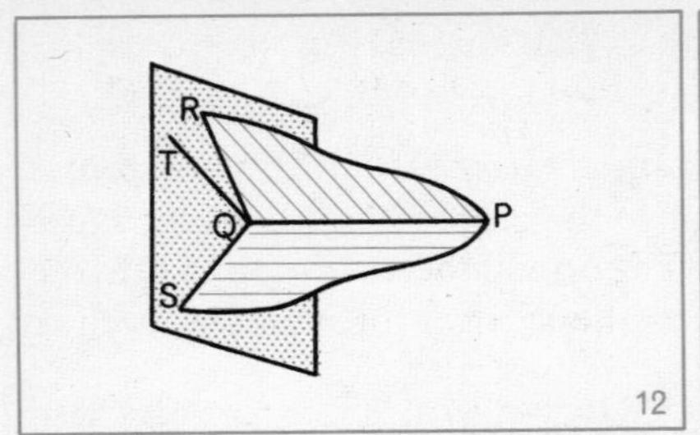

12

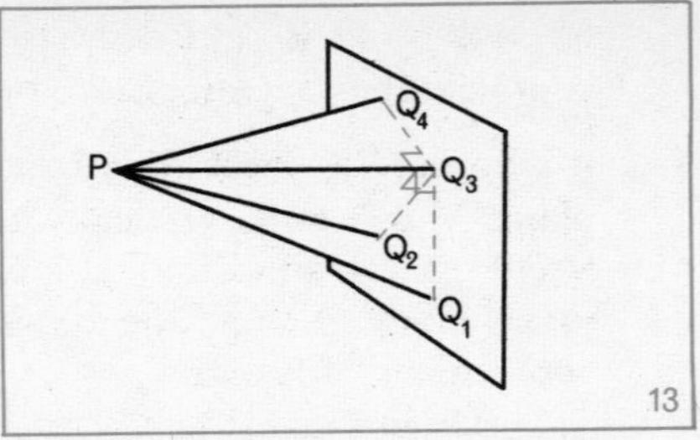

13

contact along PQ. By rotating PQS about PQ, keeping PQR fixed in position, it can be seen that all the positions QS takes are in the plane. This suggests that

if PQ is perpendicular to two lines in the plane it is perpendicular to all lines in the plane passing through Q

a theorem which can in fact be proved to be true.

In Figure 13, Q_1, Q_2, Q_3 and Q_4 are points in the plane, and Q_3P is perpendicular to the plane. Why is $PQ_1 > PQ_3$? Why is PQ_3 the shortest distance from P to the plane?

From the set of all possible distances from P to the plane we select the smallest possible distance as the definition of the distance from P to the plane.

Definition. The distance from a point to a plane is the perpendicular distance from the point to the plane.

Exercise 3

1 In Figure 14, ABCD,PQRS is a cuboid with AB = 8 cm, BC = 7 cm and CR = 6 cm.

a Copy the diagram and mark in all the right angles at B and at D.
b What is the distance from A to plane PQRS, to plane DCRS, and to plane BCRQ?

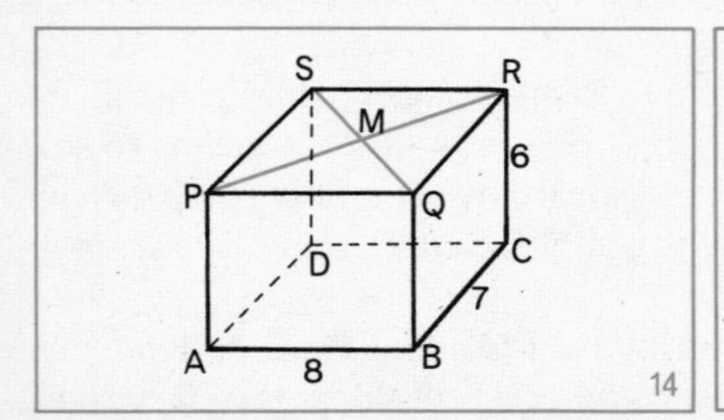

14

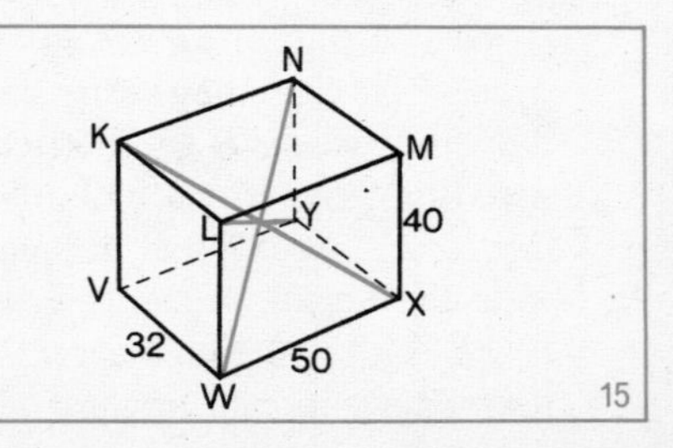

15

c What is the distance from M to the planes BCRQ, ABCD and ABQP?

2 In Figure 15, KLMN,VWXY is a cuboid with edges 32 mm, 40 mm and 50 mm long as shown.

a What are the distances from K to the planes VWXY,LMXW and MNYX?
b What are the distances from the point of intersection of the space diagonals to the faces of the cuboid?

3 Which of the following are true and which are false for Figure 14?

a There are three right angles at A.
b The distance from P to QR is PR.
c The faces of the cuboid lie in 3 pairs of parallel planes.
d PR is longer than PQ.
e There are more than three right angles shown at P.

4 In Figure 16, A,BCDE is a right pyramid on a rectangular base with BC = 8 m and BE = 6 m. AB = AC = AD = AE = 13 m.
Copy the diagram and calculate:

a the length of BD *b* the distance from A to the plane BCDE.

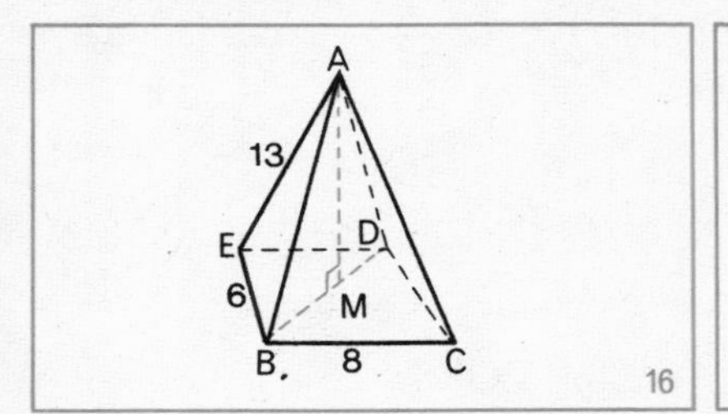

16

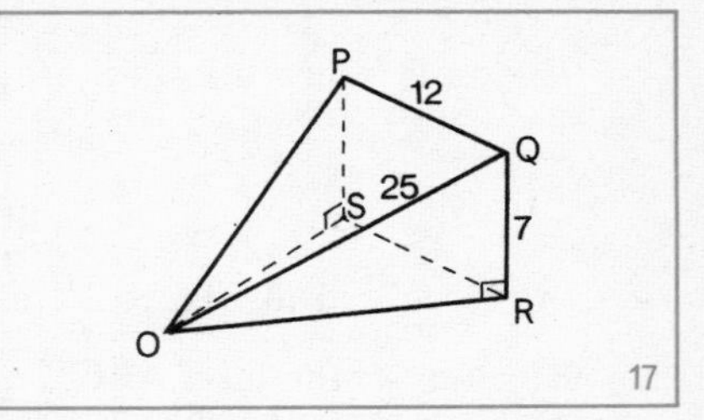

17

5 In Figure 17, the isosceles triangle ORS, with OR = OS, lies in a horizontal plane, and the rectangle PQRS in a vertical plane. OQ = 25 mm, QR = 7 mm and PQ = 12 mm. Calculate:

a the length of OR *b* the distance from O to plane PQRS.

4 *The angle between a line and a plane*

In Figure 18, RQ_1, RQ_2, RQ_3 and RQ_4 are lines in the plane, and RP is a line not in the plane. Make a model of the configuration shown in

Figure 18 using a pencil held firmly in position with the point against a plane to represent RP. Then place another pencil on the plane in positions RQ_1, RQ_2, etc., in turn. Estimate the sizes of some of the angles you form between the two pencils. Which is the greatest possible angle? Which is the smallest?

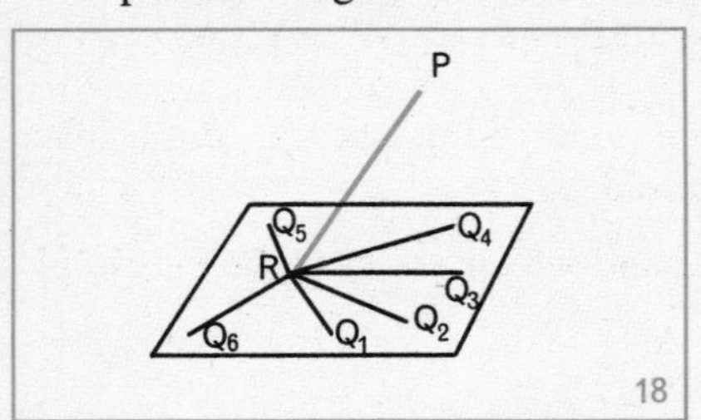

18

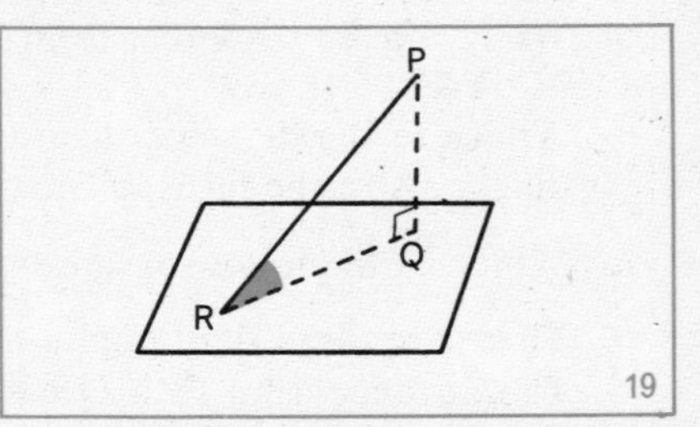

19

From the set of all such possible angles at R we have to select one as *the* angle between PR and the plane. We select the smallest possible angle. Notice that when this angle occurs RQ is the line through the feet of all the perpendiculars to the plane from points on RP. This line is called the *projection* of RP on the plane. It can be found by taking a line from P perpendicular to the plane to meet it at Q, and then joining QR.

Definition. The angle between a line PR and a plane containing R is the angle between PR and its projection QR on the plane. This angle is $\angle PRQ$ in Figure 19.

Exercise 4

1 In the cuboid in Figure 20, name the angles between:

a BR and (*1*) the base ABCD (*2*) the top PQRS
b AS and (*1*) the base ABCD (*2*) the top PQRS
c BS and (*1*) the base ABCD (*2*) the top PQRS
d BS and (*1*) the side BCRQ (*2*) the side ADSP.

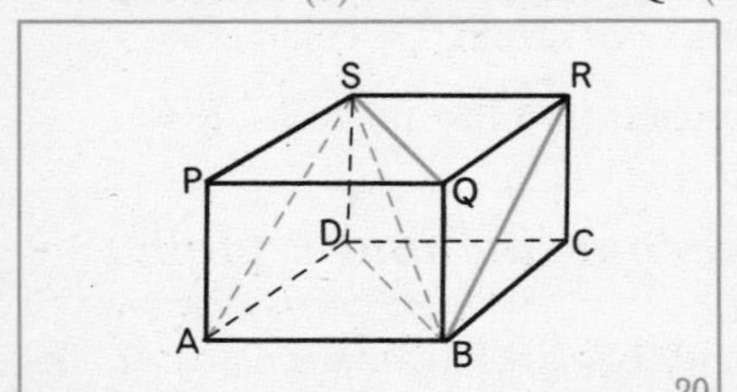

20

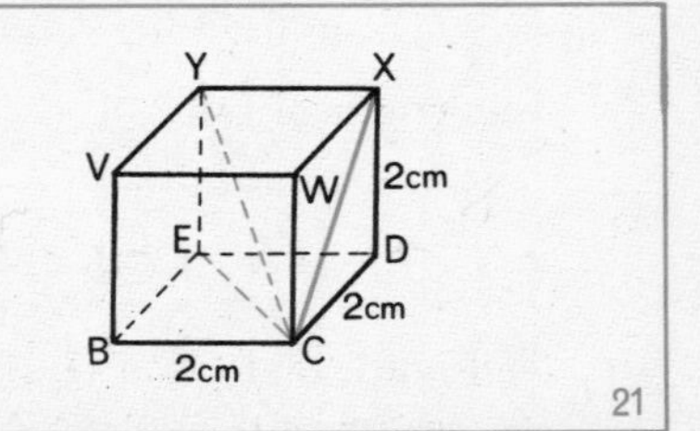

21

2 Figure 21 shows a cube BCDE, VWXY. Copy the cube (but not the coloured lines) and name the angles between:

a BW and the base *b* BX and the base *c* BY and the base
d EW and the base *e* EW and the top *f* EW and plane EBVY.

3 If the length of each side of the cube in Figure 21 is 2 cm, calculate the angle between:

a CX and the base *b* CY and the base *c* CY and the plane CDXW.

4 A spacecraft has its position fixed as shown in Figure 22. A, B and C are three points on a horizontal plane. The craft D is 15 km east and 20 km north of its base A, from which its angle of elevation is 23°. Calculate:

a the distance AC *b* the height of the spacecraft
c the angle of elevation of the craft from B.

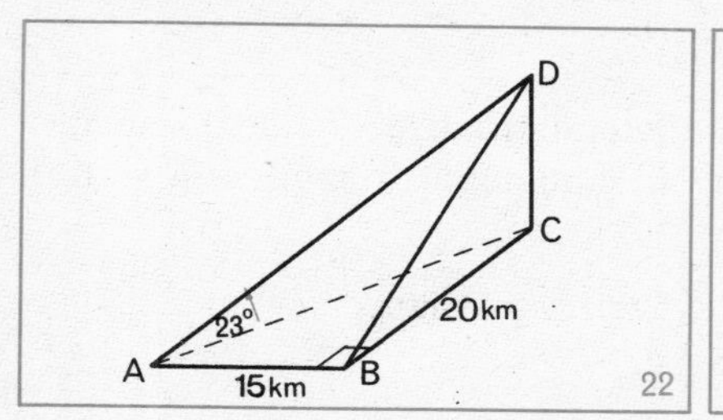

22

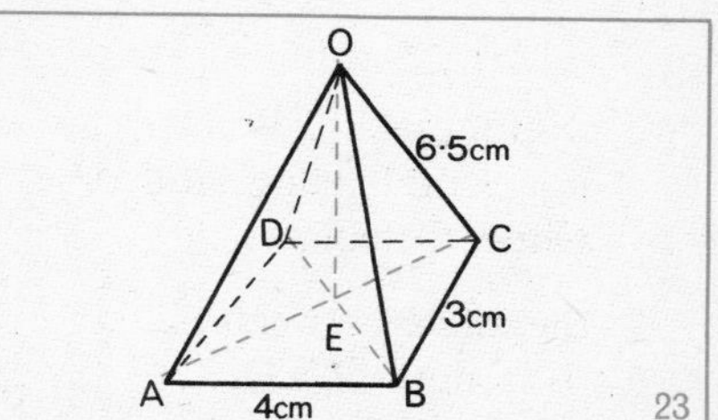

23

5 O, ABCD is a right pyramid on a rectangular base with dimensions as shown in Figure 23. AB = 4 cm, BC = 3 cm, OC = 6·5 cm, and OE is perpendicular to plane ABCD. Calculate:

a the length of AC *b* the height of O above ABCD
c the angle between OA and the base
d the size of angle AOC.

6 In Figure 23, F is the midpoint of BC. Calculate the angle between OF and the plane ABCD.

7 A right circular cone has a height of 6 cm and a base of diameter 6 cm. Calculate:

a the angle between the slanting edge of the cone and the base
b the length of the slanting edge.

8 With respect to three axes OX, OY, OZ at right angles to each other the coordinates of a point P in space are (4, 3, 6). Calculate:

a the angle between OP and the plane XOY *b* the length of OP.

9 In Figure 24, ABCD is a rectangle in a horizontal plane and ADEF is a rectangle in a vertical plane. BC = 30 m, CE = 40 m and AF = 20 m. Calculate:

a the length of BE
b the angle between BE and rectangle ABCD
c the angle between BE and rectangle ADEF.

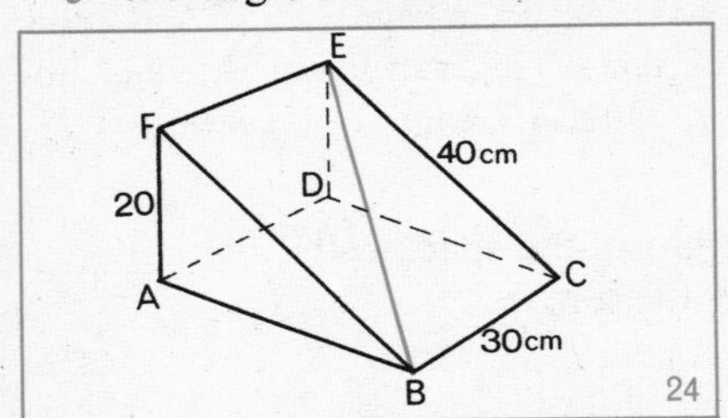

24

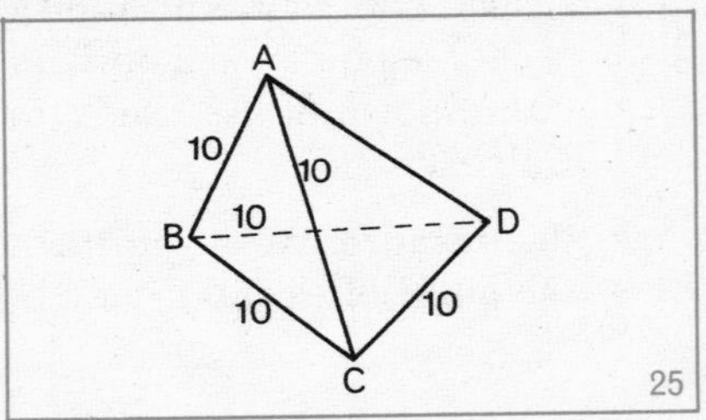

25

10 In Figure 25, BCD is an equilateral triangle of side 10 cm in a horizontal plane, and ABC is an equilateral triangle in a vertical plane. Calculate:

a the angle between AD and the horizontal plane
b the length of AD.

11 Triangle ABC lies in a horizontal plane. AB = 3 cm, BC = 4 cm and ∠ABC = 90°. CD is a vertical line 12 cm long.

a Use the converse of Pythagoras' theorem to show that ∠ABD = 90°.
b If M is the midpoint of BD, calculate the length of AM and the size of the angle between AM and the plane ABC.

5 *The angle between two planes*

So far we have defined:
(i) the angle between two lines which intersect, as in plane geometry
(ii) the angle between a line and a plane, as the angle between the line and its projection in the plane.

How can we define the angle between two planes? Evidently we must do it in terms of (i) or (ii). It is customary to use (i).

Definition. The angle between two planes which meet on a line AB is the angle between two lines, one in each plane, taken perpendicular to AB and meeting at a point on AB.

In Figure 26, CD and CE are perpendicular to AB, and so angle DCE gives a measure of the angle between the planes. Angle HBF would give the same measure if HB and BF were perpendicular to AB.

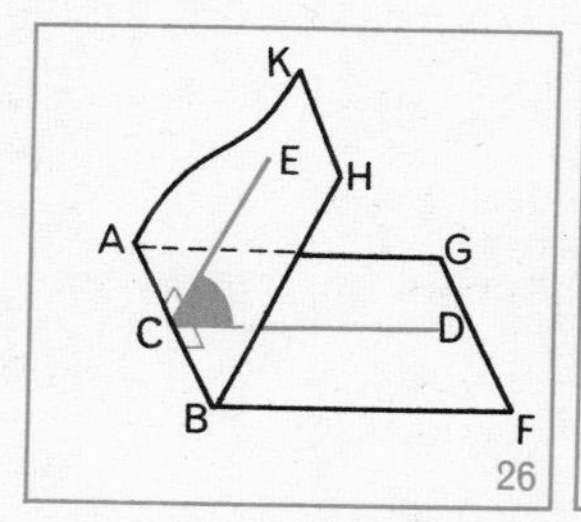

26

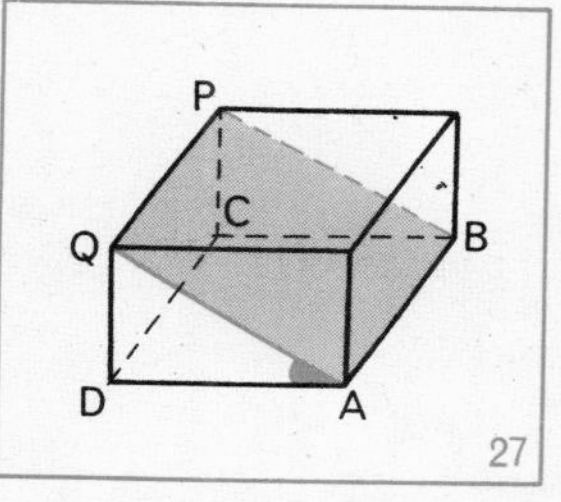

27

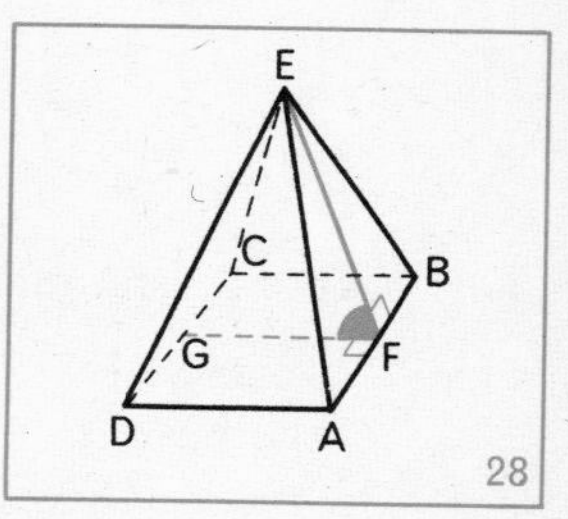

28

In the cuboid in Figure 27, ∠DAQ is the angle between the planes ABCD and ABPQ, since AB is the line of intersection of the planes, and AQ and AD are both perpendicular to AB.

In Figure 28, ∠EFG is the angle between the planes ABCD and ABE, since AB is the line of intersection of the planes, and FE and FG are both perpendicular to AB.

Example. Figure 29(i) shows a model of a bungalow 36 cm long, 36 cm wide, 15 cm high to the eaves, and rising to a central point 27 cm above the floor level. Calculate the angle of slope of the roof.

We have to find the angle between the planes ACD and BCDE. The line of intersection of these planes is CD and we require two lines, one in each plane, meeting on CD and perpendicular to CD.

We draw AP and PQ perpendicular to CD, giving ∠APQ as the required angle. The simplest way to calculate this angle is to draw AR perpendicular to PQ; it is often helpful to sketch a part of the diagram in two dimensions as shown in Figure 29(ii).

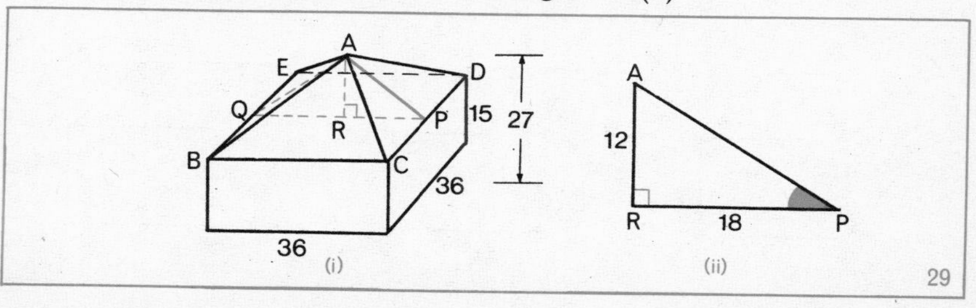

29

AR = 27 cm − 15 cm = 12 cm
PR = $\frac{1}{2}$ of 36 cm (since △APQ is isosceles) = 18 cm
tan ∠APR = $\frac{12}{18} = \frac{2}{3} \doteqdot 0{\cdot}667$
⇒ ∠APR ≑ 33·7°

Note. The use of elementary trigonometry and Pythagoras' theorem will enable you to calculate most of the required angles and lengths.

Exercise 5

1 Figure 30 shows a cube EFGH, WXYZ. Name an angle between:

a The plane EFYZ and
(*1*) the base (*2*) the top (*3*) the front (*4*) the back.

b The plane HFXZ and
(*1*) the back (*2*) the front (*3*) plane EHZW (*4*) plane FGYX.

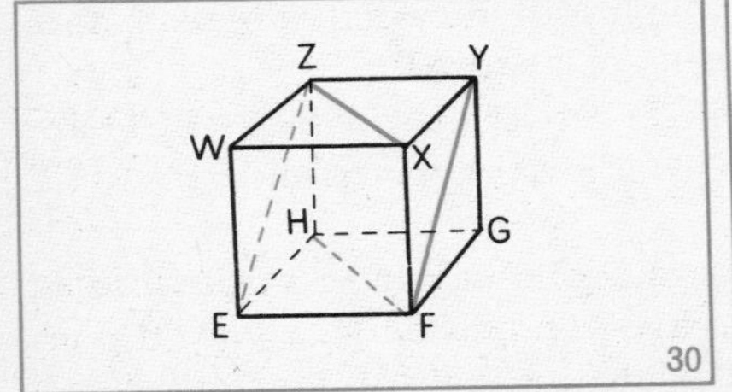

30

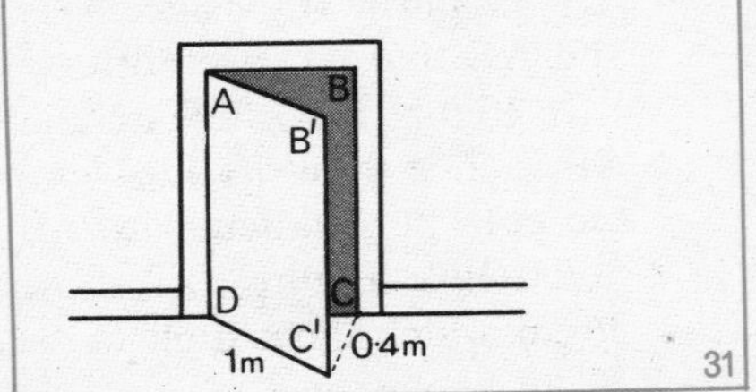

31

2 Figure 31 shows a sketch of a door which fits the rectangular space ABCD when closed. DC′ = 1 metre and CC′ = 0·4 metre.

a Which angle could be taken as the angle between the closed and open positions of the door?

b Calculate the size of this angle by sketching △CDC′ and drawing DP perpendicular to CC′.

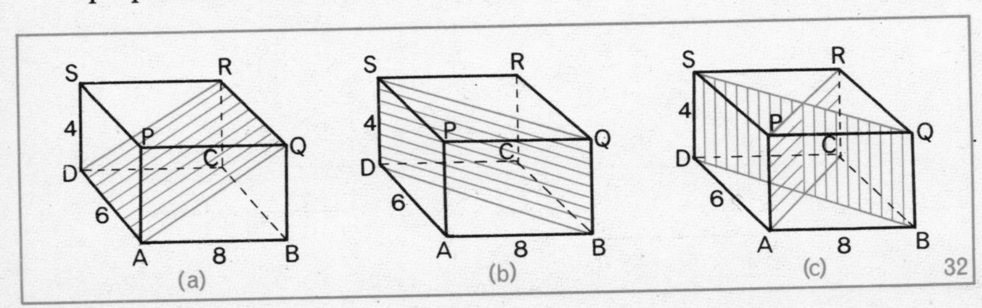

32

3 In Figure 32, ABCD, PQRS is a rectangular solid measuring 8 cm by 6 cm by 4 cm. Calculate:

a the angle between diagonal plane ADRQ and base ABCD
b the angle between diagonal plane SDBQ and face ABQP
c the acute angle between planes SDBQ and APRC.

4 Figure 33 shows a shed with a sloping roof, set on a horizontal base. Calculate the angle between the plane of the roof and the horizontal plane.

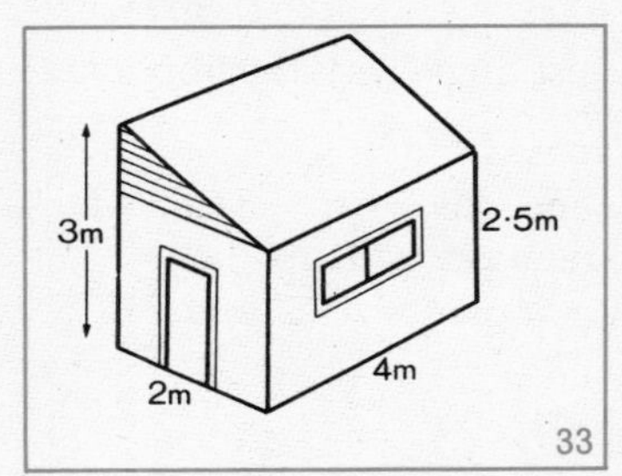

33

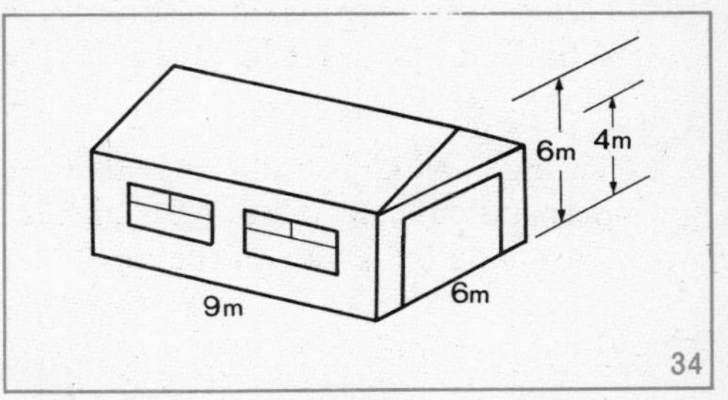

34

5 The dimensions of a large garage are shown in Figure 34. Calculate the angle of slope of the roof, and the angle between the two roof sections.

6 PQRS is a rhombus in a horizontal plane, as shown in Figure 35. Its diagonals PR and QS are 16 cm and 12 cm long respectively, and RT is a vertical line 12 cm long. Calculate the angles between the planes:

a TQS and PQRS *b* TRQ and TRS.

7 In Figure 36, ABC, DEF is an equilateral prism, three of its faces being squares of side 4 cm. (The cross-section of an equilateral prism is an equilateral triangle.)

a What is the angle between two of the square faces?

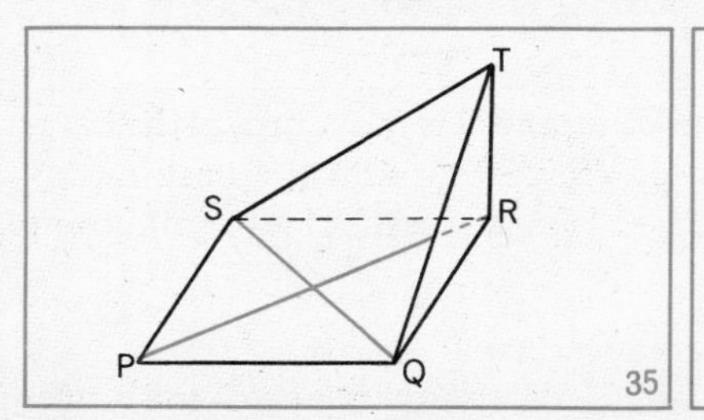

35

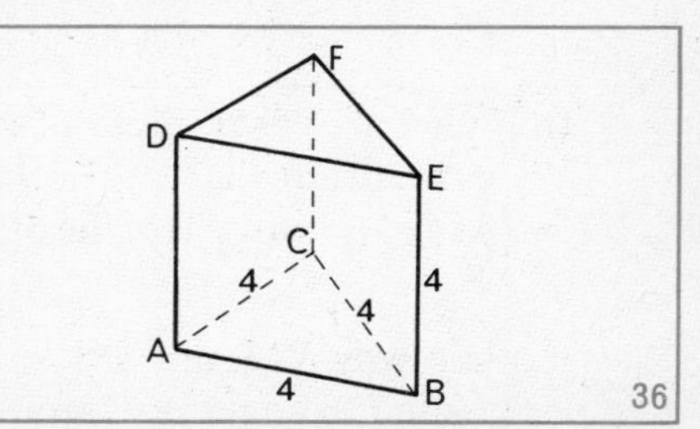

36

b What is the angle between BE and the median of triangle ABC from B? (i.e. from B to the midpoint of AC.)

c Find the angle between the planes ABC and AEC.

8 A rectangular plate HKLM rests with HK on a horizontal plane, and the plate makes an angle 60° with this plane. HK = 24 m and KL = 20 m.

a Sketch the plate as a diagonal plane of a cuboid and find the lengths of the other two edges of the cuboid. (Experimenting with a book will help you to visualize this situation.)

b What lines on HKLM make the greatest angle with the horizontal plane?

9 Triangle ABC is right-angled at A, and AB = AC = 8 cm. AP is perpendicular to plane ABC and is 16 cm long.

Calculate the size of the angle between planes ABC and PBC.

Exercise 5B

1 ABCD is a regular tetrahedron of side 2 cm, and E is the midpoint of BC, as shown in Figure 37.

a Show that AE is $\sqrt{3}$ cm long, and that $\triangle$AED is isosceles.

b Calculate the angle between:
(*1*) faces ABC and BCD (*2*) AD and face BCD.

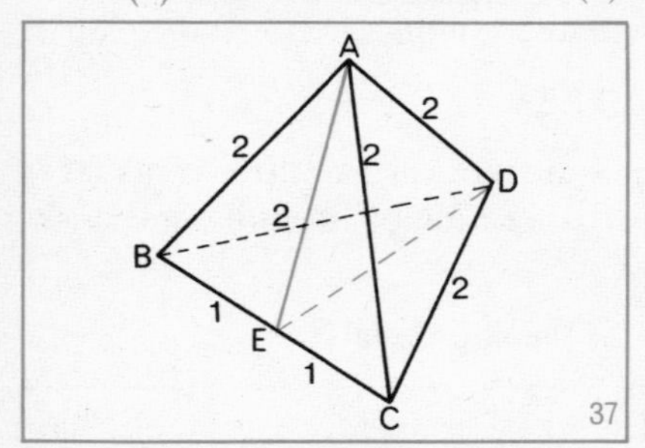

37

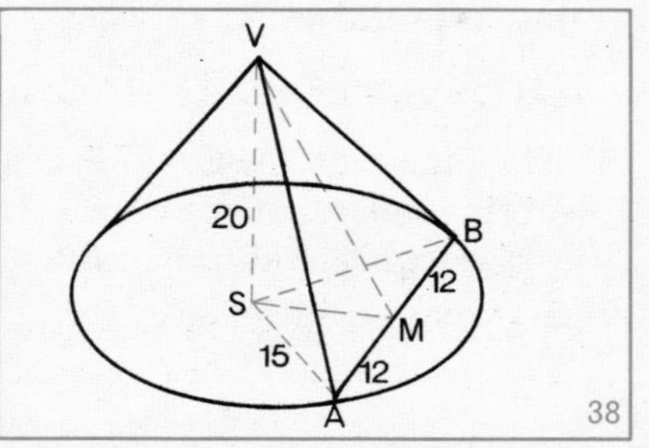

38

2 In Figure 38, V is the apex of the cone, and S is the centre of the base. VS is perpendicular to the base. VS = 20 cm and the base radius = 15 cm. A chord AB of the base is 24 cm in length. M is the midpoint of AB.

a Explain why VM and SM are perpendicular to AB.

b Calculate the angle between plane VAB and the plane of the circle.

3 A circle in a horizontal plane has centre O and radius 17 cm. AB is a chord whose midpoint M is 8 cm from O. BP is a vertical line 20 cm long.

a Calculate the length of MP, and show that $\angle OMP = 90°$.
b Calculate also the size of the angle between the planes OMP and OMB.

4 In Figure 39, OABC, PQRS is a rectangular box with sides 21, 16 and 12 cm long, as shown. Calculate the length of OR.
OR makes angles $a°$, $b°$ and $c°$ with the x, y and z-axes respectively. Show that $\cos^2 a° + \cos^2 b° + \cos^2 c° = 1$.

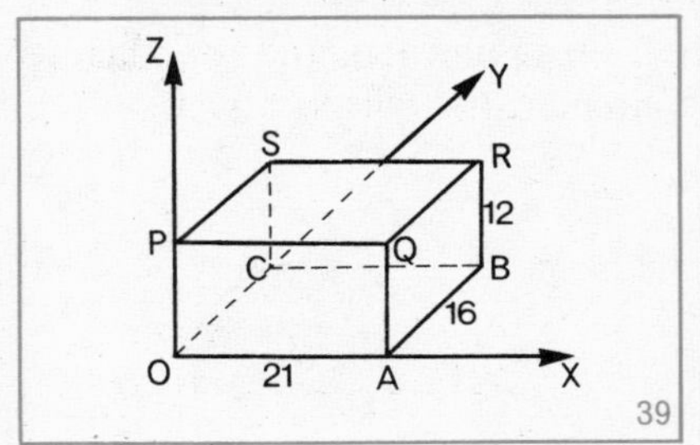

39

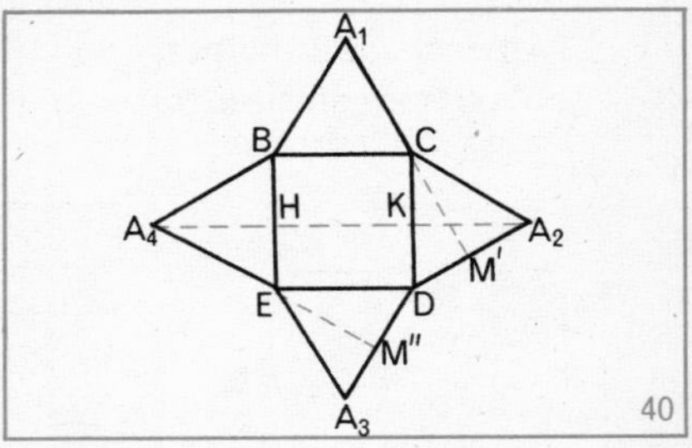

40

5 Figure 40 shows the net of a square pyramid all of whose edges are 10 cm long. When folded the pyramid becomes A, BCDE. Sketch the pyramid.

a Calculate the length of A_4H.
b Calculate the angle between the planes ABC and BCDE on the pyramid.
c M′ and M″, the midpoints of DA_2 and DA_3, become M when the net is folded. Verify that AM is perpendicular to MC and to ME. What is the size of the angle between AD and the plane EMC?
d Calculate the angle between the planes ACD and ADE.

6 The top and bottom of a prism are regular hexagons ABCDEF and PQRSTU, and the sides are congruent rectangles, one of which is ABQP. Calculate the angles between the planes:

a APTE and APUF *b* APTE and APRC *c* APTE and BQUF.

6 *The earth as a sphere; latitude and longitude*

Solids of revolution

The rotation of a plane surface through one turn about an axis generates a solid of revolution.

The solids formed by rotating the shaded areas shown in Figure 41 about LM are:

(i) a cylinder (ii) a vase shape (iii) a sphere.

In each case the perpendicular from a point P on the boundary to the axis of rotation traces out a circular disc with centre R.

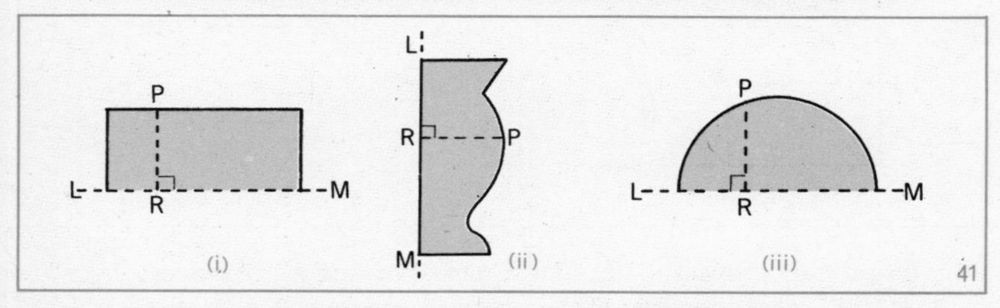

The earth is very nearly a sphere. It is slightly flattened at the poles and its surface has mountains and valleys, but in relation to its radius of about 6400 km these surface irregularities are of little significance. In this Section we assume the earth to be spherical.

To make a mathematical model of the earth we choose the diameter NS passing through the north and south poles as the axis of rotation. If O is the midpoint of NS, and OQ is perpendicular to NS, then the locus of Q is the equator (see Figure 42).

Notice that a circle seen at an angle appears as an ellipse. To make a two-dimensional diagram of a sphere, the 'equator' and the 'paral-

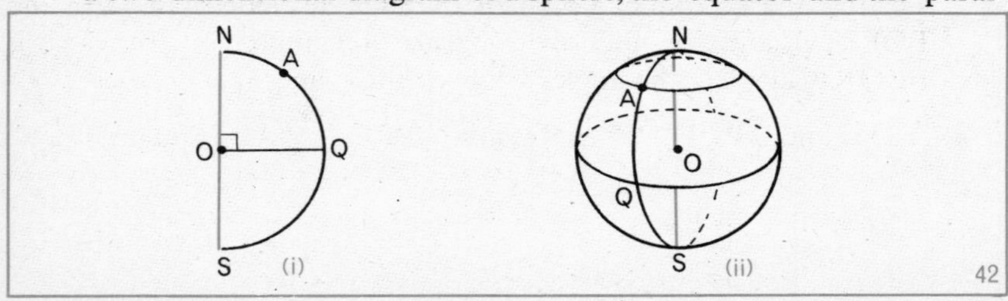

lels of latitude' and the 'meridians' must therefore be drawn as ellipses. It is a curious fact that the equatorial ellipse can be quite surprisingly round without losing sight of the 'south pole'. (Experiment with a football for sphere, and a matchstick for south pole. Stand well away from the person holding the ball and matchstick.)

Latitude and longitude

We can determine the position of a point A on the surface by stating:

(i) which circular section perpendicular to NS contains A
(ii) which position of the semicircle rotating about NS contains A.

These pieces of information are given by specifying (i) the latitude and (ii) the longitude of A.

(i) Figure 43(i) shows the equator and the *circles*, or *parallels*, *of latitude* 60°N and 50°S. The range of latitude is from 0° to 90° north or south of the equator.

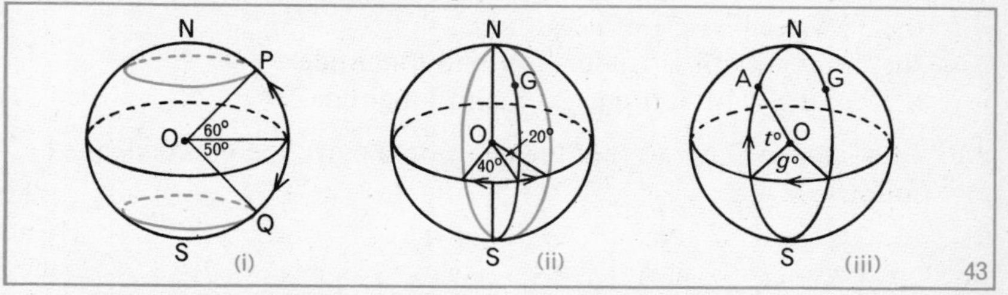

(ii) Figure 43(ii) shows the equator, the Greenwich meridian NGS and the *semicircles*, or *meridians*, *of longitude* 40°W and 20°E. The range of longitude is from 0° to 180° east or west of Greenwich.

In Figure 43(iii), point A has latitude t°N, longitude g°W.

Example 1. In Figure 44, P has latitude 60°N and longitude 50°W.

Example 2. In Figure 45, Q has latitude 30°S and longitude 50°E.

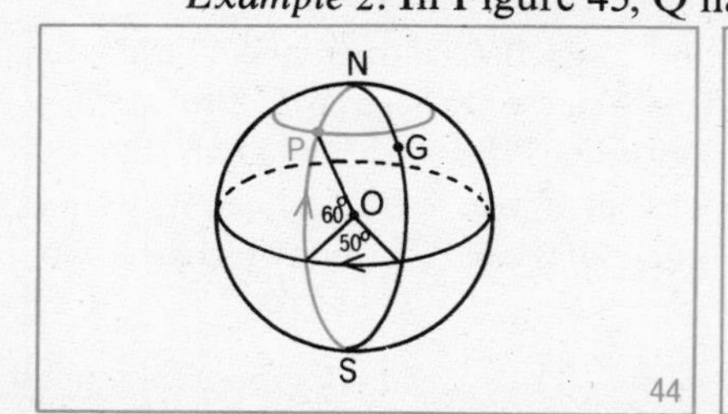

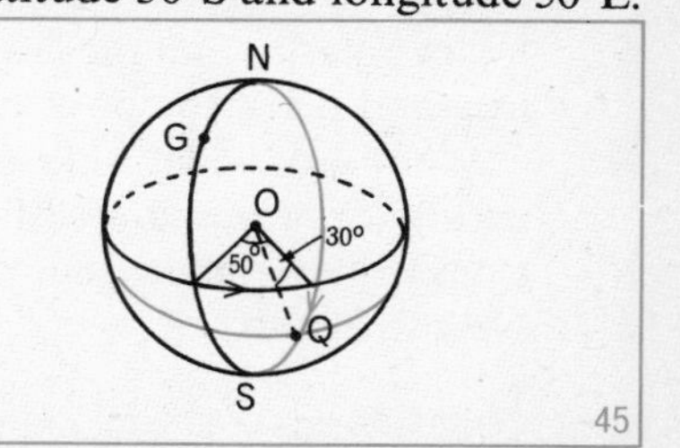

Exercise 6

1 Sketch the solids of revolution formed by rotating the shaded areas in Figure 46 about the axes shown.

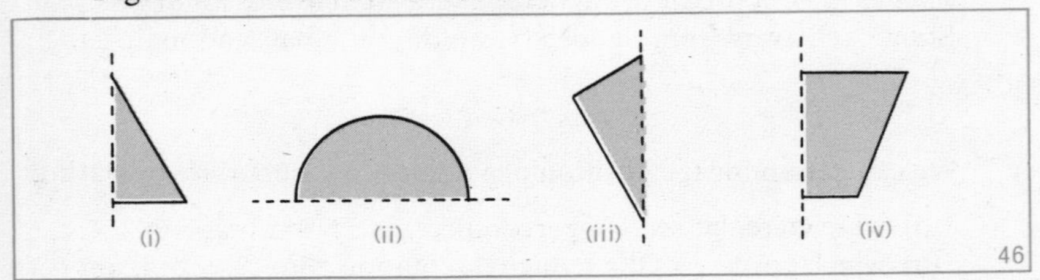

2 Draw a large sketch of the earth, marking in the centre, the north and south poles, the equator, Greenwich (about 51°N) and the Greenwich meridian.

3 In your sketch for question **2** mark:

a the 40°N circle of latitude
b the 30°W semicircle of longitude
c the point A with latitude 40°N and longitude 30°W
d the point B with latitude 20°S and longitude 10°E.

4 In a similar figure to that for question **2** mark the points defined as follows:

	C	D	E	F	H
Latitude	40°S	0°N	90°N	0°N	75°S
Longitude	30°E	45°W	100°W	100°E	120°W

5 What is the difference in latitude and the difference in longitude between the following pairs of points?

a P(60°N, 10°W) and Q(80°N, 50°W)
b R(10°N, 90°W) and S(10°S, 10°E)
c T(5°N, 125°E) and U(72°N, 125°W)

6 From an atlas find the latitude and longitude of the place where you live, and mark the position of this place on a sketch of the earth.

What are the latitude and longitude of the place on earth diametrically opposite yours? Is it true that if you were to bore a tunnel from Britain right through the centre of the earth you would arrive in New Zealand?

The radius of a circle of latitude

In Figure 47 the latitude of A is $t°$, i.e. $\angle QOA = t°$.
So $\angle OAC = t°$. (Why?)

In $\triangle OAC$, $\cos t° = \dfrac{AC}{AO}$.

So $AC = AO \cos t°$.

The radius of a circle of latitude $t° = R \cos t°$, where R = radius of earth.

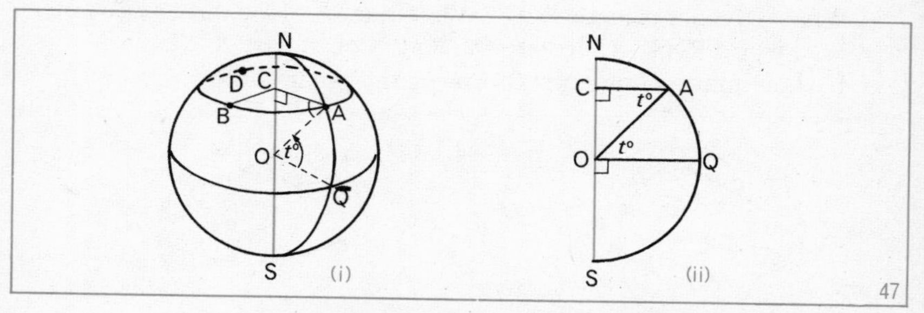

The distance along a circle of latitude

Suppose that in Figure 47(i) A has latitude 60°N, longitude 55°E, and B has latitude 60°N, longitude 13°W. We can calculate the length of arc AB as follows.

$$\frac{\text{length of arc AB}}{\text{circumference of latitude circle ABD}} = \frac{\text{size of } \angle ACB}{360°}$$

$$\text{So arc AB} = \frac{\angle ACB}{360°} \times 2\pi(R \cos t°)$$

$$\doteqdot \frac{68}{360} \times 2 \times 3{\cdot}14 \times 6400 \times \cos 60°$$

$$\doteqdot 3800\,\text{km}$$

nos	*logs*
68	1·833
6·28	0·798
6400	3·806
cos 60°	$\bar{1}$·699
	6·136
360	2·556
3800	3·580

Exercise 7B

(Assume that the radius of the earth is 6380 km to three significant figures.)

1 Show in a sketch of the earth the circles of latitude 60°N and 35°S. Calculate the radius of each of these circles.

2 Calculate the radius and circumference of the circle of latitude 80°N.

3 What is the latitude of the circle whose circumference is half that of the earth?

4 Figure 48 illustrates the path of an aircraft which flies from R to T along half of the circle of latitude 45°N, and another which flies 'over the pole' along the arc RNT.

a Calculate the length of the flight RPT from R to T along the circle of latitude 45°N (to the nearest 100 km).
b What is the size of angle ROT? What fraction of the earth's circumference is arc RNT? Calculate the length of the arc RNT.
c By how many kilometres shorter is the polar route?

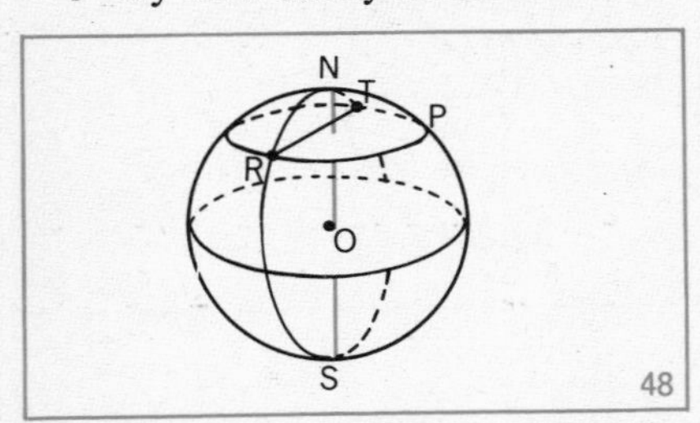

48

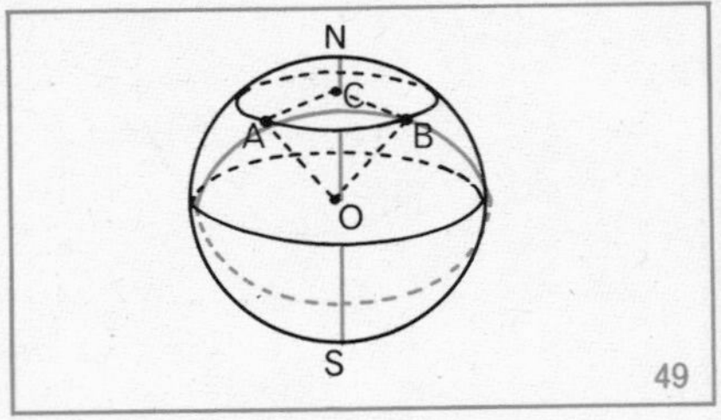

49

5 Kirkwall in the Orkney Islands has latitude 59°N and longitude 3°W, and Uranium City in Canada has latitude 59°N and longitude 109°W. Calculate the length of the arc of the circle of latitude between them.

6 In Figure 49, A has latitude 60°N and longitude 85°W, and B has latitude 60°N and longitude 25°W. Calculate:

a the length AC of the radius of the circle of latitude 60°N
b the length of AB, a side of triangle ABC
c the size of $\angle$AOB, an angle of triangle AOB
d the length of the great circle arc AB in which the plane through A, O and B cuts the surface of the earth (which would be the shortest flight path of an aircraft from A to B.)

7 When the metre was originally defined in 1799 it was intended to be one ten-millionth part of the arc of a meridian from pole to equator. Use this information to calculate the radius of the earth according to the ideas of those days, in kilometres to the nearest 10 km.

Summary

1 *The position of a point* can be defined in one, two or three dimensions as follows:

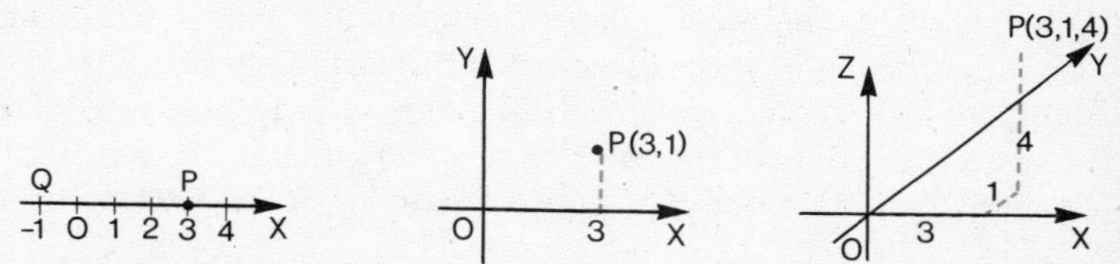

2 *The distance from a point to a plane* is the perpendicular distance from the point to the plane. A line perpendicular to a plane is perpendicular to all lines in the plane that it meets.

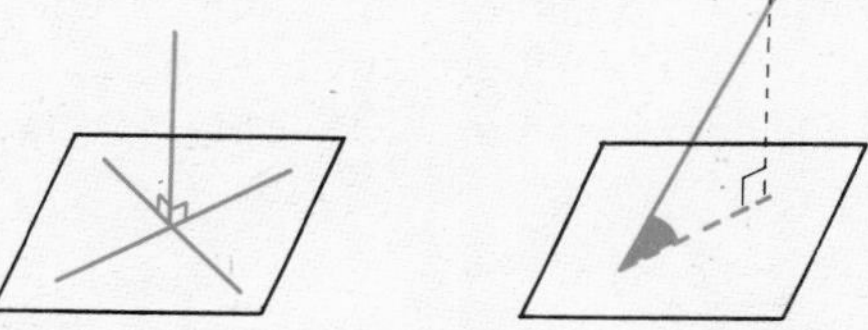

3 *The angle between a line and a plane* is the angle between the line and its projection in the plane.

4 *The angle between two planes* is the angle between two lines, one in each plane, which are perpendicular to the line of intersection of the planes and meet at a point on that line.

5 *The position of a point on the earth's surface* can be defined by its latitude $t°$(N or S) from the equator to its circle of latitude, and its longitude $g°$(W or E) from the Greenwich meridian to its meridian of longitude.

Topics to explore

1 Nautical miles

Navigators use nautical miles in their calculations at sea or in the air. A nautical mile is the distance on the earth's surface between two points on the equator whose difference in longitude is $\frac{1}{60}$ degree. Taking the radius of the earth to be 6380 kilometres, show that the length of a nautical mile is approximately 1·86 kilometres.

2 Measuring your latitude

Figure 1 shows a cross-section of the earth. Suppose your latitude at A is $a°$, and HH′ represents the horizontal at A. The earth's axis SN points close to the Pole Star, which is so far away (more than 10^{12} km) that the line from A to the Pole Star is very nearly parallel to SN.

$a = b$. Why? $b = c$. Why? So $a = c$.

In order to measure your latitude, therefore, which angle can you measure? What instrument is used for this purpose?

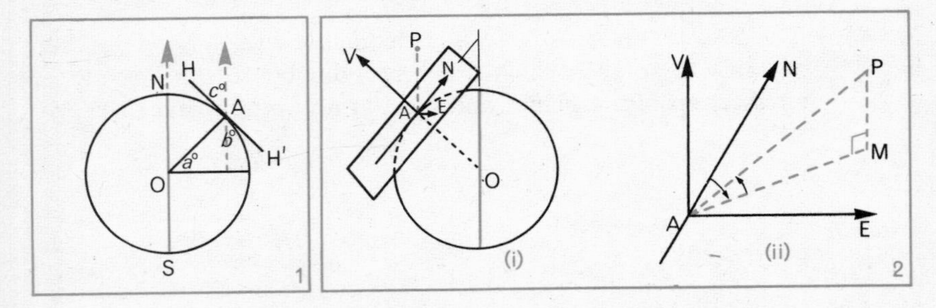

3 Determination of the position of a point in space

Figures 2(i) and (ii) suggest a way of determining the position of an object in space. Describe how the position of P would be given. (∠PAM is called the *altitude*, ∠NAM the *azimuth*.)

Revision Exercises

Revision Exercise on Chapter 1
Three Dimensions

Revision Exercise 1

1 In Figure 1, OABC, DEFG is a cuboid with OA = 8 units, AB = 6 units, BF = 4 units.

a With axes as shown, write down the coordinates in three dimensions of the eight vertices of the cuboid and of the point of intersection of the space diagonals.

b Calculate the length of OB exactly, and OF to three significant figures.

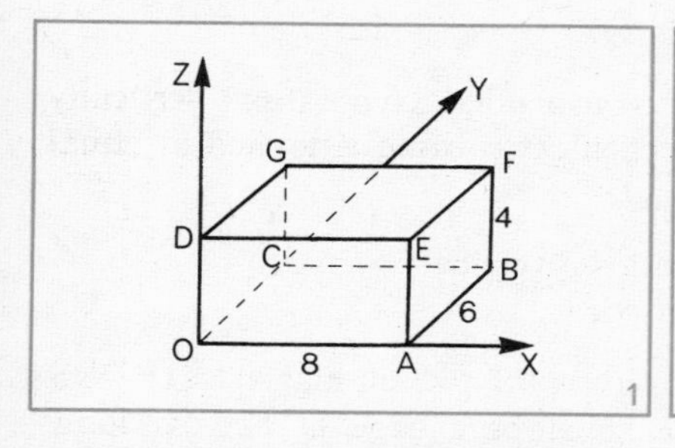

1

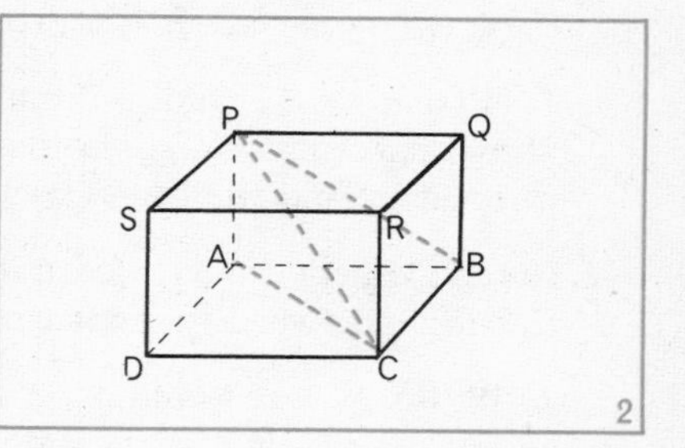

2

2 In Figure 2, ABCD, PQRS is a cuboid.

a Name four right angles at A.

b Name the line through P perpendicular to plane:
(*1*) ABCD (*2*) ADSP.

c Name the angle between PC and plane:
(*1*) ABCD (*2*) ABQP.

3 Which of the following are true and which are false for Figure 2?

a DC is the hypotenuse in triangle DAC.

b Triangle PCB is isosceles.

c PC = AR *d* PC > PB

e R lies in the plane of triangle PAC

f The distance from P to plane BCRQ is PQ.

4 *a* Sketch the solid given by the net in Figure 3. Lengths are in cm.

b Name the points which coincide with E and with J.

c Calculate: (*1*) the lengths of RQ and BS

(*2*) the distance between the planes PQSR and BCHG.

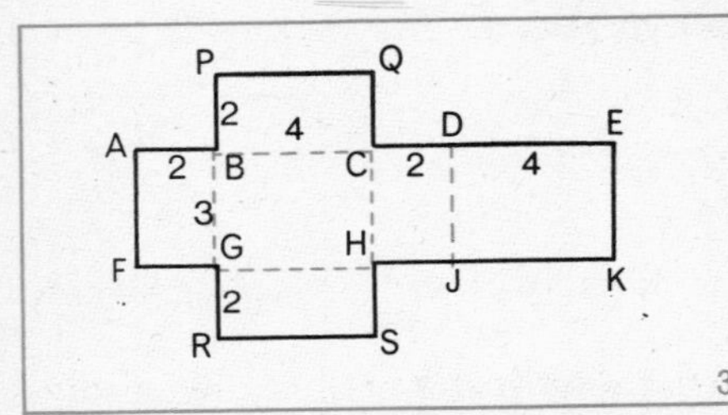

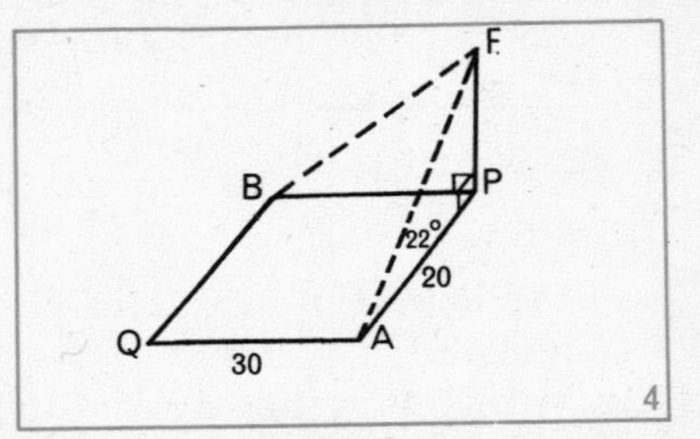

5 In Figure 4, APBQ represents a horizontal rectangular piece of ground, and FP a flagpole. AP = 20 m, AQ = 30 m and ∠PAF = 22°. Calculate:

a the length of FP *b* ∠PBF

c the angle of elevation of F from Q.

6 A circular ring of radius 20 cm is suspended from a point vertically above its centre by six cords, each 80 cm long, attached at equal intervals round the ring. Calculate:

a the angle between each cord and the vertical

b the angle between two adjacent cords.

7 O,PQRS is a pyramid on a horizontal rectangular base PQRS. PQ = 12 cm, QR = 9 cm, and each sloping edge is 12·5 cm long. Calculate:

a the length of PR *b* the height OH of the pyramid

c the angle between:

(*1*) OR and the base (*2*) plane OQR and the base.

8 In Figure 5, ABCD, PQRS is a cuboid with BC = 8 cm, CD = 10 cm and BQ = 6 cm. Calculate:

a the lengths of QC and PC

b the angle between (*1*) PC and CD (*2*) PC and plane CBQR.

9 Taking the edges of the floor of a room which meet in a corner as axes,

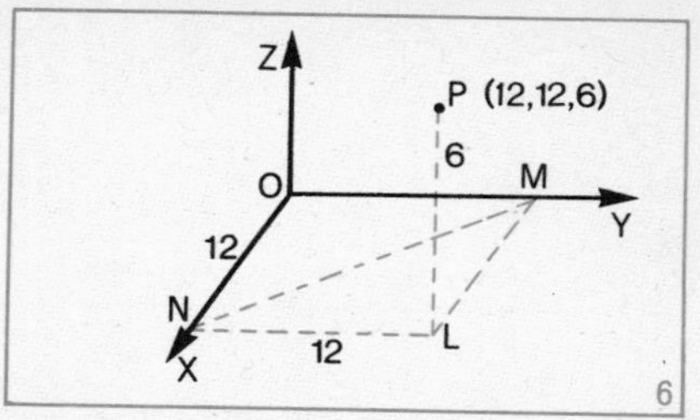

and 1 metre as unit, the position of a point P is (12, 12, 6) as shown in Figure 6. Calculate the angles between the following planes:

a PLN and PLM *b* PLN and the floor *c* PMN and the floor.

10 In Figure 7, ABC is a horizontal triangle, right-angled at A, in which AB = 6 cm, and AC = 4·5 cm. AX = 3 cm, BY = 5·5 cm and CZ = 9 cm. Calculate:

a the lengths of BC, XY and XZ exactly, and YZ to two significant figures

b ∠YXZ *c* the area of △XYZ.

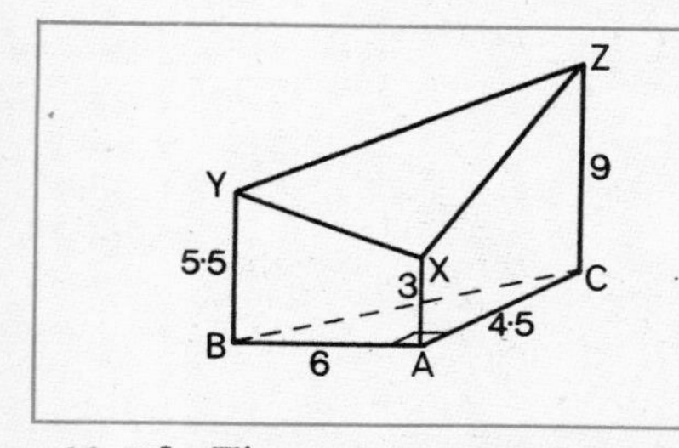

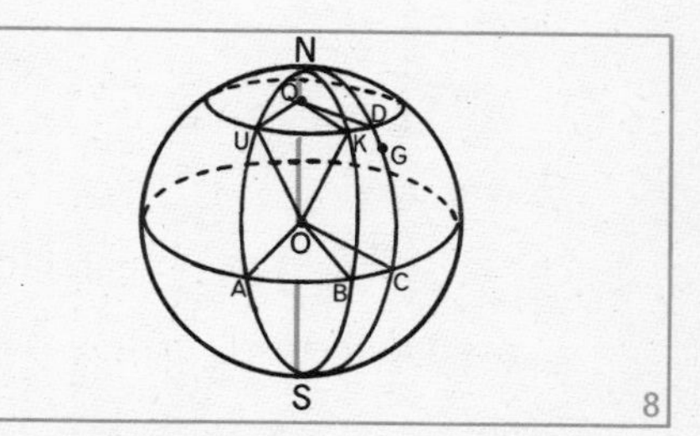

11 In Figure 8, O represents the centre of the earth, N and S the poles, and circle ABC the equator. G is Greenwich, U is Uranium City (Canada) with latitude 59°N, longitude 109°W, and K is Kirkwall in the Orkney Islands with latitude 59°N, longitude 3°W.

a Name an angle which gives the latitude of U, and two angles for the longitude of U.

b Find the difference in latitude and longitude of U and K.

c Give the sizes of angles BOC, AOB, UQD, OQK and UON.

12 Lhasa in Tibet has latitude 30°N, longitude 91°E. Tripoli in Libya has latitude 30°N, longitude 20°E. How long would it take an aircraft flying at 500 km/h to travel due east from Tripoli to Lhasa? (Take the radius of the earth to be 6380 km.)

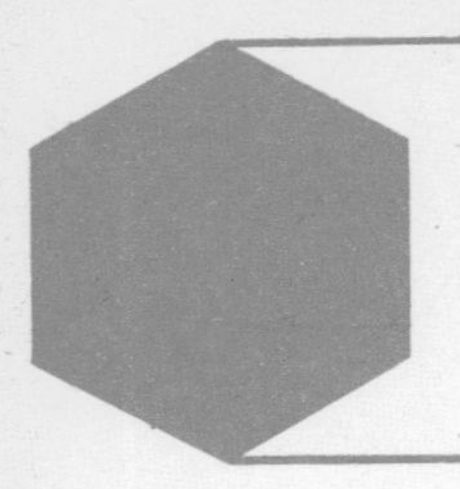

Cumulative Revision Section (Books 5-7)

Revision Topic 1
The Cosine, Sine and Tangent Functions

Reminders

1 *Definitions.* P has Cartesian coordinates (x, y) and polar coordinates $(r, a°)$.

$\cos a° = \frac{x}{r}$, $\sin a° = \frac{y}{r}$, $\tan a° = \frac{y}{x}$; P is the point $(r\cos a°, r\sin a°)$

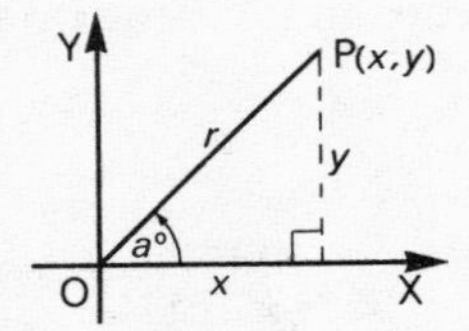

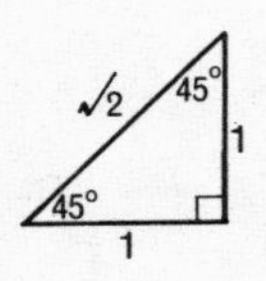

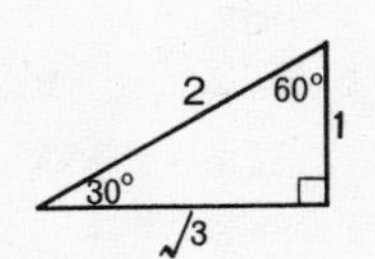

2 *Special angles.*

$\cos 45° = \frac{1}{\sqrt{2}}$, $\tan 45° = 1$, etc.; $\sin 30° = \frac{1}{2}$, $\cos 60° = \frac{1}{2}$, etc.

3 *Extending the use of the trigonometrical tables: $a > 90$.*

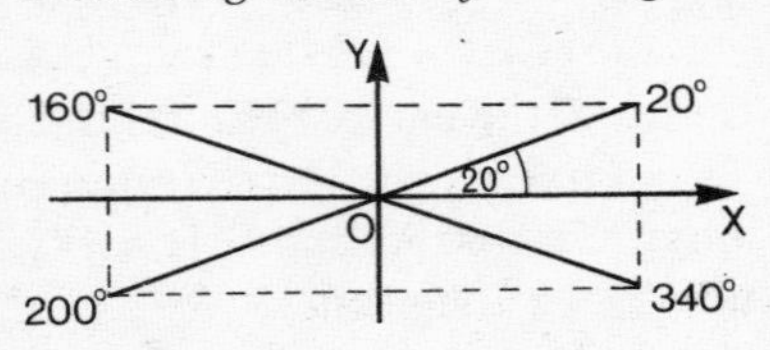

sin +	all +
tan +	cos +

e.g. $\sin 160° = \sin 20° = 0{\cdot}342$
$\cos 200° = -\cos 20° = -0{\cdot}940$
$\tan 340° = -\tan 20° = -0{\cdot}364$

4 *The graphs of the sine and cosine functions*

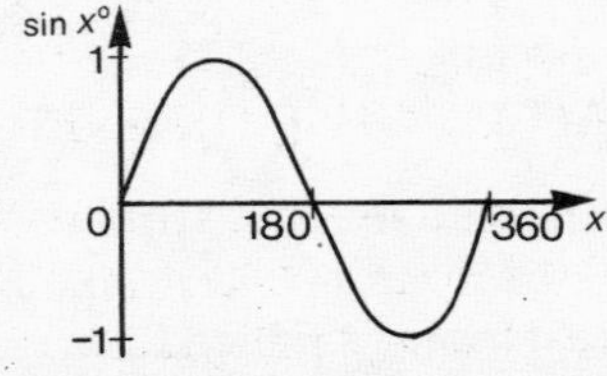

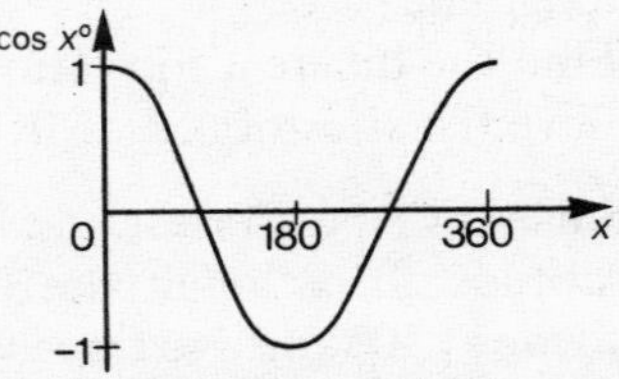

Period of graphs = 360°, $\sin 90° = 1$, $\sin 180° = 0$, $\cos 180° = -1$, etc.

5 *Formulae*: For all A, $\cos^2 A + \sin^2 A = 1$ and $\tan A = \dfrac{\sin A}{\cos A}$

Exercise 1

1 A is the point (8, 6). Write down the values of cos XOA, sin XOA and tan XOA, and find the size of ∠XOA.

2 If $\cos a° = 0{\cdot}636$, $0 < a < 90$, find the values of $\sin a°$ and $\tan a°$.

3 Sketch a triangle with angles of 30°, 60° and 90°. Hence simplify:

a $\sin 30° + \cos 60°$ *b* $\tan^2 30° + \tan^2 60°$ *c* $\sin 60° + \cos 30°$.

4 Calculate the length of side a in △ABC in which:

a ∠A = 25°, ∠B = 90°, $b = 10$ cm *b* ∠B = 41·2°, ∠C = 90°, $c = 69{\cdot}8$ cm

5 ABCD is a parallelogram with angle A = 70°, AD = 3·5 cm and AB = 6·5 cm. DK is perpendicular to AB. Calculate:

a the length of DK *b* the area of the parallelogram.

6 In Figure 1, O is the centre of each circle, and the lengths are in millimetres. PQ is a tangent in (ii).

a Name a right angle in each diagram, and explain why it is right.
b Calculate k in each part.

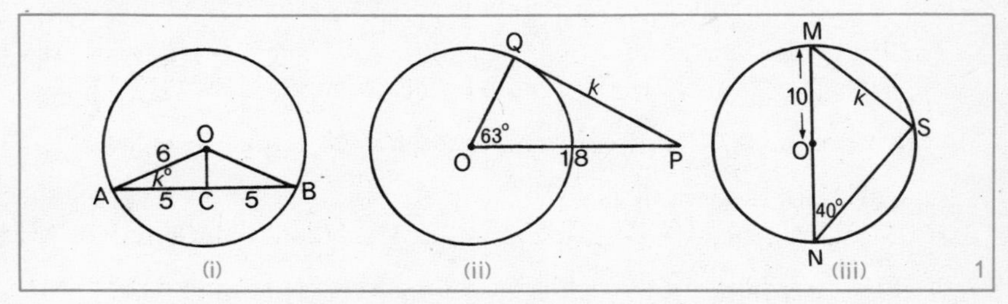

7 Use tables and the sine and cosine graphs to find the values of:

a sin 200° *b* cos 300° *c* tan 400° *d* sin 180°
e sin 450° *f* tan 111° *g* cos 270° *h* cos (−45°)

8 Find the solution sets of the following equations for $0 \leqslant x \leqslant 360$

a $\sin x° = 0{\cdot}942$ *b* $\cos x° = 0{\cdot}733$ *c* $\tan x° = 3{\cdot}420$
d $\cos x° = -0{\cdot}423$ *e* $2 \sin x° = 1$ *f* $3 \cos x° = 2$

Revision Topic 2 Triangle formulae

Reminders

For every triangle ABC we have:

1 *The Sine Rule:* $\dfrac{a}{\sin A} = \dfrac{b}{\sin B} = \dfrac{c}{\sin C}$

2 *The Cosine Rule:* $a^2 = b^2 + c^2 - 2bc\cos A$, or $\cos A = \dfrac{b^2+c^2-a^2}{2bc}$

3 *The Area of a Triangle*: $\triangle = \frac{1}{2}bc\sin A$

Note. For obtuse angles in triangles it is worth remembering that:
the sine of an angle = the sine of its supplement;
the cosine of an angle = −the cosine of its supplement.

Exercise 2

1 Calculate the smallest angle of a triangle with sides 4 m, 6 m and 8 m.

2 The lengths of the sides of a triangle are 3 cm, 5 cm and 7 cm.

a Show that the largest angle in the triangle is 120°.
b Calculate the area of the triangle (to 1 decimal place).

3 The longest side in a triangle is 14·5 cm long. Two of the angles are 45° and 60°.

a Calculate the third angle and the length of the shortest side.
b Calculate the area of the triangle.

4 Given $a = 4$ cm and $b = 5$ cm, and the area of triangle ABC $= 7{\cdot}85\ \text{cm}^2$, calculate the possible sizes of angle C.

5 In triangle ABC, AB = 3·6 cm, angle B = 24°, angle C = 54°.

a Calculate the length of AC.
b If the bisector of ∠BAC meets BC at D, calculate the length of AD.

6 The hour hand of a clock is 5 cm long and the minute hand is 6 cm long. Find the distance between the ends of the hands at 5 o'clock.

7 In triangle ABC, A = 51°, B = 68°, $c = 4{\cdot}5$ m. Calculate:
a side b *b* the area of triangle ABC.

8 In triangle ABC, B = 60°, $a = 2$ and $b = 2\sqrt{3}$.

a Write down the value of sin 60° from a '30°, 60°, 90°' triangle.
b Calculate A and c in triangle ABC.

9 In triangle XYZ, XY = 17·2 cm, YZ = 21·3 cm, ZX = 16·0 cm. Find the three angles, and the area of the triangle.

10 In triangle ABC, $a = 9{\cdot}5$ cm, $b = 8{\cdot}4$ cm, C = 73°. Calculate:

a the area of the triangle *b* side c *c* angles A and B.

11 An aircraft flies 100 km from A to B on a course 075°, then 200 km from B to C on a course 343°. How far north and how far east are B and C from A?

12 An object Q is 6 km from P on a bearing N 20°E, and object R is 7·5 km from P on a bearing N 75°E. Calculate the distance QR.

13 In Figure 2, rectangle ABCD is tilted in a vertical plane so that B is 6 cm above the horizontal line HAK. Calculate:

a the height of C above HK
b the angle through which the rectangle would have to be turned about A to bring C vertically above A.

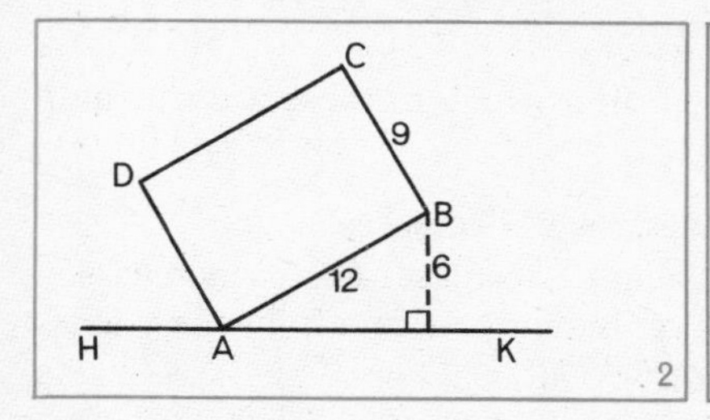

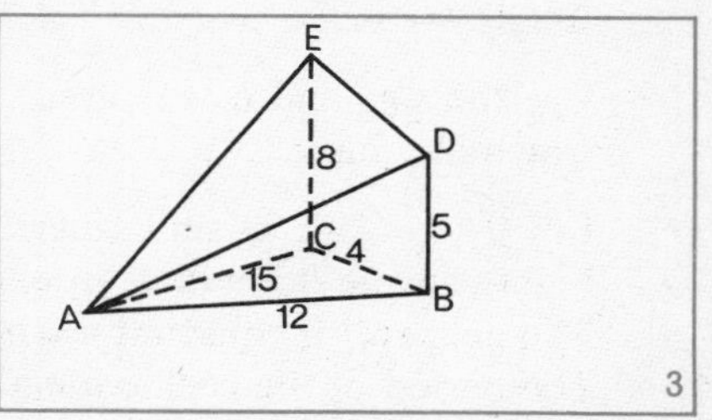

14 In Figure 3, △ABC is in a horizontal plane, and BD and CE are vertical lines. The lengths are in centimetres. Calculate:

a the lengths of AD, AE and DE
b the size of ∠ADE *c* the area of △ADE

15 An aircraft leaves airfield F and flies 120 km on course 032° to a point G. It then changes course to 143° and flies 90 km to a point H. Find the distance and bearing of H from F.

16 Balloch, Balmaha and Rowardennan are villages on the shores of Loch Lomond. Balmaha pier is 8690 m from Balloch pier on a bearing 020°. Rowardennan pier is 9600 m from Balmaha pier on a bearing 324°. Find the distance and bearing of Rowardennan pier from Balloch pier by calculation, and check by scale drawing.

Cumulative Revision Exercises

Exercise 1A

1 Find the Cartesian coordinates of the points with polar coordinates $(5, 53{\cdot}1^\circ)$ and $(10, 110^\circ)$.

2 Find two replacements for x, where $0 \leqslant x \leqslant 360$, in each of the following:

a $\cos x^\circ = 0{\cdot}075$ *b* $\tan x^\circ = -10{\cdot}02$ *c* $4\sin x^\circ = 1$
d $\cos x^\circ - 1 = 0$ *e* $\frac{1}{2}\sin x^\circ = \frac{1}{3}$ *f* $8\tan x^\circ - 3 = 0$

3 A rhombus of side 10 cm has one of its angles of size 54°. Calculate the lengths of its diagonals.

4 Using the '30°, 60°, 90°' and '45°, 45°, 90°' triangles, state which of the following are true and which are false:

a $\sin 30^\circ + \cos 60^\circ = \sin 90^\circ$ *b* $\sin 45^\circ \cos 45^\circ = \frac{1}{2}$
c $\cos 30^\circ + \cos 330^\circ = \sqrt{3}$ *d* $\tan 45^\circ + \tan 135^\circ = 0$

5 Use the sine and cosine graphs to find which of these are true and which are false:

a For $0 < x < 180$, as $\sin x^\circ$ increases $\cos x^\circ$ decreases.
b For the same domain as in ***a***, as $\sin x^\circ$ decreases $\cos x^\circ$ increases.
c The curve with equation $y = \sin x^\circ$ is symmetrical about the y-axis.
d The period of the cosine function is 360°.

6 A ship touring between islands sails 25 km from A to B on a course N 60°E, then 20 km from B to C on a course N 43°E, then 10 km from C to D on a course S 35°E. Calculate:

a how far D is east and north of A *b* the bearing of D from A.

7 For the cuboid in Figure 4, with lengths in cm, calculate:

a the lengths of BG, AG and AF
b the angles which BG and AG make with the base
c the angle between the plane EBCH and the base.

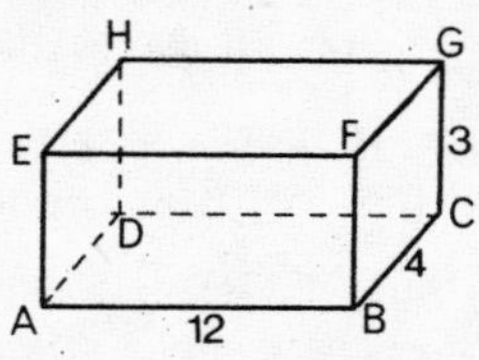

4

8 Calculate the largest angle and the area of the triangle with sides 12 cm, 15 cm and 24 cm long.

Exercise 1B

1 A is the point $(x, 10)$, and $\angle XOA = 68{\cdot}2°$. Calculate x.

2 Solve the following equations, where $0 \leqslant x \leqslant 720$:

a $2\sin x° + 1 = 0$ *b* $\sqrt{3}\tan x° = 1$ *c* $4\cos x° = 0$

3 A regular pentagon ABCDE is inscribed in a circle, centre O, with radius 4 cm. Draw OP perpendicular to AB, and calculate:

a the length of AB *b* the area of the pentagon.

4 $p = \cos\theta + \sin\theta$ and $q = \cos\theta - \sin\theta$. Prove that:

a $p^2 + q^2 = 2$ *b* $pq = 2\cos^2\theta - 1$ *c* $\sin^2\theta = \frac{1}{2}(1 - pq)$

5 The lines shown in Figure 5 are in the same vertical plane. AC is v units long. Give expressions in terms of a, b and v for the lengths of the projections of AC on:

a AH *b* AM *c* a line perpendicular to AM, and in the same plane.

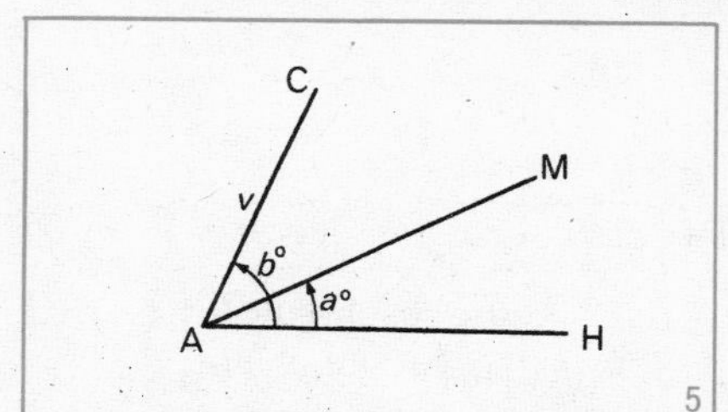

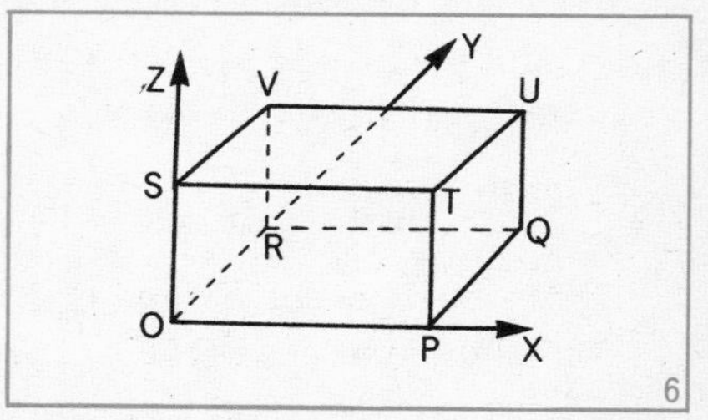

6 In the cuboid in Figure 6, P is the point $(5, 0, 0)$, $Q(5, 4, 0)$, $U(5, 4, 3)$.

a Write down the coordinates of T, S and V.
b Calculate the angles between OV and plane OPQR, and OU and plane OPQR.
c Give the point of intersection of the space diagonals.
d Calculate the angle between OU and VP.
e Calculate the angle between the planes SRQT and OPTS.

7 a A is the smallest angle in a triangle. Show that $\cos A \geqslant 0{\cdot}5$.

b Show that if in a $\triangle ABC$, $\angle B = \angle C$, then $\cos A = 1 - \dfrac{a^2}{2b^2}$.

8 V, ABCD is a pyramid on a square base of side 1·6 m. Each slant edge is 2·2 m long. Calculate: *a* the height of the pyramid
b the angle each sloping face makes with the base.

Exercise 2B

1 Find the polar coordinates, in the form $(r, a°)$ for each of the points $P(3, 4)$, $Q(-8, 15)$ and $R(-5, -12)$.

2 In triangle ABC, $\angle B = 27°$, $\angle C = 31°$ and $BC = 9{\cdot}8$ cm. Calculate:

a the length of AB *b* the area of the triangle.

3 M is the point $(a\cos\theta,\ -b\sin\theta)$ and N is $(-b\cos\theta,\ a\sin\theta)$, where a and b are positive. Prove that the length of MN is $a+b$.

4 O is the centre of a horizontal circle of radius 10 cm. OV is a vertical line of length 12 cm. PQ is a chord of length 16 cm in the circle. Calculate:

a the angle which VQ makes with the horizontal
b the angle between the planes VPQ and OPQ.

5 Prove that:

a $3\sin^2\theta + 2\cos^2\theta = 3 - \cos^2\theta = 2 + \sin^2\theta$
b $\sin\theta\cos\theta\tan\theta = 1 - \cos^2\theta$
c $\tan\theta + \dfrac{1}{\tan\theta} = \dfrac{1}{\sin\theta\cos\theta}$ *d* $\dfrac{\cos\theta}{1+\sin\theta} + \dfrac{1+\sin\theta}{\cos\theta} = \dfrac{2}{\cos\theta}$

6 P is the point (a_1, b_1) and Q is (a_2, b_2). Use the cosine rule for $\triangle OPQ$ to show that $\cos POQ = (a_1 a_2 + b_1 b_2)/\sqrt{[(a_1^2+b_1^2)(a_2^2+b_2^2)]}$.

7 Two destroyers leave from a point A at the same time. The first steams at 18 km/h on a course 073° and the second on course 101°. If at the end of an hour the ships are 10 kilometres apart, show with the help of a sketch that there are two possible positions for the second destroyer, and calculate its two possible speeds.

8 Show that $(2\cos A + 5\sin A)^2 + (5\cos A - 2\sin A)^2 = 29$

9 Triangle ABC has area 30 cm². AC = 9 cm and BC = 7 cm. Calculate the possible sizes of angle ACB, and the larger of the possible lengths of AB.

10 London has latitude 51·5°N and longitude 0°, and Cardiff has latitude 51·5°N and longitude 3·25°W. Assuming that the radius of the earth is 6370 km, calculate the length of the arc of the circle of latitude from London to Cardiff.

Computer Studies

More Programs in BASIC

1 *Revision of BASIC programs*

In Book 5 we had the following program for finding the volume of a cuboid, given its length, breadth and height.

```
10 REM VOLUME OF CUBOID          Remark statement
20 INPUT L, B, H                 Input statement
30 LET V = L * B * H             Assignment statement
40 PRINT V, L, B, H              Print statement
50 STOP                          Stop statement
```

Data 3, 4, 5

The *output* would be 60 3 4 5

In BASIC programs all statements have to be numbered in order, and all writing must be in the form of capital letters.

We saw that the following symbols were available·

BASIC *operational symbols*

\+ − * / ↑

Equality and inequality symbols

< > = LE GE NE

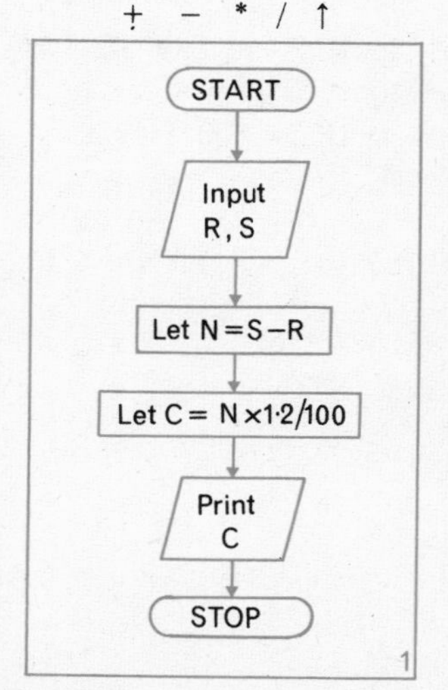

Example 1. Draw a flow chart and write a program to calculate the cost of electricity at 1·2p per unit between meter readings R and S.

```
10 REM ELECTRICITY ACCOUNT
20 INPUT R, S
30 LET N = S − R
40 LET C = N * 1·2/100
50 PRINT C
60 STOP
```

Data 500, 1500

The output would be 12.

Exercise 1 (Revision)

Draw flow charts and write programs:

1 To find the mean of three numbers.

2 To find the area of a circle, given the radius as data.

3 To calculate the volume of a sphere, given the radius as data.

4 To find the length of wire required for a skeleton cuboid.

Write BASIC programs:

5 To calculate the insurance premium payable on houses at the rate of 60p per £100.

6 To calculate the price paid for goods in a shop during a sale when a discount of 15% is given on all goods.

7 To calculate the rates payable by householders at the rate of 111p in the £.

In Book 5 we also used GOTO and conditional IF, THEN statements, as shown in Example 2 below.

Example 2. Write a program to calculate the income tax payable at the rate of 30% for incomes of £800 or more. No tax is paid on incomes or parts of incomes of less than £800.

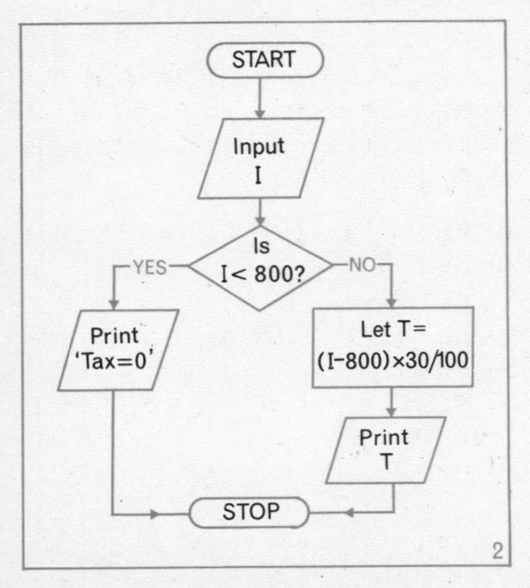


```
10 REM INCOME TAX
20 INPUT I
30 IF I<800 THEN 70
40 LET T
   = (I-800)*30/100
50 PRINT 'TAX = ', T
60 GOTO 80
70 PRINT 'TAX = 0'
80 STOP
```

Example 3. Write a program to input pairs of numbers, each pair consisting of an examination candidate's number followed by his mark. In each case output the candidate's number followed by 'PASS' if the mark is 50 or more, and by 'FAIL' if the mark is less than 50. The data are terminated by two negative numbers.

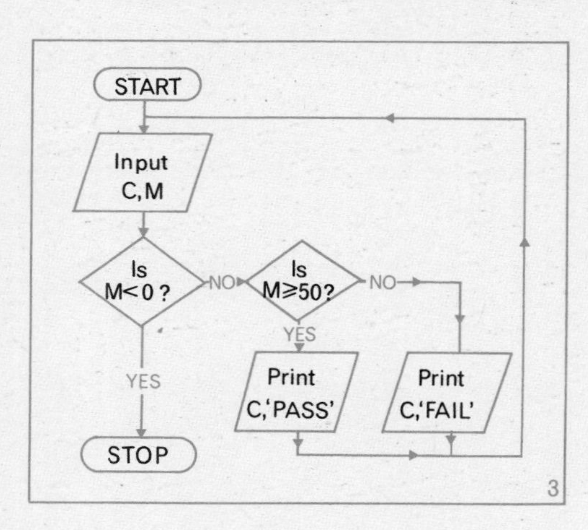


```
10 REM EXAM RESULTS
20 INPUT C, M
30 IF M<0 THEN 90
40 IF M GE 50 THEN 70
50 PRINT C, 'FAIL'
60 GOTO 20
70 PRINT C, 'PASS'
80 GOTO 20
90 STOP
```

Data 1, 80, 2, 49, 3,
75,..., −1, −1
The output would be
1 PASS 2 FAIL 3 PASS ...

Looping

The process of carrying out part of a program over and over again is called *looping*. This is illustrated in Example 4 by means of the *counter* technique provided by statements 20 and 60.

Example 4. Write a program to read 100 numbers one at a time, and to print out each number followed by its square.

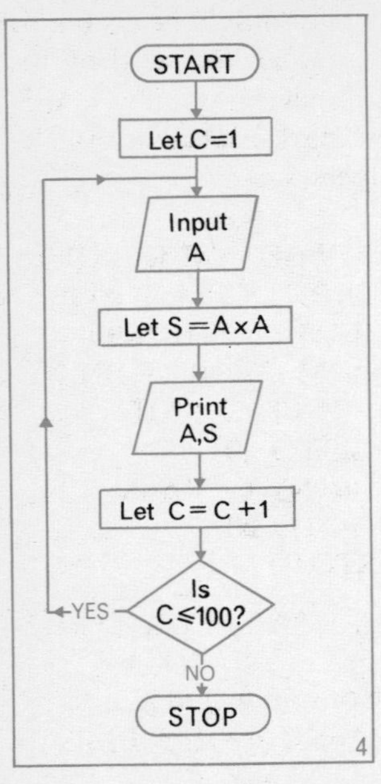


```
10 REM 100 NOS AND SQUARES
20 LET C = 1
30 INPUT A
40 LET S = A * A
50 PRINT A, S
60 LET C = C+1
70 IF C LE 100 THEN 30
80 STOP
```

In the above example, statements 30 to 70 are being repeated over and over again. The statement

```
70 IF C LE 100 THEN 30
```

is used to test the count which controls the number of times the loop is performed.

Exercise 2 (Revision)

1 Write programs for the flow charts in Figures 5 and 6.

a reads two unequal numbers and prints out the positive difference between them.

b reads the passmark for an examination and a list of marks, then prints out the number of passes.

2 A salesman receives commission of 10% of his annual sales up to £5000 and 5% for the amount over £5000. Write a program for finding his commission.

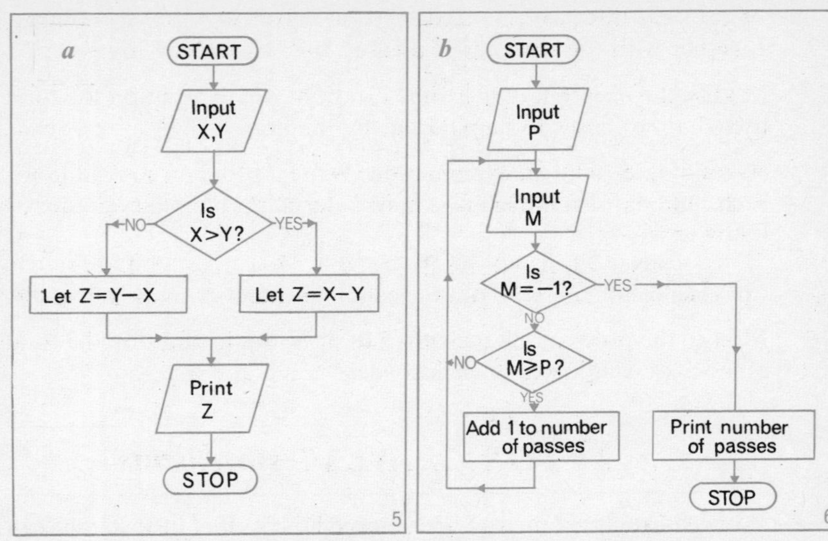

3 Write a program to calculate the income tax payable at the rate of 35% for incomes of £1000 or more. No tax is paid on incomes or parts of income of less than £1000.

4 Write a program for calculating the areas of 100 rectangles, and printing out the length, breadth and area in each case.

5 In a car rally, competitors must complete the course within a certain time to qualify. Write a program to input the qualifying time, followed by the time taken by each competitor, and to output the number of competitors qualifying. The data are terminated by a negative number.

6 Income tax allowance on life assurance premiums was calculated as follows:

For premiums of £25 or more the allowance was $\frac{2}{5}$ of the premium; for premiums of less than £25 the allowance was £10. Write a program to read a number of premiums and print them out along with their tax allowance. The data are terminated by −1.

7 A firm employing 11 persons allows travelling expenses based on the engine capacities of their cars as follows: Up to 1100 cc, 6p per km; 1101–1500 cc, 7p per km; over 1500 cc, 8p per km. Write a program to read as data for each employee an employee number, engine capacity

of car, and distance travelled. Output a list of employee numbers together with the travelling expenses due to each employee.

8 Modify the program of question *7* to print out in addition the total travelling expenses to be paid out by the firm.

9 30 candidates take an examination. Write a program to read in for each candidate a number and a mark. The marks have to be graded as follows:

Over 79, A; 60–79, B; 45–59, C; less than 45, D.

For each candidate print out the candidate number, mark and grade.

10 Modify the program of question ***9*** to print out in addition the total number of candidates in each grade.

2 *FOR...... NEXT statements*

Worked Example 4 on page 273 showed how a set of instructions can be repeated over and over again. Example 1 below shows a simpler way to do this.

Example 1. Write a program to read 100 numbers one at a time, and to print out each number followed by its square.

```
10 REM 100 NUMBERS AND SQUARES
20 FOR C = 1 TO 100
30 INPUT A
40 LET S = A * A
50 PRINT A, S
60 NEXT C
70 STOP
```

The loop is achieved by means of these two statements.

The statement 20 FOR C=1 TO 100 means 'Assign to C the value 1, and count in 1s up to 100'.

The statement 60 NEXT C adds 1 to the current value of C and tests whether C is less than or equal to 100. If $C \leqslant 100$, the loop in the program is repeated; otherwise the loop is terminated.

Compare the above with the program and flow chart in Figure 4 for Worked Example 4.

Example 2. Write a program to find the mean of 50 numbers. *See below, left.*

```
10 REM MEAN OF 50 NOS
20 LET S = 0
30 FOR C = 1 TO 50
40 INPUT A
50 LET S = S + A
60 NEXT C
70 LET M = S/50
80 PRINT 'MEAN = ', M
90 STOP
```

```
10 REM YEARS WAGES
20 LET T = 0
30 FOR C = 1 TO 52
40 INPUT W
50 LET T = T + W
60 NEXT C
70 PRINT 'TOTAL = ', T
80 STOP
```

Example 3. Write a program to calculate a man's total weekly wages for a year. *See above, right.*

Example 4. Write a program for the values of the function f defined by $f(x) = 2x^3$ for elements of the domain $-8, -7, -6, \ldots, 8$.

```
10 REM VALUES OF FUNCTION
20 FOR C = -8 TO 8
30 PRINT C, 2*C↑3
40 NEXT C
50 STOP
```

FOR statements with STEPS

The statement

FOR C = 1 TO 50 STEP 2

means that C takes 1 for its first value, and the values then increase in steps of 2 at a time, i.e. C = 1, 3, 5, 7, ..., 49.

Examples of other possible steps are:

FOR C = 50 TO 0 STEP −1

FOR C = 1 TO 5 STEP 0·5

Example 5. Write a program to find the sum of all the odd numbers between 0 and 1000.

```
10 REM SUM OF ODD NUMBERS
20 LET S = 0
30 FOR C = 1 TO 999 STEP 2
40 LET S = S + C
50 NEXT C
60 PRINT S
70 STOP
```

Worked Examples 1–5 show how easy it is to perform a loop using a FOR statement.

Exercise 3

Write BASIC programs including a FOR statement in each.

1 To find the values of the function f defined by $f(x) = 10x^2$ for elements of the domain $-4, -3, -2, \ldots, 4$.

2 To calculate a class's average daily attendance over a period of 200 school days.

3 To find the sum of the first 100 natural numbers.

4 To read 50 numbers one at a time, and to print out each number and its cube.

5 To convert temperatures in Fahrenheit to Celsius using the formula $C = \frac{5}{9}(F - 32)$, taking values of F at 1° intervals from 0° to 212°.

Use FOR statements with STEPS in programs for the following:

6 To find the values of the function g defined by $g(x) = x^3 - 3x^2 + 5$ for elements of the domain 0, 2, 4, ..., 10.

7 To find the sum of all the even numbers between 501 and 1001.

8 To convert pounds to dollars at the rate of 2·8 dollars to the £, in steps of £5 up to £100.

9 To find areas of squares with sides from 10 cm to 1000 cm in length, in steps of 10 cm.

10–12 Rewrite programs for questions ***4***, ***7*** and ***9*** of Exercise 2, including FOR statements in your programs.

3 *Standard functions*

The values of certain functions such as square roots, sines and cosines are often required in mathematics. The computer does not store tables of values of these functions, but calculates them as required by means of stored programs that it possesses; special BASIC codes are available for obtaining them.

BASIC functions consisting of three-letter codes, followed by the variable in brackets, include:

SQR(X) for the positive square root of X, where $X \geqslant 0$.

INT(X) for the integral (whole number) part of X.

ABS(X) for the absolute value of X, i.e. the value of X when its sign is ignored.
COS(X) where X is measured in *radians*. (A number
SIN(X) of degrees must be multiplied by the factor $\pi/180$
TAN(X) to convert it to radians.)

Example 1. Write a program to calculate the length of the hypotenuse in a right-angled triangle.

```
10 REM LENGTH OF HYPOTENUSE
20 INPUT B, C
30 LET S = B↑2+C↑2
40 LET A = SQR(S)
50 PRINT 'HYPOTENUSE = ', A
60 STOP
Data 6, 8
Output HYPOTENUSE = 10
```

Example 2. Write a program for the roots of the quadratic equation $ax^2+bx+c=0$ using the formula $x = \dfrac{-b \pm \sqrt{(b^2-4ac)}}{2a}$

```
10 REM ROOTS OF QUADRATIC EQUATION
20 INPUT A, B, C
30 LET X1 = (-B+SQR(B↑2-4*A*C))/(2*A)
40 LET X2 = (-B-SQR(B↑2-4*A*C))/(2*A)
50 PRINT X1, X2
60 STOP
```

Example 3. Write a program for converting a number of centimetres to metres and centimetres.

```
10 REM CM TO M AND CM
20 INPUT C                  (e.g. 234)
30 LET M = INT(C/100)       (2)
40 LET C = C-M*100          (234-200 = 34)
50 PRINT M, 'METRES', C, 'CENTIMETRES'
60 STOP
Data 234
The output would be 2 METRES 34 CENTIMETRES
```

Example 4. Write a program to calculate side a of $\triangle$ABC, being given b, $\angle$A and $\angle$B. We use the sine rule, noting that we cannot use 'small' letters in BASIC programs, and remembering to multiply the number of degrees in angles by $\pi/180$.

$$\frac{a}{\sin A} = \frac{b}{\sin B}$$

$$\Leftrightarrow \quad a = \frac{b \sin A}{\sin B}$$

or A1 = B1 sin A2/sin B2

We use A1, B1 for sides and A2, B2 for angles in radians.

```
10 REM SINE RULE
20 INPUT A, B, B1
30 LET A2 = A * 3·14/180
40 LET B2 = B * 3·14/180
50 LET A1 = B1 * SIN(A2)/SIN(B2)
60 PRINT A1
70 STOP
```

Exercise 4

Write BASIC programs:

1 To find the length of side of a square, given its area.

2 To calculate the radius of a circle, given its area and 3·14 for π.

3 To find the length of one side of a right-angled triangle, given the lengths of the hypotenuse and the other side.

4 To find the larger root of the quadratic equation $px^2 + qx + r = 0$.

5 To convert metres to kilometres and metres.

6 To convert pence to pounds and pence.

7 To convert minutes to days, hours and minutes.

Write BASIC programs for the following, the angles being given in degrees:

8 To calculate side b in $\triangle$ABC, given c, $\angle$B and $\angle$C.

9 To calculate side a in $\triangle$ABC, given b, c and $\angle$A.

10 To calculate the area of $\triangle$ABC, given a, b and $\angle$C.

11 To calculate tan A, using the formula tan A = sin A/cos A.

4 *DIMENSION statements*

It is sometimes useful to reserve a set of storage locations for a list of numbers or other data, and this can be done by means of a DIMENSION statement. For example,

```
5 DIM A(10)
```

reserves storage locations A(1), A(2), A(3), ..., A(10).

The following program inputs 10 items of data into a list A.

```
5  DIM A(10)
10 INPUT A(1), A(2), ..., A(10)
15 STOP
```

Data 2, 3, 7, 5, 9, 6, 12, 3, 8, 7

The effect of the program would be to assign the given data to storage locations as follows:

2	3	7	5	9	6	12	3	8	7
A(1)	A(2)	A(3)	A(4)	A(5)	A(6)	A(7)	A(8)	A(9)	A(10)

For a large number of items it is better to use a loop to read in the data. The above program can be written:

```
5  DIM A(10)
10 FOR C = 1 TO 10
15 INPUT A(C)
20 NEXT C
25 STOP
```

The first time round the loop, C = 1, and the statement 15 INPUT A(C) is read INPUT A(1), which means 'Store the first item of data in storage location A(1)'; and so on.

We can input data in several lists as shown in the example below.

Example 1. Write a program to read 12 items of data into lists X, Y and Z which hold 3, 4 and 5 items respectively.

```
5  DIM X(3), Y(4), Z(5)
10 FOR C = 1 to 3   }
15 INPUT X(C)       } Inputs first 3 items of data into list X.
20 NEXT C           }
25 FOR C = 1 to 4   }
30 INPUT Y(C)       } Inputs next 4 items of data into list Y.
35 NEXT C           }
```

```
40 FOR C = 1 to 5     }
45 INPUT Z(C)         } Inputs next 5 items of data into list Z.
50 NEXT C             }
55 STOP
```

The *output* of data stored in a list is achieved as follows.

Example 2. Write a program to read 30 numbers into a list B and then to print out the contents of the list.

```
10 DIM B(30)
20 FOR C = 1 TO 30    }
30 INPUT B(C)         } Loop to read in data to list B.
40 NEXT C             }
50 FOR C = 1 TO 30    }
60 PRINT B(C)         } Loop to print out contents of list B.
70 NEXT C             }
80 STOP
```

Arithmetic with lists

We use assignment (LET) statements, as before.

Example 3. Write a program to read 10 items of data into list X and 10 items into list Y, to add corresponding elements of X and Y as a list Z, and to print out list Z.

```
5  DIM X(10), Y(10), Z(10)
10 FOR C = 1 TO 10      }
15 INPUT X(C)           }
20 NEXT C               }
25 FOR C = 1 TO 10      }
30 INPUT Y(C)           }
35 NEXT C               }
40 FOR C = 1 TO 10      }
45 LET Z(C) = X(C)+Y(C) }
50 NEXT C               }
55 FOR C = 1 TO 10      }
60 PRINT Z(C)           }
65 NEXT C               }
70 STOP
```

Data 0, 2, 4, 6, 8, 10, 12, 14, 16, 18, 1, 3, 5, 7, 9, 11, 13, 15, 17, 19
The output (in list form) would be 1 5 9 13 17 21 25 29 33 37

Example 4. Write a program to find the mean of 1000 numbers.

```
5  REM MEAN OF 1000 NOS
10 DIM A(1000)
15 LET S = 0
20 FOR C = 1 TO 1000
25 INPUT A(C)
30 LET S = S + A(C)
35 NEXT C
40 LET M = S/1000
45 PRINT M
50 STOP
```

Exercise 5

1. Write a program to read 50 numbers into a list Y, and to find the sum of their squares.

2. Write a program to find the mean of 500 numbers.

3. The maximum number of points that can be scored in a competition is 70. A list of scores of 83 competitors is obtained. Write a program which will output each score as a percentage.

4. Write a program to read 20 positive numbers into a list Z, and to print out the largest member of this list.

5. Write a program to find values of the function f defined by $f(x) = x^3 - x^2 - 6$ for replacements of x $-4, -3, -2, -1, 0, 1, 2, 3, 4$. Store the values in a list A, and print them out.

6. Rewrite question **5** in such a way that the program will print out any pair of consecutive replacements of x for which the values of f have opposite signs.

Answers

Answers

Algebra—Answers to Chapter 1

Page 4 Exercise 1

1 not a surd *2* surd *3* not a surd *4* surd

5 surd *6* not a surd *7* not a surd *8* surd

9 not a surd *10* surd *11* $\{3\}$ *12* $\{-1\}$

13 $\{\frac{26}{3}\}$ *14* $\{-\frac{7}{4}\}$ *15* $\{-\sqrt{2}, \sqrt{2}\}$ *16* $\{-\sqrt{5}, \sqrt{5}\}$

17 $\{\frac{1}{3}\}; Q$ *18* $\{4\}; W$ *19* $\{-3\}; Z$ *20* $\{-4, 4\}; Z$

21 $\{-\sqrt{22}, \sqrt{22}\}; R$ *22* $\{3\}; W$ *23* surd *24* not a surd

25 surd *26* surd *27* not a surd *28* not a surd

29a 2·15 *b* 2·71 *c* −2·52 *d* −2·80

Page 6 Exercise 2

1 $2\sqrt{2}$ *2* $2\sqrt{3}$ *3* $3\sqrt{3}$ *4* $5\sqrt{2}$ *5* $2\sqrt{5}$

6 $3\sqrt{2}$ *7* $2\sqrt{7}$ *8* $10\sqrt{2}$ *9* $2\sqrt{6}$ *10* $3\sqrt{5}$

11 $5\sqrt{3}$ *12* $10\sqrt{3}$ *13* $6\sqrt{2}$ *14* $3\sqrt{6}$ *15* $7\sqrt{3}$

16 $10\sqrt{2}$ *17* $12\sqrt{2}$ *18* $20\sqrt{10}$ *19* $14\sqrt{2}$ *20* $100\sqrt{10}$

21 $8\sqrt{2}$ *22* $2\sqrt{5}$ *23* 0 *24* $-5\sqrt{3}$ *25* 0

26 $3\sqrt{7}$ *27* $3\sqrt{5}$ *28* $4\sqrt{2}$ *29* $\sqrt{3}$ *30* $2\sqrt{11}$

31 $20\sqrt{2}$ *32* $6\sqrt{5}$ *33* $AC = \sqrt{5}$ cm, $AD = \sqrt{6}$ cm

34a $2\sqrt{2}$ cm *b* $2\sqrt{3}$ cm *35a* $3\sqrt{5}, \sqrt{13}, 2\sqrt{10}$ cm *b* 7 cm

36 $3\sqrt{3}, \sqrt{3}, 8\sqrt{3}, \sqrt{30}$ m

Page 7 Exercise 3

1 5 *2* 2 *3* 6 *4* 1 *5* a

6 $\sqrt{6}$ *7* $2\sqrt{3}$ *8* $3\sqrt{a}$ *9* $\sqrt{(2a)}$ *10* $\sqrt{(ab)}$

11 4 *12* 6 *13* $3\sqrt{2}$ *14* $10\sqrt{2}$ *15* 4

16 $6\sqrt{6}$ *17* $\sqrt{2}-2$ *18* $3+\sqrt{3}$ *19* $5-\sqrt{5}$ *20* $\sqrt{7}-7$

21 $5\sqrt{2}+2$ *22* $3\sqrt{2}+8$ *23* 1 *24* 2 *25* 1

26 1 *27* 2 *28* -3 *29* $3+2\sqrt{2}$ *30* $5+2\sqrt{6}$

31 $8-2\sqrt{15}$ *32* 10 *33* -2 *34* 6 *35* $4\sqrt{3}$

36 8 *37* $4\sqrt{15}$ *38* $2\text{ cm}^2, 2\sqrt{3}$ cm *40* $k = 1$, $OP = \sqrt{3}$ units

Page 9 Exercise 4

1 0·707 *2* 0·577 *3* 0·447 *4* 3·46 *5* 4·47

6 1·16 *7* 1·34 *8* 14·1 *9* 0·671 *10* 0·566

11 $\frac{\sqrt{5}}{10}$ *12* $\frac{\sqrt{2}}{10}$ *13* $\frac{5\sqrt{3}}{3}$ *14* $\sqrt{2}$ *15* $\frac{1}{2}$

16 $\frac{2\sqrt{3}}{3}$ *17* $\frac{\sqrt{10}}{2}$ *18* $\frac{3\sqrt{10}}{10}$ *19* $\frac{\sqrt{7}}{7}$ *20* $\frac{\sqrt{15}}{5}$

Page 10 Exercise 5B

1 a $\sqrt{3}-1, 2$ *b* $\sqrt{5}+2, 1$ *c* $\sqrt{7}+3, -2$ *d* $3+\sqrt{2}, 7$

2 a $2\pm\sqrt{3}$ *b* $-1\pm3\sqrt{2}$ *c* $\frac{1}{2}(3\pm\sqrt{7})$

3 a 4, 1 *b* $-2, -17$ *c* $3, \frac{1}{2}$

4 a $\sqrt{2}+1$ *b* $\sqrt{5}-1$ *c* $12(2+\sqrt{3})$ *d* $7-4\sqrt{3}$

5 a 1 *b* 2 *c* 3 *d* 2

6 a $\sqrt{3}+\sqrt{2}$ *b* $\sqrt{5}-\sqrt{3}$ *c* $3\sqrt{5}+3\sqrt{2}$ *d* $\sqrt{7}-\sqrt{5}$

7 0·4142 *9* $\sqrt{7}-1, 8-2\sqrt{7}$

Algebra—Answers to Chapter 2

Page 13 Exercise 1

1 a $3^4\times3^2=(3\times3\times3\times3)\times(3\times3)=3\times3\times3\ldots$ to 6 factors $=3^6$, etc.

4 a 2^7 *b* 3^8 *c* 7^6 *d* 2^9 *e* 10^{20}

f a^9 *g* x^8 *h* f^{19} *i* z^7 *j* p^{13}

5 a x^{10} *b* a^{11} *c* 3^{14} *d* 5^8 *e* 2^{16}

f c^{12} *g* z^{10} *h* a^{10} *i* x^{30} *j* y^{12}

6 $a^p\times a^q=(a\times a\times a\ldots$ to p factors$)\times(a\times a\times a\ldots$ to q factors$)$
$=a\times a\times a\ldots$ to $(p+q)$ factors $=a^{p+q}$, by definition of a power.

Page 14 Exercise 2

1 a $5^4/5^2=(5\times5\times5\times5)/(5\times5)=5\times5=5^2$, etc

4 a 2^2 *b* 3^3 *c* 5^{11} *d* x^6 *e* 10^6

f c^1, i.e. c *g* z^6 *h* k^{11} *i* a^0(or 1) *j* x^7

5 a x^3 *b* y^2 *c* z^5 *d* a^7 *e* c^4

f 2^6 *g* 3^0(or 1) *h* 4 *i* 5^{10} *j* a

6 $a^p \div a^q = \frac{a^p}{a^q} = \frac{a \times a \times a \ldots \text{to } p \text{ factors}}{a \times a \times a \ldots \text{to } q \text{ factors}}$

$= a \times a \times a \ldots$ to $(p-q)$ factors $= a^{p-q}$, by definition of a power.

Page 15 Exercise 3

1 a $(2^2)^3 = 2^2 \times 2^2 \times 2^2 = (2 \times 2) \times (2 \times 2) \times (2 \times 2) = 2 \times 2 \times 2 \ldots$ to 6 factors $= 2^6$ *or* $(2^2)^3 = 2^2 \times 2^2 \times 2^2 = 2^{2+2+2} = 2^6$, etc.

3 a 5^6 *b* 3^{20} *c* 4^{60} *d* 2^6 *e* 10^{100}

f a^{30} *g* d^4 *h* x^{21} *i* x^6 *j* p^6

4 a a^6 *b* x^{16} *c* y^{16} *d* z^{16} *e* c^{25}

f 2^{12} *g* 2^{49} *h* 3^{12} *i* 4^{10} *j* a^{80}

5 $(a^p)^q = a^p \times a^p \times a^p \ldots$ to q factors $= a^{p+p+p \ldots \text{to } q \text{ terms}} = a^{pq}$

Page 15 Exercise 4

3 a a^4b^4 *b* a^3b^6 *c* $x^{10}y^{10}$ *d* x^6y^8 *e* $m^{15}n^6$

f $9m^2$ *g* $4m^6$ *h* $3m^2n^2$ *i* $9m^2n^2$ *j* $16m^8n^4$

Page 16 Exercise 5

1 a^5 *2* x^{16} *3* p^4 *4* y^{200} *5* k^5 *6* a^5

7 b *8* x^3 *9* a^6 *10* x^{100} *11* z^{12} *12* x^3y^6

13 2^7 *14* 4^8 *15* 7^5 *16* $(\frac{2}{3})^7$ *17* $(\frac{1}{2})^7$ *18* $(0{\cdot}375)^6$

19 3^{12} *20* 2^{20} *21* 13^9 *22* 15^8 *23* 2^2 *24* 3^3

25 5 *26* 10^{99} *27* 3^{13} *28* 2^{22} *29a* 2^{12} *b* 2^{12} *30a* 3^{13} *b* 10^9

31 No; $6^5 = (2 \times 3)^5 = 2^5 \times 3^5$ *32* ***b, c, d, f, g*** true; ***a, e, h, i*** false

33a 6×10^8 *b* $5{\cdot}6 \times 10^3$ *c* $9{\cdot}6 \times 10^5$ *d* 2×10^2 *e* 5×10^3 *f* $1{\cdot}86 \times 10^4$

34a 3 *b* 4 *c* 3 *d* 5 *e* 4 *f* 1 *g* 3 *h* 2

Page 17 Exercise 6

1 $10^0, 10^{-1}, 10^{-2}, 10^{-3}$ *2* $1, \frac{1}{10}, \frac{1}{10^2}, \frac{1}{10^3}$

3 a 1 *b* $\frac{1}{10}$ *c* $\frac{1}{10^2}$ *d* $\frac{1}{10^3}$

4 a $\frac{1}{10^4}$ *b* $\frac{1}{10^{10}}$ *c* $\frac{1}{2^5}$ *d* $\frac{1}{6^3}$

5 a 1 *b* 1 *c* $\frac{1}{2^3}$ *d* $\frac{1}{2^p}, \frac{1}{a^p}$

6 a $\frac{1}{2^5}$ *b* $\frac{1}{3^2}$ *c* $\frac{1}{a^4}$ *d* $\frac{1}{b^{10}}$ *e* $\frac{1}{c}$

f $\frac{2}{c^2}$ *g* $\frac{a}{b^3}$ *h* x^2 *i* $\frac{1}{2x^2}$ *j* $\frac{x^3}{4}$

7 $a^p \times a^q = a^{-m} \times a^{-n} = \frac{1}{a^m} \times \frac{1}{a^n} = \frac{1}{a^{m+n}} = a^{-(m+n)} = a^{-(-p-q)} = a^{p+q}$

Page 18 Exercise 7

1 $2^4, 2^3, 2^2, 2^1, 2^0$ *2* $2^{-1}, 2^{-2}, 2^{-3}, 2^{-4}, 2^{-8}$ *3* $3^4, 3^3, 3^2, 3^1, 3^0$

4 $3^{-1}, 3^{-2}, 3^{-3}, 3^{-4}, 3^{-6}$ *5* $1, \frac{1}{4}, \frac{1}{16}, \frac{1}{64}$ *6* $1, \frac{1}{6}, \frac{1}{36}$

7 $1, \frac{1}{16}, 4, 8$ *8* $1, \frac{3}{2}, \frac{9}{4}, \frac{27}{8}$ *9* $\frac{1}{3^2}$ *10* $\frac{1}{7^7}$ *11* 2^2

12 $\frac{5^2}{4^3}$ *13* $\frac{3^4}{2^3}$ *14* $\frac{1}{6^3 \times 5^2}$ *15* $\frac{3^3}{x^4}$ *16* $\frac{2}{x^2}$

17 $\frac{5^2}{y}$ *18* $\frac{25}{y}$ *19* 2^2 *20* 5×3^3 *21* $\frac{1}{3^6}$

22 $\frac{1}{2^{15}}$ *23* 5^6 *24* m^6 *25a* 3×10^{-2} *b* $2{\cdot}8 \times 10^9$ *c* $1{\cdot}27 \times 10^{-8}$

Page 19 Exercise 8

1 a a *b* $\sqrt{a}$ *2* $a, \sqrt[3]{a}$ *3* $x, \sqrt[4]{x}$ *4* $a^2, (\sqrt[3]{a})^2$ or $\sqrt[3]{a^2}$

5 a $\sqrt[5]{a^3}$ *b* $\sqrt{b^5}$ *c* $\sqrt[3]{c^4}$ *d* $\sqrt{d^3}$ *e* $\sqrt[6]{e^5}$ *f* $\sqrt{x}$

6 a $c^{1/2}$ *b* $a^{1/3}$ *c* $x^{3/4}$ *d* $x^{3/2}$ *e* $x^{5/3}$ *f* $x^{6/3}$ or x^2

7 a $\frac{1}{\sqrt{a}}$ *b* $\frac{1}{\sqrt[3]{b^2}}$ *c* $\frac{1}{\sqrt[4]{c^3}}$ *d* $\frac{1}{\sqrt[3]{d}}$ *e* $\frac{1}{\sqrt[3]{e^4}}$

8 a $p^{-1/2}$ *b* $q^{-1/3}$ *c* $x^{-3/2}$ *d* y^{-2} *e* $z^{-1/2}$

Page 20 Exercise 9

1 2 *2* 10 *3* 3 *4* 4 *5* 27

6 4 *7* 8 *8* 343 *9* $\frac{1}{6}$ *10* $\frac{1}{5}$

11 $\frac{1}{125}$ *12* 27 *13* $\frac{1}{27}$ *14* 2 *15* $\frac{1}{32}$

16 $\sqrt{a}$ *17* $\sqrt[3]{b}$ *18* $\sqrt[3]{c^2}$ *19* $\sqrt[5]{d^4}$ *20* $\frac{1}{\sqrt[4]{e^3}}$

21 a^2 *22* a^2 *23* a^4 *24* a^{-2}, or $\dfrac{1}{a^2}$ *25* $a^{-1/2}$, or $\dfrac{1}{\sqrt{a}}$

26 a^2 *27* a^{-5} or $\dfrac{1}{a^5}$ *28* y^{-2} or $\dfrac{1}{y^2}$ *29* y^8

30 y^3 *31* $\sqrt[3]{x}$ *32* x *33* 1 *34* $\sqrt[5]{a^2}$

35 a^2 *36* 5 *37* 13 *38* 0·1 *39* 6

Page 21 Exercise 10

1 4 *2* $12a$ *3* $6b$ *4* $2c$ *5* $2x$

6 $4z$ *7* $2y^{-1}$ or $\dfrac{2}{y}$ *8* $x^{-1/4}$ or $\dfrac{1}{x^{1/4}}$ *9* 1 *10* 8

11 24 *12* 36 *13* $\frac{2}{9}$ *14* $\frac{3}{2}$ *15* 1

16 $a^{1/3}$ *17* $x^{3/2}$ *18* $y^{5/3}$ *19* $x^{3/5}$ *20* $k^{1/4}$

21 $\sqrt{x}$ *22* $\dfrac{1}{\sqrt[3]{y}}$ *23* $\dfrac{1}{\sqrt{z^3}}$ *24* $\sqrt[5]{p^4}$ *25* $\dfrac{1}{\sqrt[4]{q^7}}$

26 $p^{7/2}$ *27* $q^{7/2}$ *28* $\dfrac{1}{r}$ *29* $\dfrac{1}{x^6}$ *30* $s^{3/2}$

31 $\dfrac{4}{t}$ *32a* -4 *b* $\frac{3}{2}$ *c* -3 *d* $\frac{4}{3}$ *33* $a+a^{1/2}$ *34* $1-b^{1/3}$

35 $3+\dfrac{3}{a^2}$ *36* $1·08\times10^{12}$ *37* 10^{15} *38* 400 *39* 5

40a 3×10^{-3} *b* $4·6\times10^4$ *c* $7·89\times10^2$ *41* $1·08\times10^6$ g

42a 4 *b* $\frac{1}{4}$ *c* 27 *d* $p^{1/n}$ *43* $d=\sqrt[3]{\dfrac{Q}{h}}$ *44* $D=5\sqrt{\dfrac{C}{L}}$

Page 23 Exercise 10B

1 x^{-1} *2* x^{-2} *3* $5x^{-2}$ *4* $2x^{-3}$ *5* $\frac{1}{2}x^{-1}$

6 $\frac{3}{4}x^{-1}$ *7* $\frac{5}{2}x^{-2}$ *8* $x^{-1/2}$ *9* $\frac{3}{2}x^{-1/2}$ *10* $\frac{2}{5}x^3$

11 $x^{-2}+x^{-3}$ *12* $2x^{-1}-3x^{-4}$ *13* $x-x^{-1}$ *14* $2x+\frac{1}{2}x^{-1}$

15 $x^{1/2}+x^{-1/2}$ *16* x^3+x^2 *17* $3x^3+\frac{1}{2}x^{-2}$ *18* $3x^{1/3}-\frac{1}{3}x^{-1/3}$

19 $x+x^{1/2}$ *20* $x-x^{-1/2}$ *21* $x+1$ *22* $x+x^{-1}$

23 $1+x^{-2}$ *24* x^2-x *25* $1+x^{-1}$ *26* $x^{3/2}-2x^{1/2}$

27	$u+u^{-1/3}$	*28*	$1-v^2$	*29*	$1-x^{-2}$	*30*	$p^2+2pq+q^2$
31	r^2-2r+1	*32*	$x+2x^{1/2}+1$	*33*	$x-2x^{1/2}+1$	*34*	$x^2+2x^{3/2}+x$
35	x^3-2x^2+x	*36*	x^2+2+x^{-2}	*37*	$1-2x^{-2}+x^{-4}$		
38	$x^{3/2}+2x+x^{1/2}$	*39*	$x^{1/2}-2+x^{-1/2}$	*40*	x^2+2+x^{-2}		
41	$x-2+x^{-1}$	*42*	$4x^2+2+\frac{1}{4}x^{-2}$	*43*	$\frac{1}{9}x^{-1}-2+9x$		
44	$x+1+x^{-1}$	*45*	$2-x^{-1}$	*46*	$\frac{1}{2}x^2-1+2x^{-2}$		
47	$x+6+9x^{-1}$	*48*	$\frac{1}{3}+x^{-2}-2x^{-3}$	*49*	$x^{-1}-x^{-2}$		

Algebra—Answers to Chapter 3

Page 28 Exercise 1

1 Any six of (0, 0), (0, 1), (0, 2), (0, 3), (0, 4), (1, 0), (1, 1), (1, 2), (1, 3), (2, 0), (2, 1), (2, 2), (3, 0), (3, 1)

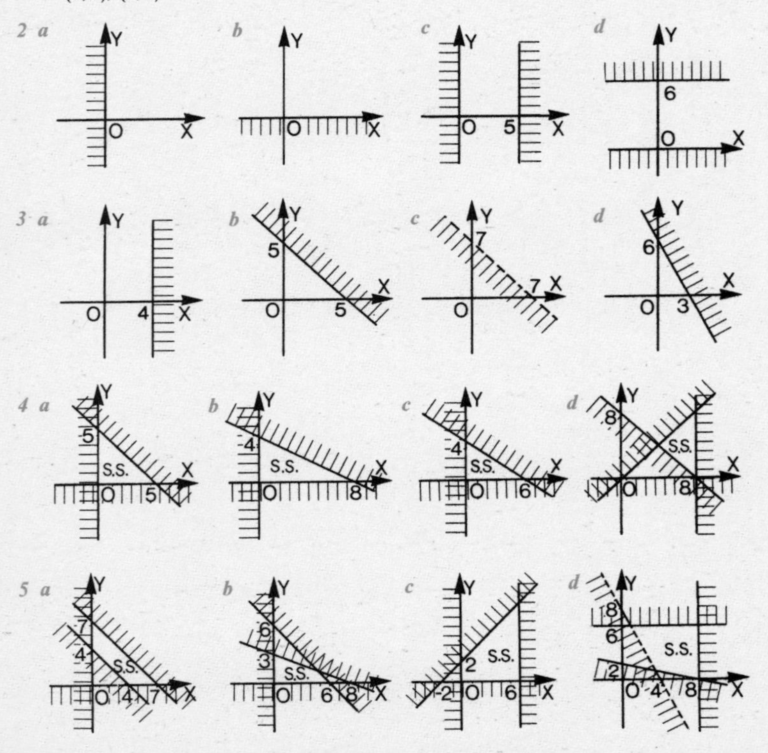

6 b 8 at (4, 4) *c* 2 at (1, 1) *d* They lie in a straight line.
e {(1, 1), (1, 2), (1, 3), (2, 1), (2, 2), (3, 1)}

7 a 1, 2, 3, 4, 5; 3, 4, 5, 6, 7; 5, 6, 7, 8, 9; 7, 8, 9, 10, 11.
b (1, 3), (3, 2), (5, 1) *c* {(1, 0), (2, 0), (3, 0), (4, 0), (1, 1), (2, 1)}
d maximum 11 when $x=5$ and $y=3$ *e* minimum 1 when $x=1$ and $y=0$

8 a 0, 14, 36, 20 *b* (1, 1), 7; (6, 1), 17; (5, 3), 25; (3, 5), 31; (1, 4), 22; (3, 3), 21; (3, 1), 11.
c *(1)* 0 at (0, 0) and *(2)* 36 at (3, 6) respectively.

9 a 0, 12, 19, 19, 9 *c* *(1)* 0 at (0, 0) *(2)* 19 at (5, 3), (2, 5), maximum value at two different points, (5, 3) and (2, 5).

10a $x>0,\ y>0,\ x+y<6$ *b* $x>0,\ y>0,\ x+2y<6$

Page 32 Exercise 2

1 c (3, 0) and (2, 1) *2* 3, 6, 2, 5, 4, 7, 6; 7 *3 c* (2, 2), (3, 0) *d* 6 *4 c* 7; (2, 1)

5 b 38 at (6, 20); 45 at (15, 5)

6 a $x\geqslant 0,\ y\geqslant 0;\ 150x+75y\leqslant 2250 \Leftrightarrow 2x+y\leqslant 30;\ 50x+75y\leqslant 1500 \Leftrightarrow 2x+3y\leqslant 60$
c (7, 15), (8, 14); 7 of the first kind, 15 of the second, or 8 of the first and 14 of the second, or 6 of the first and 16 of the second.

7 a $x\geqslant 0,\ y\geqslant 0,\ x+y\leqslant 25,\ 3x+4y\leqslant 84$
c $10x+12y$ maximised by (16, 9), giving maximum profit £268

8 a $x\geqslant 0,\ y\geqslant 0,\ x+y\leqslant 48,\ 3x+y\leqslant 72$ *c* 12 first class, 36 tourist class; £3000

Page 35 Exercise 3

1 a 2, 3, 4, 3, 4, 5, 4, 5, 6 *b* Points where $x+y$ has value k lie on $x+y=k$.
c maximum $x+y$ is 6, minimum $x+y$ is 2

2 c maximum 4, minimum 0

3 b 5 given by (3, 1) *5 c* maximum 22 at (1, 5); minimum 3 at $(1\frac{1}{2}, 0)$

6 c 20 cars, 10 buses, income £12·50 *7* 47 cars, 13 buses, £36·50

8 c 3 tablets, 4 capsules, $5\frac{1}{2}$p per day

9 15 semi-detached, 11 bungalows; £33 500.

10 $10\times$ £100 of A and $22\frac{1}{2}\times$ £100 of B giving maximum value £3250.

11a $x+y\geqslant 24,\ 2x+5y\geqslant 60$ *c* 4 tonnes of X, 20 tonnes of Y; £5600
d £9600.

Algebra—Answers to Chapter 4

Page 41 Exercise 1

1 $m_{AB}=\frac{1}{2}$, $m_{CD}=-2$, $m_{EF}=0$, $m_{GH}=-3$, $m_{JK}=\frac{1}{3}$, m_{MN} is not defined, $m_{PQ}=1$, $m_{RS}=\frac{7}{2}$

2 a $\frac{3}{5}$ *b* 3 *c* $-\frac{2}{3}$ *d* $\frac{3}{5}$ *e* $-\frac{6}{5}$ *f* not defined *g* 0

3 $2, \frac{2}{3}, \frac{1}{4}, 0$ *4* $-3, -1, -\frac{1}{2}, 0$ *5 a* $\frac{1}{3}$ *b* $-\frac{1}{3}$ *c* $\frac{1}{4}$ *d* $-\frac{1}{2}$

6 ***d, c***

Page 43 Exercise 2

1 b 0·4, 1·3 *2* 0, 0·1, 0·4, 0·9, 1·6, 2·5, 3·6, 4·9 *b* 0·4, 0·8, 1·2

3 0, 0·9, 1·6, 2·1, 2·4, 2·5, 2·4, 2·1, 1·6 *b* 0·8, 0·4, 0, −0·4

4 −3, −0·8, −0·2 *5* $2, -\frac{2}{3}, -6$ *6 a* 4, −2 *b* (3, 9)

7 a 6, 4, 2, 0 *b* −2, −4, −6

8 a $y_A = a^2p - 2aph + h^2p + aq - qh + r$, $y_B = a^2p + 2aph + h^2p + aq + qh + r$

Page 47 Exercise 3

1 a 12 km, 24 km, 36 km *b* 12 km/h *c* 36/3 = 12

2 200/75 = 8/3; 2·7 g/cm^3

3 44 million, 50 million; 200 000 people per year *b* 300 000 people per year

4 2 m/s, 6 m/s, 10 m/s, 14 m/s *5 a* 30 m/s *b* 30 m/s

6 a 52 km/h *b* 84 km/h *c* 18 km/h

Page 51 Exercise 4

1 a 84/4 = 21 cm^2 *b* 20 cm^2 *2 a* 156/4 = 39 cm^2 *b* 39 cm^2

3 a 43 cm^2 *b* 43·5 cm^2 *4* 52 cm^2 *5 b* 170 square units *c* too large

6 b 19·5 square units *c* less *7 a* 5·1 m^2 *b* 255 m^3/minute

8 a 73·2 square units *b* 53 square units *9* 10·6 square units *10* 11 160 m^2

Page 54 Exercise 5

1 3250 m or $3\frac{1}{4}$ km *2* 300 m *3* 2·58 km *4 a* 1·35 km *b* 40 km/h

5 a 200 m 5 m/s^2 *6* 10 m/s; $1\frac{2}{3}$ m/s^2, 5 m/s^2 *7* 336 litres

8 80·7 million units

Algebra—Answers to Revision Exercises

Page 57 Revision Exercise 1

1 a $\{\sqrt{0}, \sqrt{1}, \sqrt{4}, \sqrt{9}\}$ *b* $\{\sqrt{2}, \sqrt{3}, \sqrt{5}, \sqrt{6}, \sqrt{7}, \sqrt{8}, \sqrt{10}\}$ *c* ϕ *d* E

2 a $\{-1, 1\}$ *b* $\{-12, 12\}$ *c* $\{-\sqrt{2}, \sqrt{2}\}$ *d* $\{-0{\cdot}1, 0{\cdot}1\}$

3 a F *b* T *c* F *d* F *e* T *f* F *g* T *h* F

4 a 10 *b* $10\sqrt{5}$ *c* $4\sqrt{3}$ *d* $5\sqrt{6}$ *e* $6\sqrt{15}$ *f* $10\sqrt{10}$

5 $8\sqrt{2}$ cm *6* $a\sqrt{3}$ cm; $a^2\sqrt{3}$ cm^2

7 $5\sqrt{5}$ cm, $5\sqrt{10}$ cm, $5\sqrt{13}$ cm, $5\sqrt{14}$ cm

8 a 0 *b* $3\sqrt{2}$ *c* $2\sqrt{3}$ *d* 0 *e* $2\sqrt{2}$ *f* 50

9 a $6\sqrt{2}$ *b* 45 *c* 40 *d* 15 *e* 23 *f* 5 *g* $9-6\sqrt{2}$ *10a* 6; 7 *b* yes

11a $8-4\sqrt{2}$ *b* 4 *13a* $3\sqrt{3}$ *b* $\frac{3\sqrt{2}}{2}$ *c* $\frac{\sqrt{15}}{5}$ *d* $\frac{\sqrt{11}}{11}$ *e* $\frac{\sqrt{3}+1}{2}$

f $3(\sqrt{7}+\sqrt{3})$ *16a* $2\sqrt{3}$ cm *b* $2\sqrt{3}$ cm *c* $\sqrt{10}$ cm, $\sqrt{11}$ cm

Page 59 Revision Exercise 2

1 a a^7 *b* a^2 *c* a^6 *d* 2^{10} *2 a* T *b* F *c* T *d* F *e* T *f* T

3 a a^{-4} *b* a *c* a^{-8} *d* a^8 *4 a* $\sqrt{3}$ *b* $\sqrt[4]{a^3}$ *c* $\sqrt[3]{b^2}$ *d* $\sqrt{(a+b)}$

5 a $p^{1/3}$ *b* $p^{2/3}$ *c* $a^{1/2}$ *d* $a^{1/2}b^{1/2}$ *6 a* 8 *b* $\frac{1}{2}a$ *c* a^{-18} *d* a

7 a 10 *b* 11 *c* 8 *d* 125 *8 a* $9{\cdot}6\times10^4$ *b* $3{\cdot}4\times10^{-2}$ *c* $1{\cdot}2\times10^{-1}$

9 a -3 *b* 16 *c* $-\frac{1}{5}$ *10a* 7 *b* 0 *c* -1 *d* -5

11a x^2+1 *b* x^6+x^{-3} *c* $x+1$ *12* $d=2\sqrt[3]{\frac{T}{\pi f}}$ *13* 15; 1 *14* 1200

Page 60 Revision Exercise 2B

1 a $a^{-1/2}$ *b* $a^{-4}b^{-9}$ *c* 1 *2 a* 3×10^2 *b* 6×10^{-5} *c* 4×10^{-2} *d* 3×10^{-2}

3 a 4 *b* 9 *4* $l=\frac{gT^2}{4\pi^2}$; $r=\sqrt[3]{\frac{3V}{4\pi}}$ *5* $k=1{\cdot}38\times10^{-16}$

6 a 2 *b* $\frac{1}{2}$ *c* $-1{\cdot}5$ *7 a* 1·04 1·06 0·99

8 a x^{-3} *b* $2x^{-1}$ *c* $3x^3$ *d* $\frac{1}{2}x^{-1/3}$ *e* $2x^{1/2}-\frac{1}{2}x^{-1/2}$

9 a $a-1$ *b* $a-a^{-1}$ *c* $a^{1/2}+2+a^{-1/2}$

10a $x+2+x^{-1}$ *b* x^3-x^{-5} *c* $x^{-2}+x^{-3}$ *d* $\frac{1}{8}x^3+x+2x^{-1}$ *e* $x^{-1}-1-x$

f $\frac{1}{2}x-1-\frac{1}{2}x^{-1}$ *g* $x^{-1}-x^{-2}$ *h* $4x^{1/2}-4+x^{-1/2}$

11 $2\frac{1}{6}\doteqdot2{\cdot}167$ *13* $h=6{\cdot}62\times10^{-34}$, $m=9{\cdot}09\times10^{-31}$

14a $v=\left(\frac{k}{p}\right)^{2/3}$ *15* $\frac{3}{4}e^{2c}+\frac{1}{4}e^{-2c}$

Page 61 Revision Exercise 3

(In the answer diagrams, shading shows regions where points cannot lie, so the solution set of the system is left blank.)

1 a *b* *2 a* *b* 8, 30

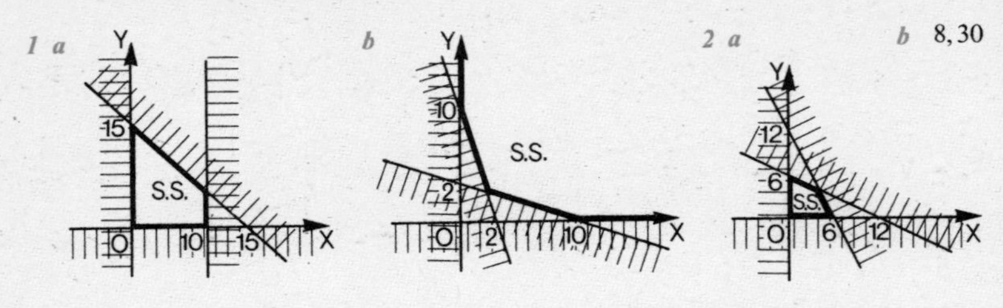

3 a

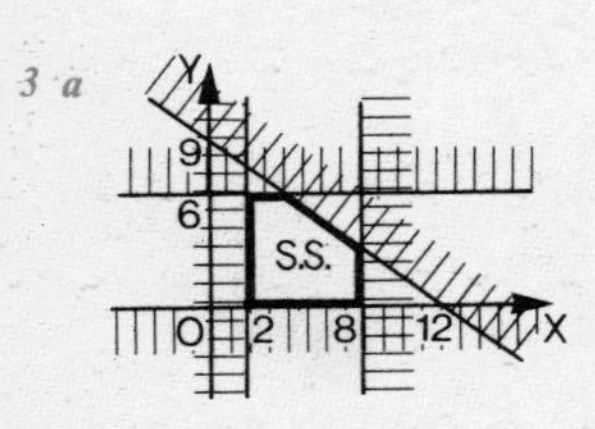

b (*1*) (8, 3) (*2*) (2, 6) (*3*) (4, 6)

c (*1*) (2, 0) (*2*) (8, 0) (*3*) (2, 0)

4 c (*1*) 210 (*2*) 130

5 a $3x+y \geqslant 20$, $x+2y \geqslant 30$, $x+y \geqslant 40$
c 2 kg of A, 14 kg of B

6 a $2x+4y \leqslant 24$, $4x+2y \leqslant 24$ *b* $P=3x+5y$ *c* 4 of each; £32

7 a (*1*) $x+y \leqslant 120$ (*2*) $2x+3y \leqslant 270$ *b* $90x+107y$ *c* 90, 30; £11 310

8 a $x \geqslant 0$, $y \geqslant 0$, $6x+8y \geqslant 132$, $y \geqslant 2x$ *b* 6, 12 *c* 48p *d* $48\frac{1}{2}$p

Page 63 Revision Exercise 4

1 b 0, undefined, $\frac{1}{2}$, $-\frac{3}{2}$, $-\frac{4}{3}$ *2 c* 0·8, 0·5, 0·25 *3 b* 0·5, 1·5, 2·5 *4 b* −4, −2, 0, 2, 4

5 b 10 m/s, 20 m/s, 30 m/s *c* 25 m/s, 43 m/s *6* 0·8 m/s, 1·6 m/s, 0·7 m/s

7 a 364 square units *b* 540 square units *8* 19 square units *9* 35 square units

10b 2100 m, 1650 m *c* 14 m/s, 11 m/s

Algebra—Answers to Cumulative Revision Section

Page 66 Exercise 1

1 a $\{p, a, r, l, e, o, g, r, m\}$ *b* $\phi, \{0\}, \{1\}, \{2\}, \{0,1\}, \{0,2\}, \{1,2\}, \{0,1,2\}$

2 a $\{11, 13, 15, 17, 19\}$ *b* $\{7, 11, 13\}$ *c* $\{11, 13\}$ *d* $\{7, 11, 13, 15, 17, 19\}$

3 a $\{x \in Z: -3 \leqslant x \leqslant 3\}$ *b* yes • *4 a* $\{4\}$ *b* $\{4\}$ *c* $\{1, 2, 3, 4, 5, 8\}$

5 a

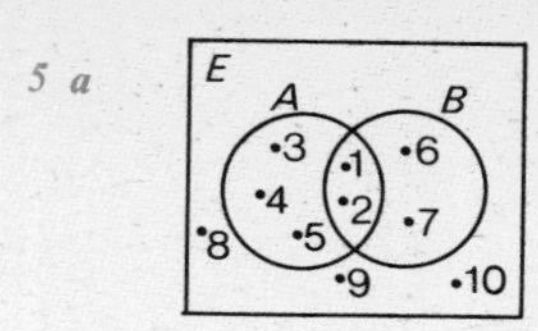

b (*1*) $\{1,2\}$ (*2*) $\{1,2,3,4,5,6,7\}$ (*3*) $\{6,7,8,9,10\}$ (*4*) $\{3,4,5,8,9,10\}$ (*5*) $\{3,4,5,6,7,8,9,10\}$ (*6*) $\{8,9,10\}$ (*7*) $\{3,4,5,6,7,8,9,10\}$ (*8*) $\{8,9,10\}$

6 a

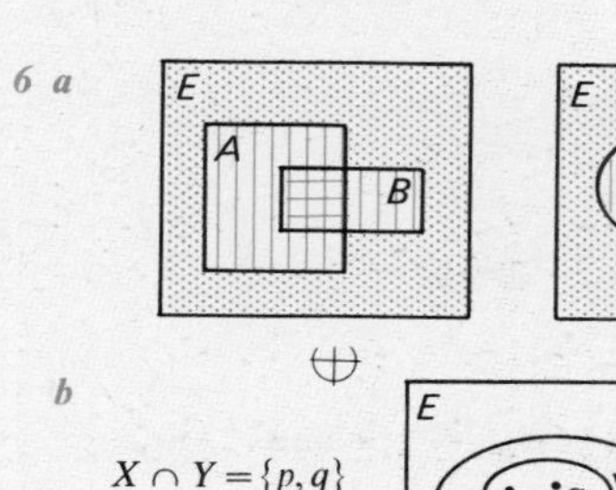

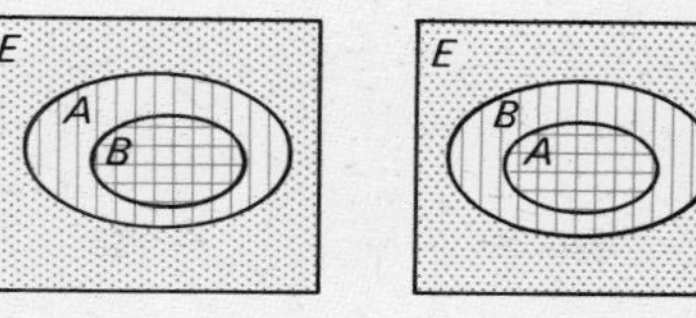

b $X \cap Y = \{p, q\}$

7

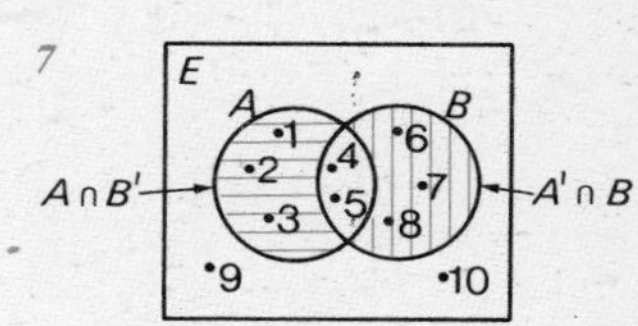

8 9; $(A \cup C)'$ or $A' \cap C'$ *9* 4 *10* 10; 0·1 *11* 2

Page 68 Exercise 2

1 a -1 *b* 3 *c* -32 *d* 104 *e* -24

2 a (*1*) 2 (*2*) $-\frac{2}{3}$ *b* (*1*) $-\frac{1}{2}$ (*2*) $\frac{3}{2}$ *c* $-\frac{2}{3} < -\frac{5}{8} < -\frac{7}{12}$

3 a 14 *b* 27 *c* -12 *d* 0 *4 a* $4n-10$ *b* $6x+y$ *c* $-2a^2-4a$

5 a $5x-y$ *b* $12x-7y$ *c* $x+9y$ *d* $5x+22y$ *6 a* T *b* F *c* F *d* T *e* T

7 a $10a^2+a-2$ *b* $6m^2-23m+21$ *c* $8p^2+2p-45$ *d* $9c^2+24c+16$
e $16x^2-8x+1$ *f* $15x^2-62xy+40y^2$ *g* e^4-4 *h* $12n^4-11n^2-5$
i $3a^2b^2-4ab-7$

8 a $24x$ cm *b* $22x^2$ cm^2 *c* $6x^3$ cm^3

9 a a *b* yes *c* yes *d* e.g. $b \circ c = c \circ b = a$

10a $4x^3-8x^2+13x-5$ *b* a^3+b^3 *11a* $10x$ mm, $6x^2-y^2$ mm^2

12a 3·8, 1·2 *b* 3·9, -1·9 *c* 0·62, -1·12 *d* -0·647, -0·953

13 $5(p^2+q^2)$; $-3p^2+8pq+3q^2$ *14a* $x^2-\dfrac{1}{x^2}$ *b* $a^2-1-\dfrac{2}{a}-\dfrac{1}{a^2}$ *15* -5

Page 70 Exercise 3

1 a $\{1,2\}$ *b* $\{7,8,9,10\}$ *c* ϕ *d* $\{3,5\}$
e $\{7\}$ *f* $\{5,6,7,8,9,10\}$ *g* $\{4,5\}$ *h* S *i* $\{1,2,3\}$

2 a

+	0	1	2	3	4
0	0	1	2	3	4
1	1	2	3	4	0
2	2	3	4	0	1
3	3	4	0	1	2
4	4	0	1	2	3

×	0	1	2	3	4
0	0	0	0	0	0
1	0	1	2	3	4
2	0	2	4	1	3
3	0	3	1	4	2
4	0	4	3	2	1

b (*1*) {2} (*2*) {4} (*3*) {4} (*4*) {2}

3 a $x=6$ *b* $x=10$ *c* $a=4$
d $x=1\frac{1}{2}$ *e* $t=-3$
f $y=-1\frac{1}{3}$ *g* $c=2$ *h* $x=2$

4 a $y=3$ *b* $a=-12$ *c* $x=11\frac{2}{3}$ *d* $x=-29$

5 a $\{24, 25, 26, \ldots\}$ *b* $\{6, 7, 8, \ldots\}$ *c* $\{s \in Q: s > -1\frac{1}{2}\}$
d $\{x \in Q: x < 8\frac{1}{2}\}$ *e* $\{s \in Q: s < 3\frac{1}{3}\}$ *f* $\{x \in Q: x \leqslant \frac{1}{6}\}$

6 a −2 *b* 3 −1 $P \cap Q = \{x \in R: -1 \leqslant x \leqslant 3\}$

7 a 55, 5 *b* 12 *c* 21 *8* 40 cm and 18 cm

9 150, 151, 152 *10* 58 and 28 years *11* 36 and 72 km/h

12a $\{-\frac{1}{3}\}$ *b* $\{x: x<0\}$ *13* 44 and 52 km/h

Page 72 Exercise 4

1 A B; 10, 14, 24 → 2, 3, 4, 5

$\{(10,2), (10,5), (14,2), (24,2), (24,3), (24,4)\}$

2 a (i), (iv) *b* (iv)

3 a $\{(0,0), (1,1), (1,-1), (4,2), (4,-2), (9,3), (9,-3)\}$ *b*
c no; e.g. 1 has two images, 1 and −1

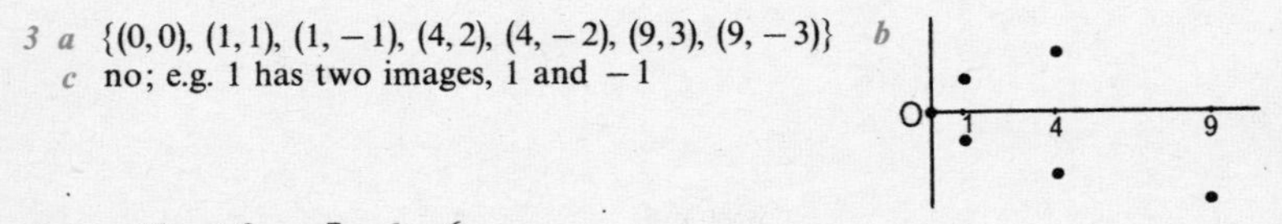

4 a −10 *b* 0 *c* 7 *d* −6

5 a (*1*) 2 (*2*) 5 *b* 1 *c* 10 or −10 *6* −60 *7* 5, −8; −13

8 a F *b* T *c* F *9* 0 has no image; $\{2, 1\frac{1}{3}, \frac{3}{4}, \frac{1}{2}\}$; yes

10a −2, −2·67, −1·27 *b* 2 *c* 1 *d*

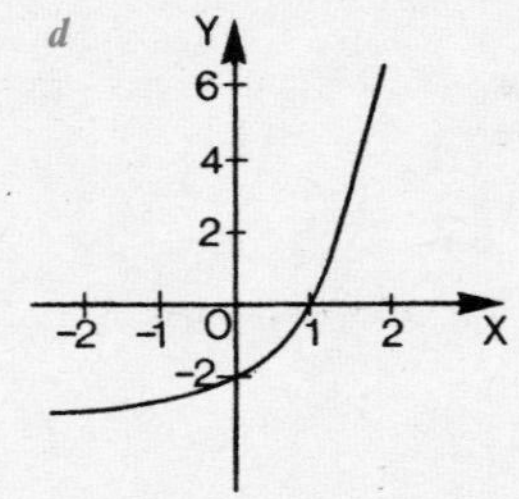

Page 74 Exercise 5

1 a $4(x-y)$ *b* $4(5x-4)$ *c* $3x(x-2y)$ *d* $x^2(x-2)$ *e* $mn(9n-16m)$
f $(a-b)(4a-4b-1)$

2 a $(4x-9)(4x+9)$ *b* $(1-3y)(1+3y)$ *c* $(a+b-5)(a+b+5)$
d $2(x-1)(x+1)$ *e* $a(b-1)(b+1)$ *f* $\frac{1}{2}m(v-u)(v+u)$

3 a $p(x^2+y^2-z^2)$ *b* $(4x-3y)(4x+3y)$ *c* $8q(2p-3q)$
d $a^2b^2(b+a)$ *e* $(x-3)(x+1)$ *f* $\pi r(2h+r)$

4 a $(x-2)(x+1)$ *b* $(a+8)(a+9)$ *c* $(y-6)^2$ *d* $(4x+3)(x+5)$
e $(2b+1)(b-3)$ *f* $(3x+2)(2x+5)$

5 $x+1$ *6* $\dfrac{x+2}{4}$ *7* $\dfrac{1}{x+2}$ *8* $\dfrac{n}{m+n}$ *9* -1 *10* $\dfrac{1+a}{2(1+b)}$

11 $\dfrac{4x^2+x-1}{x^2}$ *12* $\dfrac{20x+13}{10x}$ *13* $\dfrac{1}{6}$ *14* $\dfrac{2}{(x-1)(x+1)}$

15 $\dfrac{x}{(x-1)(x+1)}$ *16* $\dfrac{9}{(x-5)(x-2)}$

17a 12·6 *b* 9420 *18* $x-y=7; x=9, y=2$

19a $(2p+2q-3r)(2p+2q+3r)$ *b* $(1-r+s)(1+r-s)$
c $(4-3x-3y)(4+3x+3y)$ *d* $(1-2x)(1+2x)(1+4x^2)$
e $x^2(1-x)(1+x)(1+x^2)$ *f* $(a-b)(xa-xb-1)(xa-xb+1)$

20 $\left(\dfrac{20}{r}+\dfrac{30}{s}\right)$ hours *21a* $2(a+b)$ *b* $\dfrac{1}{p+3}$ *22a* x^3-1 *b* $(x-1)(x^2+x+1)$

c $(a-1)(a+1), (a-1)(a^2+a+1), (a-1)(a+1)(a^2+1), (a-1)(a^4+a^3+a^2+a+1)$

23a $\dfrac{3a+4}{3a}$ *b* $\dfrac{x(x+4)}{2(x+2)}$ *24a* (1) T (2) T (3) T *b* F *c* F *d* T

Page 76 Exercise 6

1 $x=\dfrac{P}{6}$; 16 *2* $v=\sqrt{\dfrac{R}{k}}$; 9 *3* $s=\dfrac{f^2}{16e}$; 200

4 $h=\dfrac{v^2}{196}$; 4 *5* $T=n^2+4$; 10 004 *6* $V=p^3$; 10·65

7 a $r=\dfrac{C}{2\pi}$ *b* $x=\dfrac{y-q}{2}$ *c* $r=\sqrt{\dfrac{2\pi}{s}}$ *d* $d=\dfrac{M-3}{6}$ *e* $t=\dfrac{v-u}{a}$ *f* $s=\dfrac{v^2-u^2}{2a}$

8 $x=\sqrt{\left(\dfrac{A}{4+\pi}\right)}$; 29 *9* $a=\dfrac{A-2tb-4t^2}{2t}$ *10* $a=\dfrac{Qb}{Q-3b}, Q\neq 3b$

11a $x=\dfrac{by}{1-ay}$ *b* $u=\dfrac{2xy}{x-y}$ *c* $r=\dfrac{(p-1)^2}{4}$ *d* $f=\sqrt{\dfrac{T}{2\pi LC}}$ *12a* $-\frac{2}{3}$ *b* 1 *c* $\frac{5}{6}$

13 $A=2\pi r^2+2\pi rh; h=\dfrac{A-2\pi r^2}{2\pi r}; 2\pi r^2+2\pi hr-A=0$

14a $x=\dfrac{c}{a-b}$ *b* $x=\dfrac{1}{y-1}$ *c* $x=(k-1)^2$

15a $rx^2-px+q=0$ *b* $x^2-ax+a^2=0$ *c* $x^2-(b+1)x+(b-a)=0$

16a $S = a(1+r)$ *b* $a = \dfrac{S}{1+r}$ *c* $r = \dfrac{S-a}{a}$

Page 78 Exercise 7

1 a *b* *c* *d* *e* *f*

Y, $y=4x-8$, O, X; Y, $y-7x=14$, O, X; Y, O, X, $2x+3y=12$; Y, O, X, $4x+5y+20=0$; Y, O, X, $y=-2$; Y, O, X, $y=-2x$

2 a *b* *c* *d*

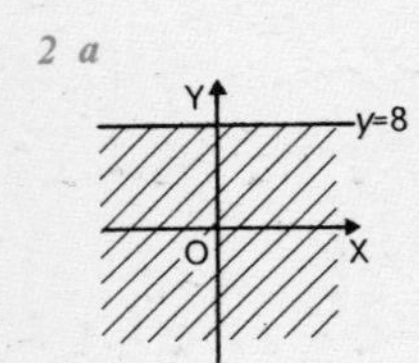

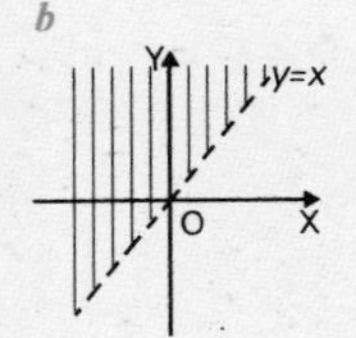

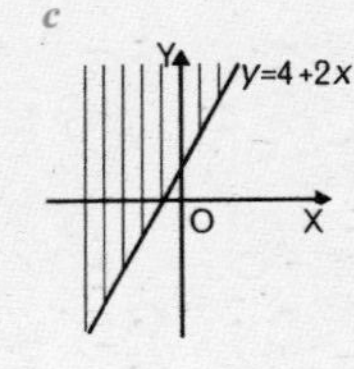

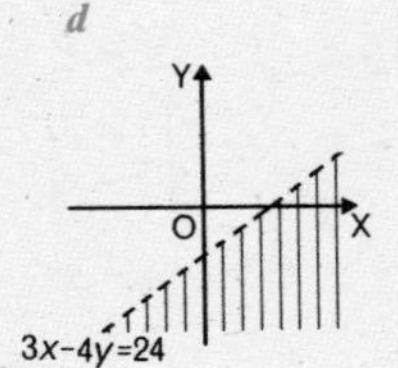

3

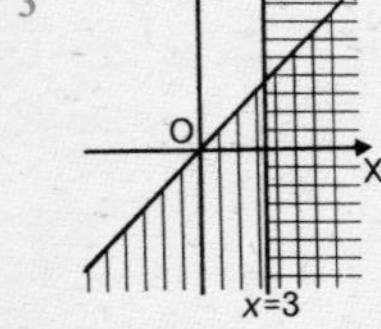

4 a $\{(2,3)\}$ *b* $\{(1,3)\}$ *5 a* $\{(6,-5)\}$ *b* $\{(1,5)\}$ *c* $\{(2,-5)\}$
d $\{(3,-1)\}$ *e* $\{(2,-3)\}$ *f* $\{(2,3)\}$ *g* $\{(7,5)\}$ *h* $\{(18,4)\}$

6 a $(1\frac{1}{2}, 0), (0, 3)$ *b* III and V *c* VI *d* II and III *e* $\{(1,1)\}$

7 a P(3, 7), Q(3, 1), R(1, 3)
b (1) $\{x, y): y \leqslant 4-x\} \cap \{(x, y): y \geqslant 2x+1\}$, $x, y \in R$
(2) $\{(x, y): y > 4-x\} \cap \{(x, y): y < 2x+1\} \cap \{(x, y): x < 3\}$, $x, y \in R$

8 $a = 4, b = -7$ *9* 22 cm, 20 cm *10* 16 of 5p and 24 of 50p

11 $(2, -1)$ *12* $-1, -2$; $y = -x-2$ *13* 15

Page 80 Exercise 8

1 a 55 *b* $W = 2S^2$ *2 a* $s = kt^3$ *b* $p = k/q^4$ *c* $m = kn^2$ *d* $u = k\sqrt{v}$

3 a 64 *b* $\frac{1}{2}$ *4 a* $1\frac{1}{2}$ ***b*** $1\frac{1}{8}$ *5 a* $Q = \dfrac{196x}{3z^2}$ *b* $1\frac{1}{3}$

6 a 4 seconds *b* 56·25 cm *7* $d = \frac{15}{2}\sqrt[3]{m}$; 64 kg

8 750 *9* $a = 2, b = 1$ *10* $11\frac{1}{4}$ seconds *11* 4·8 *12a* 320; 15

b $t = K\sqrt{d}$; '... as the square root of the distance in metres it has fallen.'

13 $\frac{8}{27}$

Page 82 Exercise 9

1 a $\{-2,-5\}$ *b* $\{2\}$ *c* $\{10,-\frac{1}{2}\}$ *d* $\{0,-10\}$

2 a $\{1,5\}$ *b* $\{-5,-4\}$ *c* $\{-\frac{3}{2},\frac{3}{2}\}$ *d* $\{-\frac{1}{2}\}$ *e* $\{0,4\}$ *f* $\{-3,11\}$

3 15 cm, 9 cm *4 a* $(-1,0),(15,0),(0,15)$ *b* 0, 14 *c* 64

5 a $\{-14,4\}$ *b* $\{-3\frac{1}{4},2\frac{3}{4}\}$ *c* $\{\frac{1}{4},\frac{3}{4}\}$

6 a $\{-3{\cdot}6,1{\cdot}6\}$ *b* $\{0{\cdot}2,2{\cdot}3\}$ *c* $\{-0{\cdot}9,0{\cdot}7\}$ *d* $\{-0{\cdot}7,2{\cdot}2\}$
e $\{-2{\cdot}8,0{\cdot}8\}$ *f* $\{-0{\cdot}7,2{\cdot}7\}$

7 a (*1*) $(0{\cdot}5,-12{\cdot}25)$ (*2*) min., $-12{\cdot}25$ (*3*) -3, 4 (*4*) $-12{\cdot}25 \leqslant f(x) \leqslant 8$ (*5*) $x=0{\cdot}5$
b (*1*) $(-0{\cdot}5, 0{\cdot}75)$ (*2*) min., $0{\cdot}75$ (*3*) none (*4*) $0{\cdot}75 \leqslant f(x) \leqslant 7$ (*5*) $x=-0{\cdot}5$
c (*1*) $(-0{\cdot}75, 3{\cdot}2)$ (*2*) max., $3{\cdot}2$ (*3*) $-2, 0{\cdot}5$ (*4*) $-12 \leqslant f(x) \leqslant 3{\cdot}2$ (*5*) $x=-0{\cdot}75$

8 5; 30 cm^2 *9 a* $\{x : x<-3 \text{ or } x>3\}$ *b* $\{x : 5 \leqslant x \leqslant 8\}$ *c* $\{x : -4<x<3\}$

10a $\{1,2\}$ *b* $\{0,2\}$ *11a* T *b* F *c* F *d* T *e* T

Page 84 Exercise 10

1 a $\begin{pmatrix}3 & 2 & 2\\1 & 1 & 0\end{pmatrix}$ *b* $\begin{pmatrix}a & 4\\2b & 6\end{pmatrix}$ *2 a* $p=3, q=10$ *b* $p=35, q=0$

3 a $\begin{pmatrix}6 & -1\\-1 & 1\end{pmatrix}$ *b* $\begin{pmatrix}4 & -6\\2 & 7\end{pmatrix}$ *4 a* $\begin{pmatrix}6 & 1 & 5\\5 & 4 & -3\end{pmatrix}$ *b* $\begin{pmatrix}-2 & 2 & -4\\-4 & -6 & 8\end{pmatrix}$

5 a (34) *b* $\begin{pmatrix}14\\6\end{pmatrix}$ *c* $\begin{pmatrix}1 & 1\\2 & 1\end{pmatrix}$ *d* (21) *e* $\begin{pmatrix}2\\1\end{pmatrix}$ *f* $\begin{pmatrix}2 & 0\\1 & 0\end{pmatrix}$

6 a $\begin{pmatrix}1 & 2\\2 & -1\end{pmatrix}\begin{pmatrix}x\\y\end{pmatrix}=\begin{pmatrix}x'\\y'\end{pmatrix}$ *b* $(5,0),(0,-5),(-5,0),(0,5)$ *c* squares

7 a $\begin{pmatrix}0 & -1\\1 & 0\end{pmatrix}$ *b* $(0,0),(0,4),(-4,4),(-4,0)$; an anti-clockwise rotation of 90° about O, followed by the dilatation [O, 2]

8 a $\begin{pmatrix}6 & -5\\-7 & 6\end{pmatrix}$ *b* $\frac{1}{4}\begin{pmatrix}-1 & -2\\4 & 4\end{pmatrix}$ *c* none *d* $-\frac{1}{2}\begin{pmatrix}-1 & 2\\2 & -2\end{pmatrix}$

9 a $\{(1,-1)\}$ *b* $\{(\frac{21}{31},\frac{25}{31})\}$ *c* $\{(\frac{66}{59},\frac{8}{59})\}$

10 $\begin{pmatrix}1 & 2\\0 & 1\end{pmatrix}\begin{pmatrix}x\\y\end{pmatrix}=\begin{pmatrix}x'\\y'\end{pmatrix}$. $(0,0),(4,0),(8,4),(4,4)$. $\begin{pmatrix}1 & -2\\0 & 1\end{pmatrix}$

11 $x=0, y=1$; $\begin{pmatrix}0 & 0\\4 & 5\end{pmatrix}$, $\frac{1}{3}\begin{pmatrix}2 & -1\\-1 & 2\end{pmatrix}$, Q^{-1} does not exist

12a $\begin{pmatrix}1 & 0\\0 & -1\end{pmatrix}$ *b* dilatation [O, 2] *c* $\begin{pmatrix}2 & 0\\0 & -2\end{pmatrix}$, $\{(0,0),(8,-4),(12,2)\}$

d $\begin{pmatrix}\frac{1}{2} & 0\\0 & \frac{1}{2}\end{pmatrix}$, dilatation [O, $\frac{1}{2}$]

Algebra—Answers to Cumulative Revision Exercises

Page 86 Cumulative Revision Exercise 1A

1 a $\frac{1}{4}$ *b* 11 *c* 8 *d* 125 *2 a* $4\sqrt{2}$ *b* $3\sqrt{2}-6$ *c* $3+2\sqrt{2}$

3 a $\{-1,0,1,2\}$ *b* $\{-3,-2,3\}$ *c* $\{2\}$ *d* $\{-1,0,1,2,3\}$

4 a $2ab, 10, 0$ *b* $b*a = 2ab = a*b$ *5 a* $1\frac{1}{2}$ *b* 16

6 a $\{0{\cdot}3, 3{\cdot}7\}$ *b* $\{-0{\cdot}9, 0{\cdot}5\}$ *7 a* $12(k^2+1)$ *b* $12(k-1)(k+1)$ *c* $(2x+1)(x-1)$

8 a $\{1\}$ *b* $\{r : r \leqslant -\frac{3}{5}\}$ *9 a* $\frac{1}{b-1}$ *b* $\frac{6-r}{6}$ *c* $\frac{7-x-x^2}{(3-x)(3+x)}$

10 4:1 *11a* 2, 3, 5, 9 *b* 5 *12*

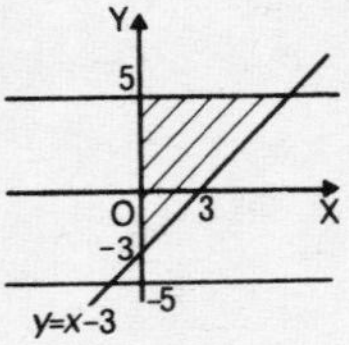

Page 87 Cumulative Revision Exercise 1B

1 a $\frac{1}{2xy^2}$ *b* $\frac{4x}{y}$ *c* $\frac{y}{4x}$ *2* $3a^2-b^2 = 10\sqrt{2} \notin Q$ *3 a* 4 *b* 4

4 a $1\frac{1}{3}$ *b* $\pm 0{\cdot}4$ *5 a* $\{0,2\}$ *b* $\{-1,0,1\}$ *c* $\{1,2\}$

6 a F *b* F *c* F *d* T *7 a* $-\frac{x+2}{2x}$ *b* $\frac{-x}{(x+2)(x-1)}$

8 a 6 *b* yes *c* 2, 8 *9* $S = \frac{1}{2}n(n+1)$ *10* $12ab-3b^2$, or $3b(4a-b)$

Page 88 Cumulative Revision Exercise 2A

1 a $x = \frac{-qy-r}{p}$ *b* $x = \pm\sqrt{(r^2-y^2)}$ *c* $x = \frac{bc-ad}{d}$

2 $\{(-1,-1)\}$ *3 a* $2, \frac{3}{4}, -8$ *b* $2, -2$ *4 a* 14 or -15 *b* 12

5 a $\frac{2x^2+x+6}{2x^2}$ *b* $\frac{2ab}{(a-b)(a+b)}$ *c* $\frac{1}{x+2}$

6 16; $x = 2, y = 4$

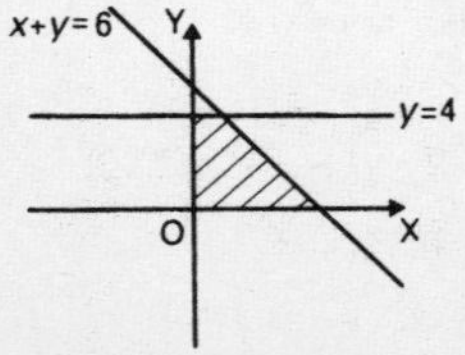

7 a $3x^2+6$ *b* $-3x^2+12x-8$ *c* $x^2-8x+11$

8 $\begin{pmatrix} -1\frac{1}{2} & \frac{1}{2} \\ -1 & 0 \end{pmatrix}$ *9* $a = 36, b = 6$ *10a* 12·25 *b* 4·0

11a 3 *b* $\frac{1}{4}$ *c* $\frac{3}{2}$ *12* 120, 3

Page 89 Cumulative Revision Exercise 2B

1 a $n=\dfrac{Ir}{E-IR}$ *b* $a=\dfrac{1}{\pm\sqrt{(1+r^2)}}$ *2* $P\cap Q=\{(-2,\frac{1}{3})\}\subset R$

3 a 4 *b* 5 *c*

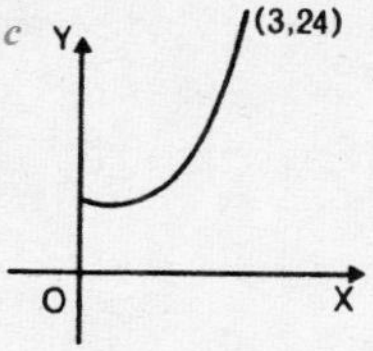

4 a $P=4a^2b^2$ *b* 16

5

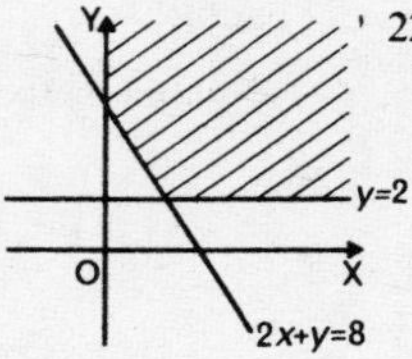

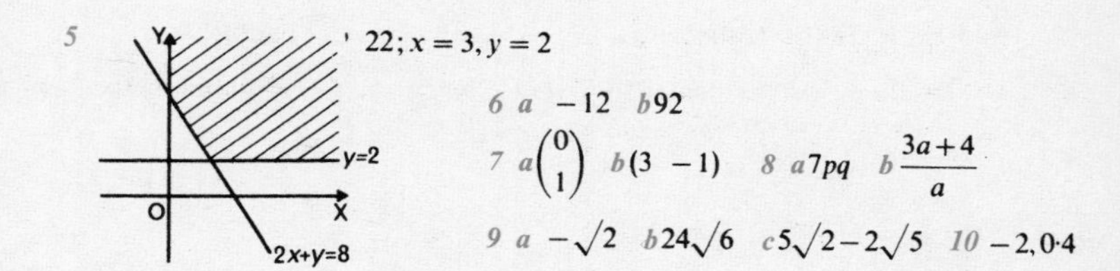

22; $x=3, y=2$

6 a -12 *b* 92

7 a $\begin{pmatrix}0\\1\end{pmatrix}$ *b* $(3\ \ -1)$ *8 a* $7pq$ *b* $\dfrac{3a+4}{a}$

9 a $-\sqrt{2}$ *b* $24\sqrt{6}$ *c* $5\sqrt{2}-2\sqrt{5}$ *10* $-2, 0{\cdot}4$

Page 90 Cumulative Revision Exercise 3A

1 2·16 *2* $\{x:1<x<2\}$ *3* 2500; 50 *4 a* F *b* F *c* F

5 a I, II, III *b* I, VI, VII, IX *c* I *d* IV, V, VIII

6 a (*1*) 0 (*2*) 12 *b* 0·35 *7* $\{x:x>8\}$

8 a $(p+q-r)(p+q+r)$ *b* $(2x-5)(x+2)$ *c* $b(a+b)(a-3b)$

9 a $\{(-1\frac{1}{2},2)\}$ *b* $\{(-1{\cdot}4, 1{\cdot}8)\}$ *10* 8 or 10

Page 91 Cumulative Revision Exercise 3B

1 a y *b* $x+2+\dfrac{1}{x}$ *c* 0 *2 a* $\{x:-5<x<5\}$ *b* $\{x:2\leqslant x\leqslant 4\}$

c $\{x:x<-2 \text{ or } x>3\}$ *3 a* $(10-x)$ cm, $(10-2x)$ cm *c* 25, 5

4 $\{-2{\cdot}6, -0{\cdot}4\}$ *5 a* 6 *b* $4\sqrt{2}$ *c* $24\sqrt{2}$ *d* 1

6 a (*1*) I, II, VI, VII (*2*) I, II *b* $\{(x,y):y\leqslant 0,\ y\leqslant x+3,\ y\leqslant -x+3\}$

7 a $d=\frac{1}{15}v^2+\frac{2}{3}v$ *b* 40, 20 *8 a* $\begin{pmatrix}0&-1\\1&0\end{pmatrix}, \begin{pmatrix}0&1\\1&0\end{pmatrix}, \begin{pmatrix}-1&0\\0&1\end{pmatrix}, \begin{pmatrix}1&0\\0&-1\end{pmatrix}$

b reflection in y-axis, and in x-axis *c* $\begin{pmatrix}-1&0\\0&-1\end{pmatrix}$, half turn about origin

Geometry—Answers to Chapter 1

Page 95 Exercise 1

1 c They are perpendicular to the axis. *d* The points get closer to each other.

2 (iv) *4* {P, Q}, {R}, ϕ *5* perpendicular to radius

6 a $x = 90$, $y = 60$ *b* $a = 50$, $b = 90$; $c = 40$ *c* $\angle OCP = \angle OCQ = 90°$, $\angle AOB = 140°$, $\angle OAB = \angle OBA = 20°$, $\angle OAC = \angle OCA = 40°$, $\angle OCB = \angle OBC = 30°$, $\angle ACP = 50°$, $\angle BCQ = 60°$.

7 DG = 10 cm, EF = 2·5 cm

8 a 13 cm *b* circle, centre Z, radius 13 cm *c* same as *b*

9 Both are perpendicular to UV. *10* two pairs of parallel sides

11 Both are perpendicular to axis of symmetry. *12* Both are perpendicular to axis of symmetry.

Page 98 Exercise 2

1 a ORT, OST *b* TR = TS, OR = OS *c* $\angle ORT = \angle OST$, $\angle ROT = \angle SOT$, $\angle RTO = \angle STO$ *d* $\angle ORT$, $\angle OST$ *2* 7·4 cm

3 CBD, ABD; $\angle CBD = \angle CDB = 70°$, $\angle ABD = \angle ADB = 20°$, $\angle BAD = 140°$

4 a OQ; P↔R *b* $\angle SRQ = \angle SOP = \angle ROS = 70°$, $\angle SQP = \angle SQR = \angle OPS = \angle ORS = 20°$; four right angles at S *c* 6·7 cm

5 a PR, QS *b* It is a rhombus.

6 a AY = AZ, BZ = BX, CX = CY *b* AB = 7 cm, BC = 9 cm, CA = 8 cm *c* AB = 14 cm, AC = 15 cm *d* 15 cm *e* 18 cm

7 b BC = 7 cm, CA = 8 cm, AB = 9 cm *c* 5 cm, 3 cm, 3 cm

Page 100 Exercise 3

1 b one *3 a* PQ is the perpendicular bisector of CD. *b* a kite

4 a H, R, S *b* (*1*) parallel to each other, perpendicular to line of centres (*2*) PQ = RS

5 a R, S *b* T lies on AB. *6 a* images of T, B, A *b* It is a common tangent. *c* AB = CD *7* as in question *6* *8 a* T, 2·5; 10 cm *b* T, −2·5; $4\frac{2}{7}$ cm

10b BC = 4 cm, DE = 4 cm, AD = 6 cm, AE = 2 cm *c* 14·9 cm

11b BC = 4 cm, DE = 4 cm, AD = 11 cm, AE = 7 cm *c* 24 cm

12b BC = 2 cm, DE = 2 cm, AD = 12 cm, AE = 10 cm *c* 24 cm

13b alternate angles equal

Page 102 Exercise 4

1 a *b* *c* $\angle AOB = 2\angle ACB$

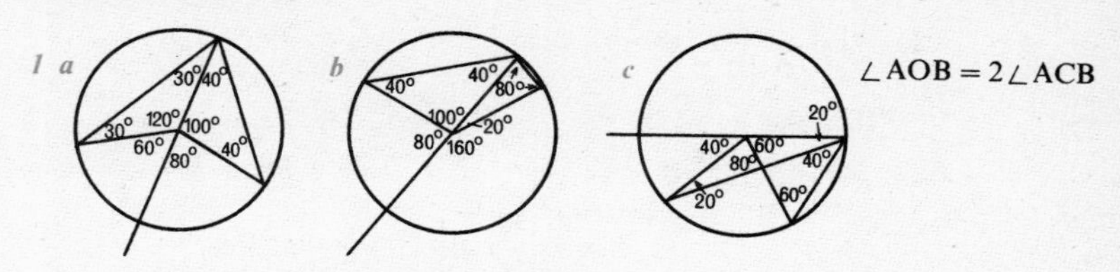

2 $\angle OAC = x°$, $\angle OBC = y°$, $\angle AOC = (180-2x)°$, $\angle BOC = (180-2y)°$, $\angle AOP = 2x°$, $\angle BOP = 2y°$; yes

3 a $\angle BAC$, $\angle BEC$, $\angle BDC$ *b* $\angle DBE$, $\angle DCE$ *c* $\angle ABD$, $\angle ACD$

4

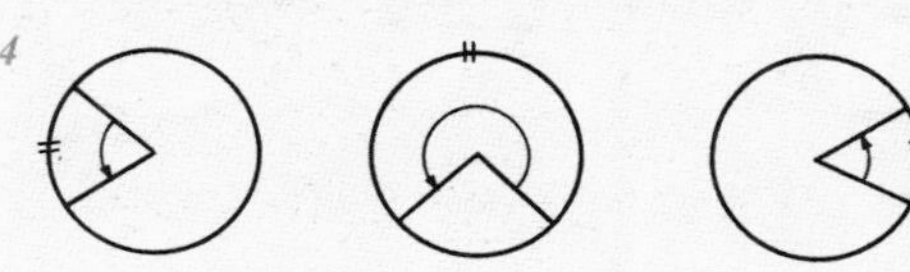

5 a 76° *b* 26° *c* 230°, 130° *6 a* 50° *b* 130° *7 a* 60°, 30° *b* 300°, 150°

8 a 80° *b* $(180-2x)°$ *9* 72°, 36° *10* 45°, $22\frac{1}{2}°$; 36°, 18°; $\frac{360°}{n}$, $\frac{180°}{n}$

11 $\angle YOZ = 120°$, $\angle OYZ = \angle OZY = 30°$

12 $\angle P = 65°$, $\angle Q = 60°$, $\angle R = 55$

Page 105 Exercise 5

1 a 180° *b* 90° *2* (i) (ii) (iii)

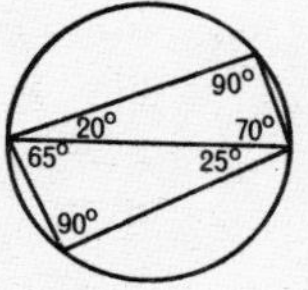

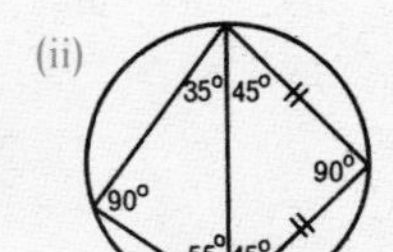

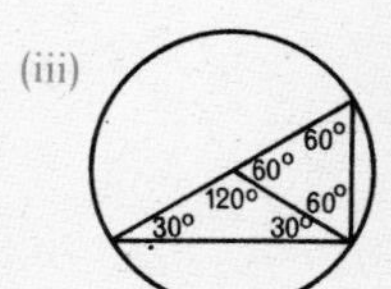

3 all angles right,
or equal diagonals bisecting each other.

4 OAT and OBT are right angles

Page 106 Exercise 6

1 a all 40° *b* all 36° *c* all $x°$ *d* all equal *2* $x = 40$, $y = 60$, $p = q = 20$

3 *4* *5*

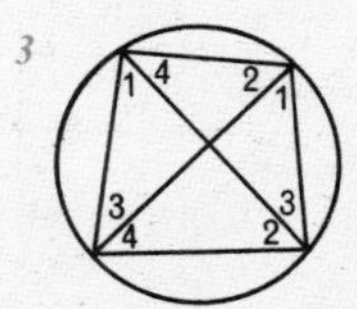

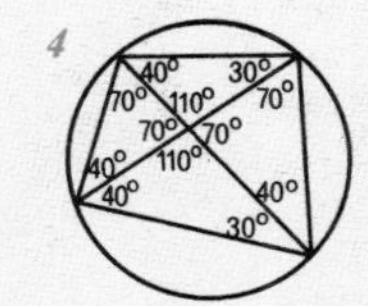

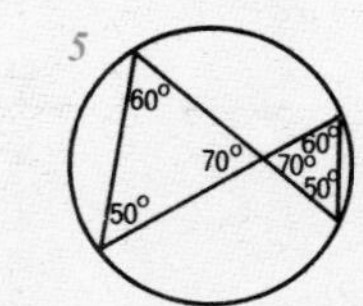

equiangular

6 $\angle XYZ = \angle YXZ = \angle ZAB = 35°$, $\angle XZY = \angle AZB = 110°$

7 CD = 9 cm, CE = 7·5 cm *8* 5 cm *9* △sKNM, HNG

Page 108 Exercise 7

1 (*i*) *a* 220° *b* 70 *c* 110 *d* 180 (*ii*) *a* 280° *b* 40 *c* 140 *d* 180

2 a $2y°, 2x°$ *b* 360° *c* $(2x+2y)°$ *d* 180

3

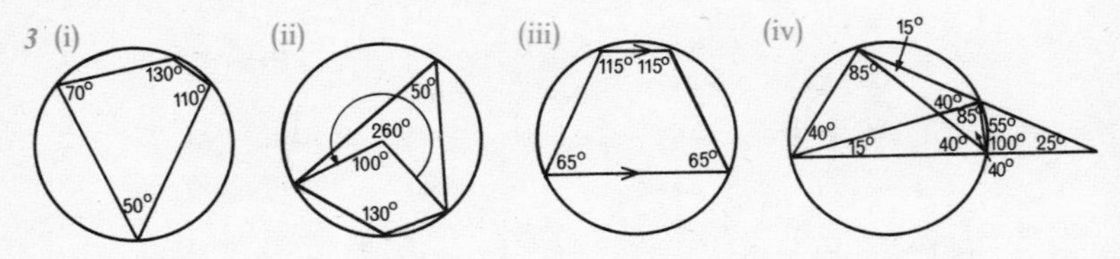

4 $\angle ABE = \angle EFC = 105°$, $\angle DEB = \angle BCF = 80°$, $\angle EBC = 75°$, $\angle BEF = 100°$ *5* 85

Page 109 Exercise 7B

1 a supplementary *b* supplementary *c* equal *2* tangent at P, $\angle RPT = \angle RSP$

3 Both are perpendicular to ST; alternate angles; under reflection in ST; angles in same segment are equal. *6* no, $\{P_1 : P_1$ lies inside the segment APB$\}$

7 $\{P_2 : \angle AP_2B < \angle APB$, P_2 on same side of AB as P$\}$

8 If $\angle APB = \angle AQB$, A, P, Q, B are concyclic; yes

10a Opposite angles are supplementary; BDHF, CEHD *b* $\angle BFC = \angle BEC$; C,D,F,A; A,E,D,B *c* at the midpoints of AH, BH, CH; BC, AC, AB

Geometry—Answers to Chapter 2

Page 114 Exercise 1

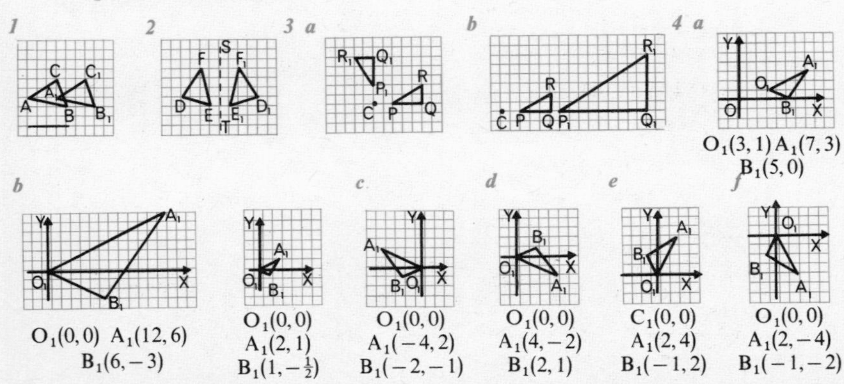

Page 116 Exercise 2

1 a $\overrightarrow{PR}$ *b* $\overrightarrow{PR}$ *c* $\overrightarrow{PT}$ *d* $\overrightarrow{PQ}$ *2 a* $\frac{2}{3}$ *b* $\frac{1}{2}$ *c* $\frac{1}{3}$ *d* $\frac{2}{3}$

3 a $\begin{pmatrix}1\\3\end{pmatrix}$ *b* $\begin{pmatrix}5\\4\end{pmatrix}$ *c* $\begin{pmatrix}\frac{5}{2}\\2\end{pmatrix}$ *4 a* (4, 0), (6, 2) *b* $\begin{pmatrix}3\\3\end{pmatrix}$, (6, 2)

5 a $(6, -1), (6, 2); (a+4, b-1), (a+4, b+2)$ *b* $\begin{pmatrix}4\\2\end{pmatrix}$; $(6, 2), (a+4, b+2)$

6 a $(8, 2), (5, -5)$ *c* commutative

Page 117 Exercise 3

1

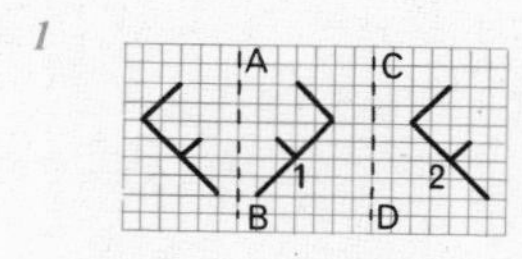

F_1 by turning over, F_2 by sliding.

translation $\begin{pmatrix}14\\0\end{pmatrix}$; 14 units; 7 units

2 in each case a translation perpendicular to AB and CD, twice distance from AB to CD

3 translation equal in magnitude but in opposite direction

4

P_1	(8, 2)	(10, 3)	(12, 0)	$(6, -1)$	$(12-a, b)$
P_2	(14, 2)	(12, 3)	(10, 0)	$(16, -1)$	$(a+10, b)$

; translation $\begin{pmatrix}10\\0\end{pmatrix}$

5 a

P_1	$(-4, 2)$	$(-2, 3)$	(0, 0)	$(-6, -1)$	$(-a, b)$
P_2	(14, 2)	(12, 3)	(10, 0)	$(16, -1)$	$(a+10, b)$

; translation $\begin{pmatrix}10\\0\end{pmatrix}$

b

P_1	(0, 2)	(2, 3)	(4, 0)	$(-2, -1)$	$(4-a, b)$
P_2	(14, 2)	(12, 3)	(10, 0)	$(16, -1)$	$(a+10, b)$

; translation $\begin{pmatrix}10\\0\end{pmatrix}$

c P_2 the same in each case *d* $x = -1, x = 4; x = 1, x = 6$, etc

6

P_1	(18, 2)	(20, 3)	(22, 0)	$(16, -1)$	$(22-a, b)$
P_2	$(-6, 2)$	$(-8, 3)$	$(-10, 0)$	$(-4, -1)$	$(a-10, b)$

; $M_2 M_1 = -M_1 M_2$

Page 120 Exercise 3B

1 b 2, 5 *c* 6, 12 *d* 3, 2; $-6, -12$ *2 a* b *b* $k-h$ *c* $2(k-h)$ *d* $[a+2(k-h), b]$

3 a (9, 4) *b* (10, 5) *c* (19, 0) *d* $(-7, -2)$ *e* $(-9, 4)$ *4 a* (2, 5) *b* $(2, -3)$

5 a $x = 270, 450$, etc.; yes *b* $x = -180, 0, 180$, etc.; yes

Page 121 Exercise 4

1 half turn about O *2* half turn about O in each case *3* half turn about O

4 a $(-2, -1)$ *b* $(-4, -3)$ *c* $(3, -2)$ *d* $(-5, 6)$ *e* $(5, -6)$ *f* $(5, 6)$

5 a $(3, -2)$ *b* $(3, -2)$ *c* $(-3, -2)$ *d* $(-3, -2)$ *e* $(-3, 2)$ *f* $(-3, 2)$

6 a H *b* H *c* X *d* X *e* I *f* I

Page 122 Exercise 4B

1 a $(a,-b)$ *b* $(-a,b)$ *c* $(-a,-b)$ *2* $(-a,-b)$ *3 a* Y *b* Y *c* X *d* I

4

$\circ$	I	H	X	Y
I	I	H	X	Y
H	H	I	Y	X
X	X	Y	I	H
Y	Y	X	H	I

5 a I *b* I *c* H *d* I

6

(x,y)	$(2,1)$	$(0,0)$	$(7,4)$	$(-1,2)$
$T(x,y)$	$(2,5)$	$(0,6)$	$(7,2)$	$(-1,4)$
$S(x,y)$	$(10,1)$	$(12,0)$	$(5,4)$	$(13,2)$
$H(x,y)$	$(10,5)$	$(12,6)$	$(5,2)$	$(13,4)$
$T\circ S(x,y)$	$(10,5)$	$(12,6)$	$(5,2)$	$(13,4)$
$H\circ T(x,y)$	$(10,1)$	$(12,0)$	$(5,4)$	$(13,2)$

$T\circ S=H;\ H\circ T=S$

Page 124 Exercise 5

1 rotation, 60° anticlockwise about O *2*

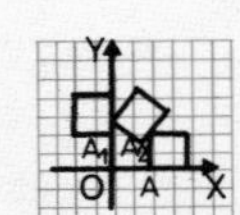

rotation 45° anticlockwise about O

3 a *b*

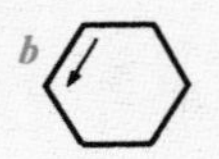

c *d*

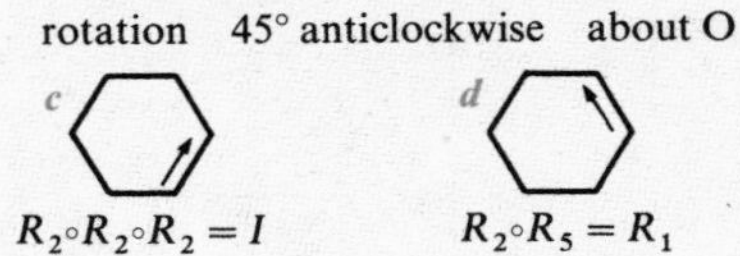

$R_2\circ R_2\circ R_2=I$ $R_2\circ R_5=R_1$

4 a OE *b* OE *c* FA *d* A *e* △OFA *f* OCDE *5 a* R_5 *b* I *c* I *d* R_2

6 a

B A O C D

b

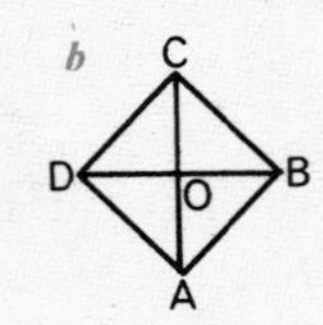

c

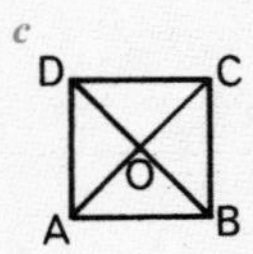

d

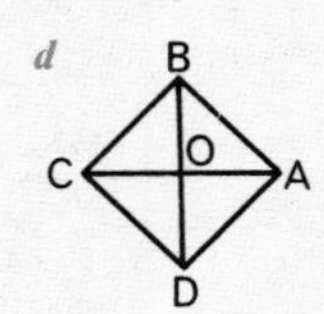

7 a $(-1,0)$ *b* $(-1,2)$ *c* $(-\sqrt{2},\sqrt{2})$

Page 126 Exercise 6

1 a

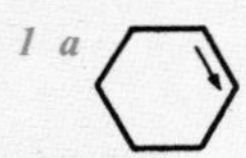

b

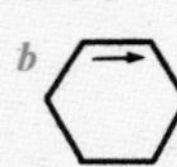

c

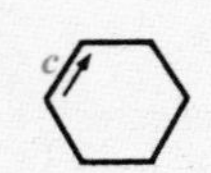

d

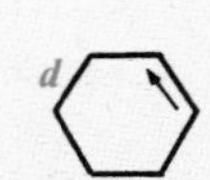

e

3 a R_2 *b* $Y\circ X, C\circ B$, etc. *4 a* R_3 *b* $C\circ X, Z\circ B$, etc.

6 (1, 3), (0, 5), (−2, 4) in both cases *7* (−3, −1), (4, −2); 90°, 270°, about O

8 (3, 1), (−4, 2); 270°, 90°, about O; yes

Page 128 Exercise 6B

1 a *b* *c*

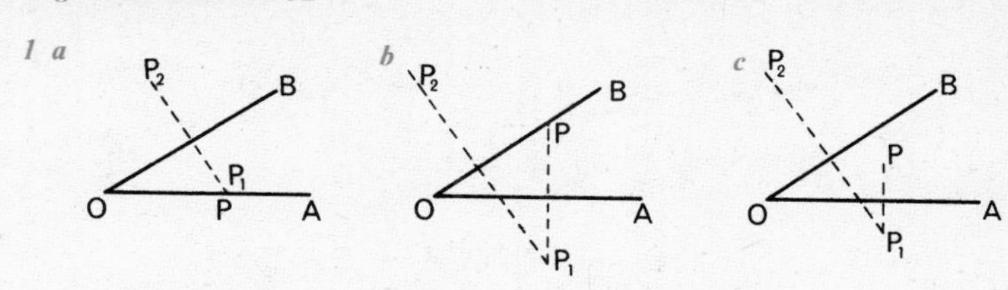

3 a rotation of 80° about A from AB to AB_1; *b* 9

4 a *b*

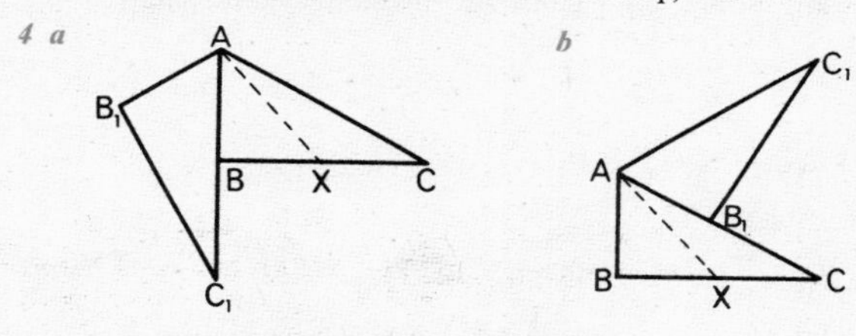

5 (2, 60°), (10, 70°), (5, 80°), (3, 95°), (4, 110°)

6 a rotation about O, equal to 2∠BOA from OB to OA
b rotation about O, equal to 2∠DOC from OD to OC
c rotation about O, equal to 2(∠BOA + ∠DOC) in sense from OB to OA
d rotation about O, equal to 2(∠AOB + ∠COD) in sense from OA to OB
e same as ***c***

Page 129 Exercise 7B

1 a $X \circ H = Y$ *b* $Y \circ Y = I$ *c* $Y \circ X = H$ *d* $Y \circ H = X$

2

$\circ$	I	H	X	Y
I	I	H	X	Y
H	H	I	Y	X
X	X	Y	I	H
Y	Y	X	H	I

3 a symmetrical about main diagonal
b no exceptions

4

$\times$	I_1	H_1	X_1	Y_1
I_1	I_1	H_1	X_1	Y_1
H_1	H_1	I_1	Y_1	X_1
X_1	X_1	Y_1	I_1	H_1
Y_1	Y_1	X_1	H_1	I_1

5 a same pattern in both tables (isomorphic) *b* yes, no

6 O(0, 0), $A_1(0, 2)$, $B_1(4, 4)$, $C_1(4, 2)$

7 O(0, 0), $A_1(0, 4)$, $B_1(-6, 2)$

Page 131 Exercise 8B

1 a M_x, M_y, H *b* H, M_x, M_y *c* M_y, I *d* R_{90}, R_{-90}, I *e* R_{90}, H
$H \circ M_x = M_y$; $M_y^2 = I$; $R_{90}\, R_{-90} = I$; $R_{90}^2 = H$; yes

2 a $\begin{pmatrix}-3\\2\end{pmatrix}$ *b* $\begin{pmatrix}5\\3\end{pmatrix}$ *c* $\begin{pmatrix}-q\\p\end{pmatrix}$; yes

3 $(\sqrt{2}, \sqrt{2})$, $(0, 2)$, $(-\sqrt{2}, \sqrt{2})$, $(-2, 0)$; $+45°$, about O

4 $\begin{pmatrix}0 & -1\\1 & 0\end{pmatrix}$; rotation of $+90°$ about O; yes

5 yes; $\begin{pmatrix}0 & -2\\2 & 0\end{pmatrix}$ *6* $(-2, 4)$, $(-2, 8)$, $(-6, 8)$, $(-6, 4)$; yes

7 square→parallelogram; no; same area *8* square→parallelogram; no; area of image is three times area of square

Geometry—Answers to Revision Exercises

Page 136 Revision Exercise 1

1 2 cm *2 b* Both are perpendicular to PQ. *c* PQ and the perpendicular diameter

3 a {P : P lies inside the circle} *b* {P : P lies on the circumference}
c {P : P lies outside the circle} *d* ϕ

4 a 15 cm *b* 14·4 cm *c* $\angle A = 28{\cdot}1°$, $\angle C = 33{\cdot}7°$, $\angle AOC = 118{\cdot}2°$

5

6 a AF = AE, BD = BF, CE = CD
b 10 cm *c* 5 cm

7 a $\angle PTQ = 54°$, $\angle POQ = 126°$ *b* $\angle PTQ = (180 - 2x)°$, $\angle POQ = 2x$
c make $\angle POQ = 140°$

8 12 cm *9 a* *b*

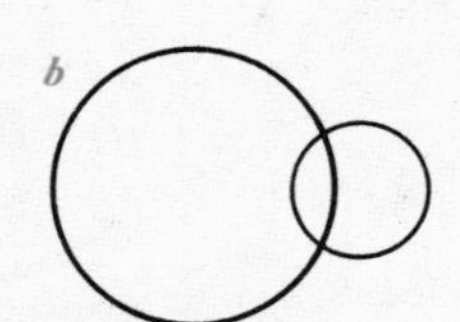

10 48 mm

11 $\angle ACB = 90°$, $\angle ABC = 45°$ *12a* 12 cm *b* 192 cm^2 *c* 192 cm^2

13 Each angle = 60°. *14* $\triangle OAB$ is equilateral. *15a* 40° $(180 - 2p)°$

16

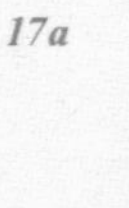

17a

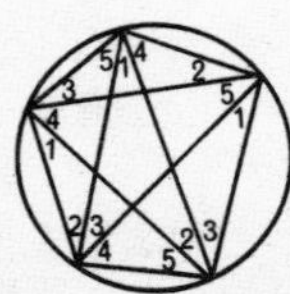

b angles in ***a*** marked 2 + 5; 2 + 3; 3 + 4; 1 + 4; 1 + 5

18 $\angle A = 40°$, $\angle B = 60°$, $\angle C = 80°$ *19b* angle in a semicircle

20 $\angle P = 75°$, $\angle R = 105°$, $\angle PQR = 100°$, $\angle PSR = 80°$

21a They are equiangular. *b* DE = 5 cm, DC = 6 cm

22a They are equiangular. *b* PS = 12 cm, TS = 3 cm

23 70°; alternate angles equal

24a corresponding angles equal

b $\angle ABE = \angle ADF = (180 - p)°$,
$\angle CAB = \angle BFD = (180 - q)°$,
$\angle ABF = \angle FDG = p°$,
$\angle BAD = q°$

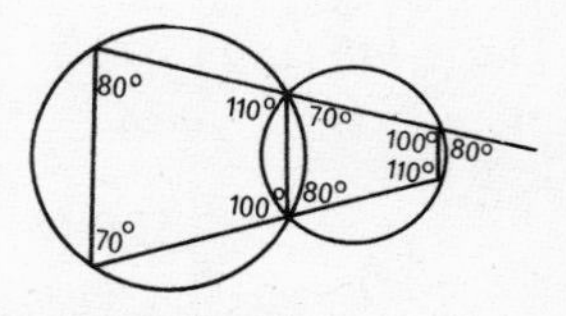

25a $\angle ECF = \angle EAB$, $\angle DCF = \angle DAB$ *b* AE bisects $\angle DAB$

26a

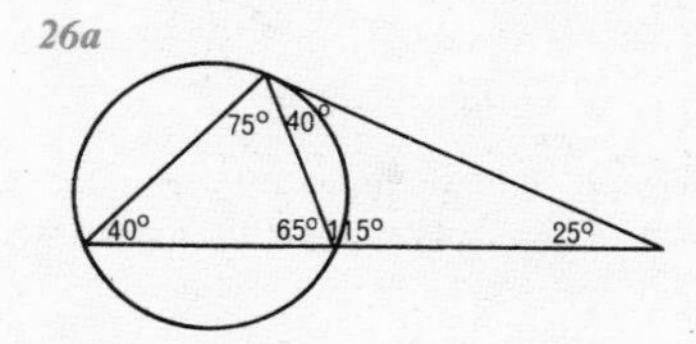

b

(180−2x−y)°
x°
y°
(x+y)°
x°
(180−x−y)°

c 6 cm

27a $\angle BFC = \angle BEC$ *b* $\angle EFC = \angle EBC$, $\angle ECF = \angle EBF$

Page 140 Revision Exercise 2

1 a $\begin{pmatrix} 5 \\ -3 \end{pmatrix}, \begin{pmatrix} -1 \\ 4 \end{pmatrix}$ *b* each $\begin{pmatrix} 4 \\ 1 \end{pmatrix}$ *c* (9, −2), (8, 2)

2 a half turn about E *b* reflection in BE *c* reflection in AE *3 a* $\overrightarrow{BG}$ *b* $\overrightarrow{DC}$ *c* $\overrightarrow{AB}$

4 a (6, 2), (12, 2) *b* translation $\begin{pmatrix} 10 \\ 0 \end{pmatrix}$ *c* $\overrightarrow{PP_2} = 2\overrightarrow{AB}$ *5 a* 10 *b* 1 *c* $k - h = 5$

6 a (4, 5), (4, −3) *b* $M_2 \circ M_1 = \begin{pmatrix} 0 \\ 4 \end{pmatrix}, M_1 \circ M_2 = \begin{pmatrix} 0 \\ -4 \end{pmatrix}$

7 a translation $\begin{pmatrix} 6 \\ 0 \end{pmatrix}$ *c* (6, 0), (0, −12) *d* $y = 2x - 12$

8 a translation $\begin{pmatrix} 0 \\ 12 \end{pmatrix}$ *c* (0, 12), (−6, 0) *d* $y = 2x + 12$

9 a (*1*) $\overrightarrow{AC}$ (*2*) half turn about L (*3*) half turn about midpoint of KL (*4*) dilatation [A, 3] *b* BK, CL, etc.; CL, LE, etc.

10a (5, −2), (−5, −2) *b* (−5, 2), (−5, −2) *c* $X \circ Y = Y \circ X$ *d* half turn about O

11a *b* *c*

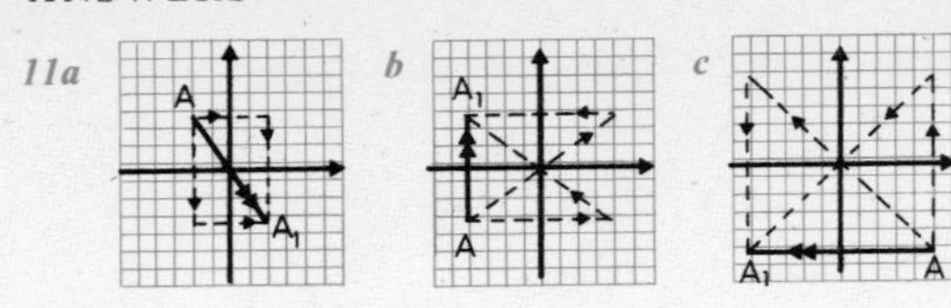

12 half turn about O; *a* $y = \frac{1}{2}x$ *b* $y = x - 2$ *c* $y = 3x + 2$

13a $\theta = 38, \theta = 53$ *b* $(2, 38°), (2, 53°)$ *c* $+27°$ about O *d* It is their sum.

14a $(2, 30°), (2, 70°)$ *b* $60°$ *c* rotation of $+60°$ about O

15

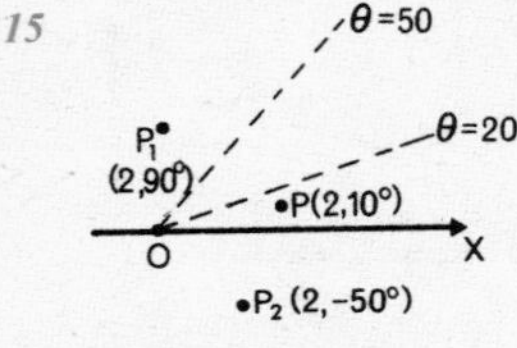

$M_2 \circ M_1 = -M_1 \circ M_2$

16a, b, c

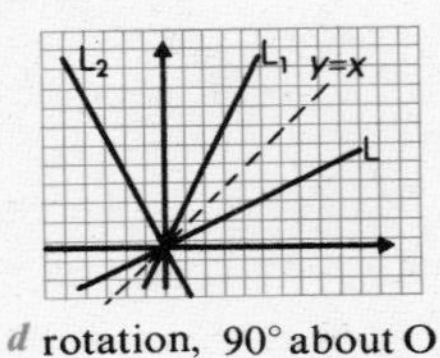

d rotation, $90°$ about O

e e.g. $(4, 2), (-2, 4)$ *f* $y = -2x$

17a translation twice distance between axes, at right angles to axes, from first axis to second *b* rotation about intersection of axes, twice angle between axes, from first axis to second

c, d

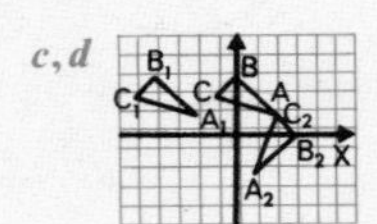

18a, b, c

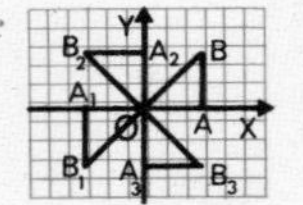

rotation, $+90°$ about O; OA_1, OB_1, etc.; no; yes; fourth

19a equiangular; $90°, 45°, 45°$ *b* $45°$ from BA to BM; $\sqrt{2}$

20a M, reflection in x-axis; N, $+90°$ about O

b $P = \begin{pmatrix} 0 & -1 \\ -1 & 0 \end{pmatrix}$; reflection in $y = -x$

21a $\begin{pmatrix} 1 & 2 \\ 1 & 1 \end{pmatrix}$ *b* $(0, 0), (1, 1), (3, 2), (2, 1)$ *c* no *d* no

22a $(-2, -4), (-5, -2)$ *b* $(-4, 2), (-2, 5)$ *c* $(-4, -2), (-2, -5)$ *d* $(4, -8), (10, -4)$

23a $(-4, 3), (3, -2)$ *b* $(-3, -4), (2, 3)$ *c* $(12, -9), (-9, 6)$

Geometry—Answers to Cumulative Revision Section

Page 146 Exercise 1

1

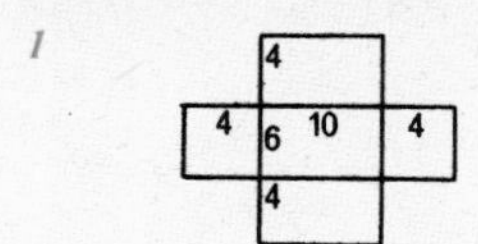

2 a 2 cm edge *b* 30

3 a 12 *b* 2 *c* 6 *d* 2

4 **a, d** *5* Equal diagonals bisect each other. *7*

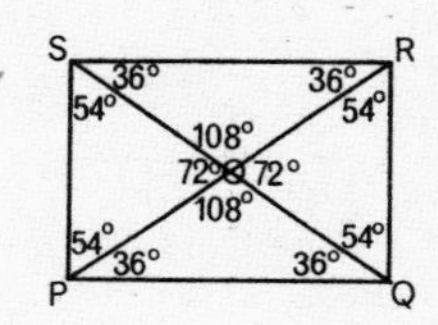

8 Adjacent side = 4 cm.

9 a (−1, 6), (2, 4) *b* (0, 5), (4, 4·5)

11 (8, 8), (8, 2), (2, 8) *12* (2, 6) and (4, −2)

13 (0, 6), (4, 7); (2, −2), (6, −1)

Page 149 Exercise 2

1 a 15 *b* 12·5 *2* 45°, 45°, 135°, 135° *3 a* T *b* F *c* T

4 a base 12 cm, height 8 cm; base 16 cm, height 6 cm *b* 2 parallelograms, 1 kite

6 10 square units *7 a* 23 km *b* 18 or 19 km *8* no; yes

Page 150 Exercise 3

1

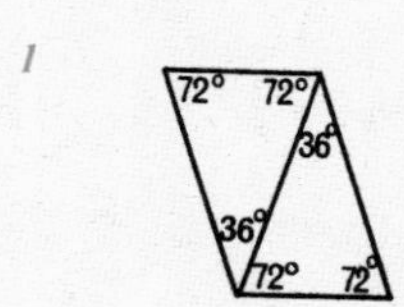

3

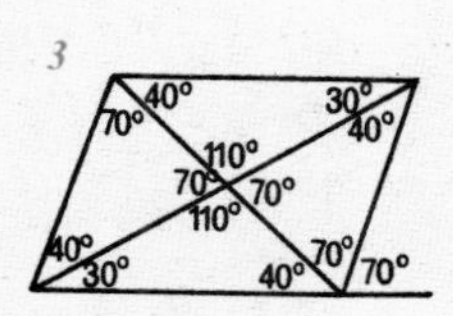

4

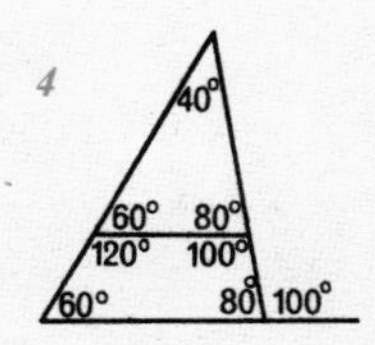

5 (0, 3), (4, 5), (0, −3); each is 2 △OPQ.

6 (4, 3), (0, 5) *7* **a, d** *8* parallelogram; 4 : 1

9

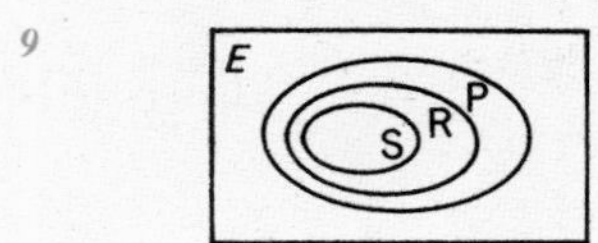

12 $A_1(7, -2), B_1(3, 4)$

Page 153 Exercise 4

1 a $x = 0$ *b* $y = 0$ *c* $y = x$ *d* $x = 2$ *e* $y = -x$ *2 b* hexagon *c* 3 *d* $x = 0, y = 0$

3 rhombuses *4* translation $\begin{pmatrix}10\\0\end{pmatrix}$; (−13, −3), (−14, −3), (−14, −5)

5 a $(-x, y)$ *b* $y=(x+1)(x+3)$

6 a (0, 0), (1, 1), (3, 3), etc. *b* $y=x$, excluding (2, 2) and (5, 5) *c* $(-1, -1)$ *d* (2, 2)

7 $\{(-4, -3), (0, 0), (1, 2)\}$

8 $C=\begin{pmatrix} 0 & 1 \\ -1 & 0 \end{pmatrix}$; reflection in x-axis, reflection in $y=x$, rotation of $-90°$ about O

Page 154 Exercise 5

1 a *b* *2*

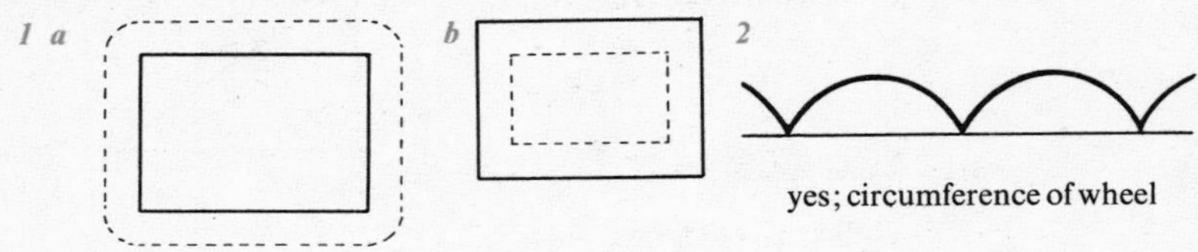

yes; circumference of wheel

3 a *b*

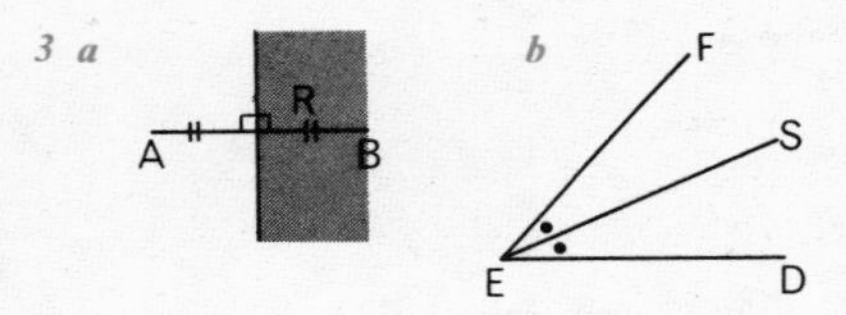

4 *5*

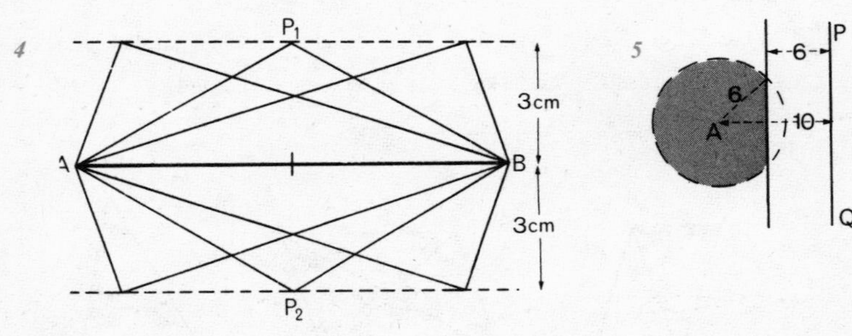

6 $3y=2x$ *7* $p=3, q=-3$ *8* $\{(x, y): y<4, y\geqslant 2, y\geqslant -2x\}$

9 a $y=\frac{1}{2}x+1$ *b* $y=-\frac{2}{3}x+3$

10a $2x-y-3=0$ *b* $x+2y-8=0$ *c* $2x+3y-3=0$

11a 2, (0. −3) *b* −3, (0, 4) *c* $-\frac{2}{3}$, (0, 2)

12a (−1, −1), (0, 0), (1, 1), etc. *b* $y=x$ *c* 70°, 70° *d* (2, 2)

13a *b* $h\geqslant 4$ *c* $k\geqslant 6, l\geqslant 8$

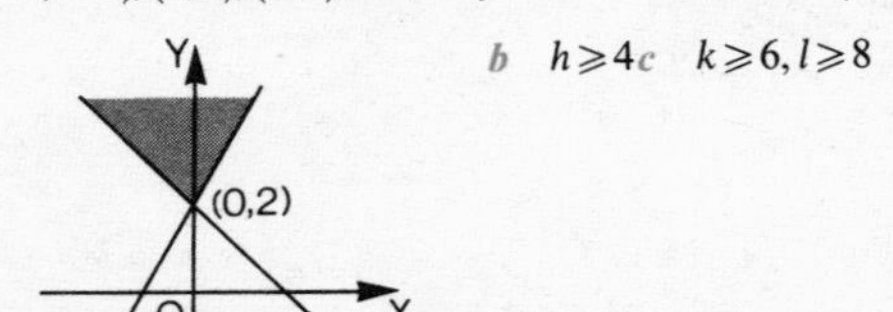

14

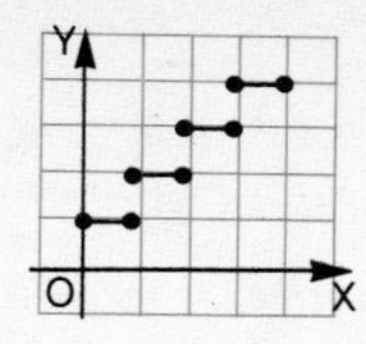

Page 156 Exercise 6

1 8 m, 4 m *2* 6·4 km; *a* 6·4 km *b* 9·1 km *3 b* 7·1 cm

4 19·5 cm, 12 cm *5* 14 square units

6 $3y = -4x$; OA = OB = 5; 10, 20 square units

7 $x - y = 3$; perpendicular bisector of AB *8 a* $x^2 + y^2 = 25$; circle, centre O, radius 5

10 $x^2 = 12y; x = 0$ *11* $AP^2 = (x-2)^2 + y^2, BP^2 = (x-8)^2 + y^2$.

12a 336 square units *b* 13·4 *13a* (3, 4) *b* (5, 0) *c* $(\frac{1}{2}, 1\frac{1}{2})$

Page 159 Exercise 7

1 a $-\boldsymbol{v}$ *b* $\boldsymbol{u}-\boldsymbol{v}$ *c* $-\boldsymbol{v}$ *d* $2\boldsymbol{v}$ *e* $\boldsymbol{v}-\frac{1}{2}\boldsymbol{u}$

2 a F *b* F *c* T *d* T *3 a* $\boldsymbol{u}$ *b* $\boldsymbol{p}$ *c* $\boldsymbol{O}$ *d* $-\boldsymbol{u}$ *e* $-\boldsymbol{t}$ *f* $\boldsymbol{u}$

4 a $\begin{pmatrix}11\\1\end{pmatrix}$ *b* $\begin{pmatrix}-5\\-2\end{pmatrix}$ *c* $\begin{pmatrix}\frac{3}{2}\\4\end{pmatrix}$ *d* $\sqrt{29}$ *5 a* $\begin{pmatrix}0\\0\end{pmatrix}$ *b* $x = 3, y = -1$

6 a 2 *b* $\begin{pmatrix}2\\-6\end{pmatrix}$ *c* 2, −4 *7 a* $\begin{pmatrix}6\\2\end{pmatrix}, \begin{pmatrix}-6\\-2\end{pmatrix}$ *b* It is a parallelogram.

Page 160 Exercise 8

1 16 cm, 3 cm *2 a* equiangular and sides proportional
b equiangular, 76°, 52°, 52° *c* (*1*) angles not equal (*2*) sides not proportional

3 3, $6\frac{2}{3}$ *4* equiangular, $\frac{AB}{DC} = \frac{BE}{CE} = \frac{EA}{ED}$; 7·5 cm, 4·5 cm

5 a T *b* F *c* F *d* F *e* T *6 a* (4, 3) *b* (11, 8) *c* (3, 2) *d* (12, 11)

7 a [A, 2] *b* [B, 2] *c* [X, 4] where X divides SV in ratio 2:1

8 a $\frac{1}{2}$ *b* $\frac{1}{2}$ *c* 0 *d* $-\frac{3}{2}; -\frac{6}{7}$ *9* $x + 2y = 16$

Page 162 Exercise 9

1 45° *2 a* 72° *b* 108° *c* 18·8 cm *d* 62·8 cm^2 *4 a* 72° about O *b* 36°

5 $\sqrt{5}$; (2, 1), 2·5 square units *6* 90° about A, 90° *7* 80 mm, 77·4 mm

9

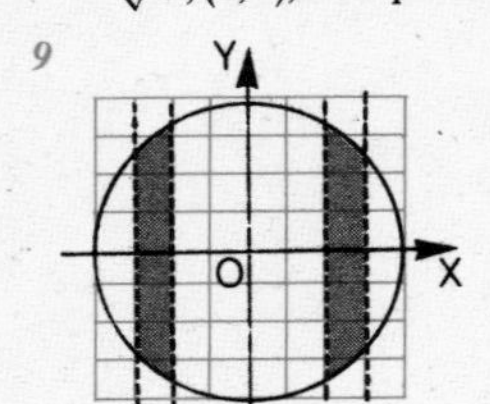

yes, O, second; yes, $x = 0, y = 0$

10a $x^2 + y^2 = 81$ *b* $4x^2 + 4y^2 = 9$ *c* $16x^2 + 16y^2 = 9$ *d* $x^2 + y^2 = 144$

Geometry—Answers to Cumulative Revision Exercises

Page 164 Cumulative Revision Exercise 1A

1 $(x+1)^2 + (y-3)^2, (x-5)^2 + (y+3)^2; x - y = 2$ *2 a* 72 cm^2 *b* $DA \leftrightarrow D_1A_1$

3 a $2\boldsymbol{v}$ *b* $2\boldsymbol{v} - \boldsymbol{u}$ *c* $\frac{2}{3}\boldsymbol{u} + \frac{1}{3}\boldsymbol{v}$ *4* $p = -2, q = 2$

5 the x-axis; rotation +45° about O, dilatation $[O, \sqrt{2}]$

6 $x = 1\frac{1}{3}, y = 6\frac{2}{3}$ *7* 7·5 mm, 4·8 mm

8 $k = \dfrac{m+n}{m}$; BC∥DE, BC = $\dfrac{m+n}{n}$ DE

Page 165 Cumulative Revision Exercise 1B

1 16, 24 units *3* 3:5

5 $3(\boldsymbol{u}+\boldsymbol{v}); 2(7\boldsymbol{u}+5\boldsymbol{v}), 10(\boldsymbol{u}+\boldsymbol{v})$; P, Q, D collinear

6 $32y = 3x^2$ *7*

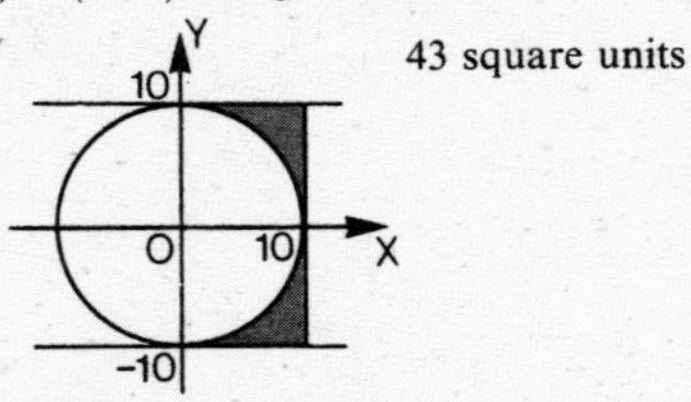

43 square units

Page 166 Cumulative Revision Exercise 2A

1

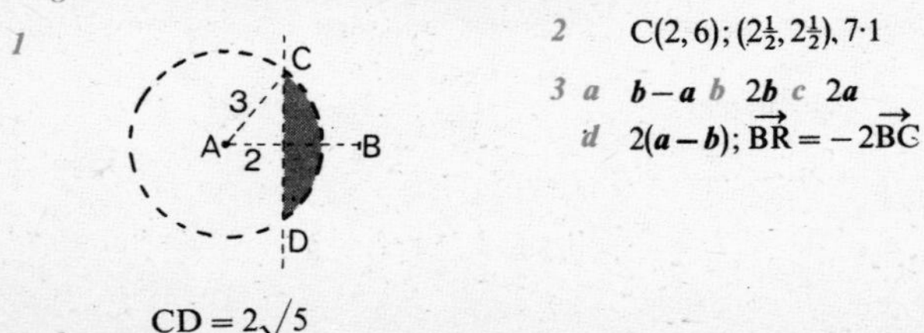

CD = $2\sqrt{5}$

2 C(2, 6); $(2\frac{1}{2}, 2\frac{1}{2})$, 7·1

3 a $\boldsymbol{b} - \boldsymbol{a}$ *b* $2\boldsymbol{b}$ *c* $2\boldsymbol{a}$
d $2(\boldsymbol{a} - \boldsymbol{b})$; $\overrightarrow{BR} = -2\overrightarrow{BC}$

4 a $\angle BOC = 120°$, $\angle OAB = \angle OBA = 35°$, $\angle OAC = \angle OCA = 25°$, $\angle OBC = \angle OCB = 30°$, $\angle XAB = 55°$, $\angle YAC = 65°$ *b* $\angle XAB = \angle ACB$, $\angle YAC = \angle ABC$

5 $A(-4, -4), c = 6$ *6* yes, 1:2; *a* 1:2 *b* 1:2 *c* 1:4

7

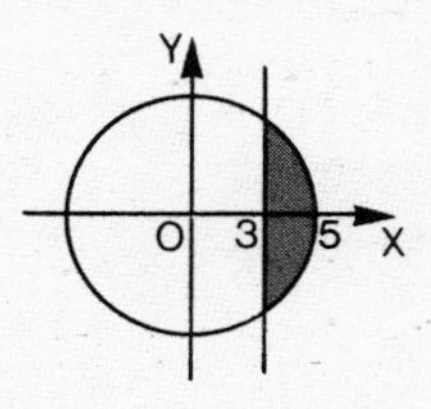

$\{(x, y): x^2 + y^2 \leqslant 25, -5 \leqslant x \leqslant -3\}$

8 $AB \parallel DC$ *a* $\boldsymbol{u} + \boldsymbol{v} - k\boldsymbol{u}$ *b* $\frac{1}{2}(\boldsymbol{u} + \boldsymbol{v} - k\boldsymbol{u})$ *c* $\boldsymbol{u} + \frac{1}{2}\boldsymbol{v}$ *d* $\frac{1}{2}(1 + k)\boldsymbol{u}$. $EF \parallel AB$ and $EF = \frac{1}{2}(AB + DC)$

Page 167 Cumulative Revision Exercise 2B

1

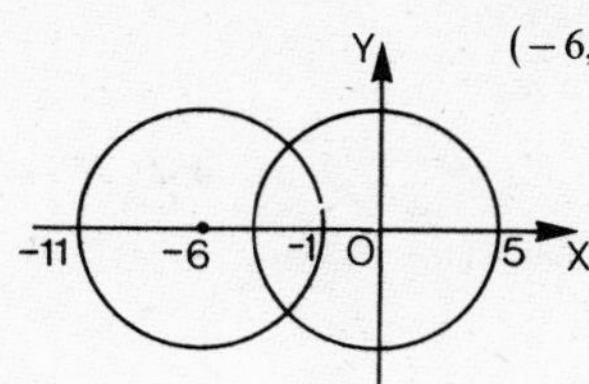

$(-6, 0)$

2 a $y = x + 8$ *b* $\{(x, y): x^2 + y^2 \leqslant 64, y \geqslant x + 8\}$ *c* $16(\pi - 2)$ square units

3 $\boldsymbol{v} + \boldsymbol{u}, \boldsymbol{v} - \boldsymbol{u}, \boldsymbol{v} - 2\boldsymbol{u}, \boldsymbol{u} - 2\boldsymbol{v}$

4 N is midpoint of AC

5 a opposite angles supplementary *b* draw circle with diameter OP

6 a equiangular *b* 2 *c* 120°

Page 168 Cumulative Revision Exercise 3A

1 a *(1)* $x = -90, 90, 270$, etc. *(2)* intersections with x-axis *b* *(1)* $x = 90$, none *(2)* none, (180, 0) *(3)* none, (0, 0) *c* $\begin{pmatrix} 360n \\ 0 \end{pmatrix}$, where n is any integer

2

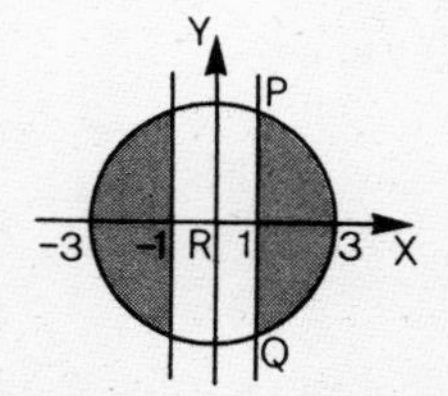

3, 1 *4* 162° *5 b* $L_1(16, 10)$ *d* (6, 8)

6 $\overrightarrow{BA}$, $\overrightarrow{CD}$; $\overrightarrow{CB}$, $\overrightarrow{DA}$; $\overrightarrow{AC}$, $\overrightarrow{DB}$; 2, $\sqrt{2}$

7 $A_1(6,6), B_1(18,9), C_1(9,15), A_2(6,-6)$,
$B_2(18,-9), C_2(9,-15)$; yes *8* 15 cm *9* 9 cm, 10 cm

Page 169 Cumulative Revision Exercise 3B

1 $\frac{3}{4}\boldsymbol{u}, \frac{1}{2}\boldsymbol{v}+\frac{1}{4}\boldsymbol{u}$ *2 a* $\begin{pmatrix}-1\\1\end{pmatrix}, \begin{pmatrix}3\\4\end{pmatrix}$ *b* $\begin{pmatrix}4\\3\end{pmatrix}$ *c* 5 *3 a* (20, 14) *b* (−8, −10)

4 a $2y = x, \frac{1}{2}$ *b* (*1*) (4, −2) (*2*) $2y = -x$ *c* translation $\begin{pmatrix}0\\4\end{pmatrix}$; $2y = x+8$

5 $\begin{pmatrix}1 & 1\\1 & -1\end{pmatrix}$ *a* line $y = x$ *b* line $y = -x$ *6* 47·2° *7* 21°

8 four pairs of angles in the same segment, etc. *9* dilatation [O, $\sqrt{2}$]

Arithmetic—Answers to Chapter 1

Page 174 Exercise 1

1 a 673 840 *b* 673 800 *c* 674 000 *d* 670 000

2 a 8·7 *b* 11·3 *c* 507·0 *d* 39·1 *e* 0·4 *f* 0·1 *g* 5·0

3 a 8·12, 8·1 *b* 16·09, 16 *c* 2·47, 2·5 *d* 0·38, 0·38 *e* 1·00, 1·0

4 a 6·1 *b* 5·01 *c* 19 000 *d* 18 900
e 0·0052 *f* 5 *g* 10·00 *h* 3·14

5 a 3 *b* 4 *c* 2 *d* 1 *e* 3

6 a 1·43 *b* 1·429 *c* 1·4 *d* 1·43

Page 175 Exercise 2

1, 4, 6, 8, 10, 11 are exact

Page 177 Exercise 3

1 1 cm, 0·5 cm, 8·5 cm, 7·5 cm

2 1 m, 0·5 m, 124·5 m, 123·5 m

3 1 km, 0·5 km, 234·5 km, 233·5 km

4 1 kg, 0·5 kg, 13·5 kg, 12·5 kg

5 0·1 cm, 0·05 cm, 7·55 cm, 7·45 cm

6 0·1 kg, 0·05 kg, 17·85 kg, 17·75 kg

7 0·1 cm, 0·05 cm, 18·25 cm, 18·15 cm

8 0·1 cm, 0·05 cm, 1·65 cm, 1·55 cm

9 0·1 litre, 0·05 litre, 3·15 litres, 3·05 litres

10 0·01 cm, 0·005 cm, 1·035 cm, 1·025 cm

11 0·1 h, 0·05 h, 51·25 h, 51·15 h

12 0·01 s, 0·005 s, 10·245 s, 10·235 s

Page 179 Exercise 4

1 0·5 m, $\frac{1}{250}$ *2* 0·5 kg, $\frac{1}{50}$

3 0·5 km, $\frac{1}{30}$ *4* 0·05 m, $\frac{1}{50}$

5 0·5 cm, 8·3% *6* 0·5 kg, 4·2%

7 0·05 litre, 1·4% *8* 0·05 m, 1·1%

9 17% *10* 1·7% *11* 0·17% *12* 2%

13 2% *14* 2% *15* 0·59% *16* 0·33%

17 4·2% *18* 0·51% *19* 0·13° *20* 0·30%

Page 180 Exercise 5

1 a 13 g, 11 g *b* 78 m, 74 m *c* 4·4 cm, 4·2 cm

d 6·4 s, 6·2 s *e* 5·3 kg, 4·3 kg *f* 2·2 cm², 1·8 cm²

g 1·45 s, 1·35 s *h* 15·25 m, 14·75 m

2 a 2 cm *b* 1 g *c* 0·2 cm *d* 0·3 kg *e* 0·14 cm² *f* 0·1 cm³

3 a (7 ± 2) mm *b* (81 ± 2) m *c* $(12{\cdot}5 \pm 1{\cdot}5)$ kg

d $(5{\cdot}6 \pm 0{\cdot}2)$ kg *e* $(4{\cdot}65 \pm 0{\cdot}05)$ kg *f* $(1{\cdot}27 \pm 0{\cdot}01)$ cm

g $(0{\cdot}90 \pm 0{\cdot}05)$ s *h* $(1{\cdot}35 \pm 0{\cdot}05)$ cm

4 ***a, b, d, e*** *5* ***c, d, e*** accepted

6 *(i)* 130 mm, 126 mm *(ii)* 127·5 mm, 122·5 mm

(iii) 421·5 mm, 418·5 mm

Page 182 Exercise 6

1 a 15 cm, 13 cm *b* 30 g, 28 g *c* 9·1 m, 8·9 m

d 16·5 cm, 16·3 cm *e* 2·33 kg, 2·31 kg *f* 4·2 g, 4·0 g

2 a 1 cm *b* 1 g *c* 0·1 mm

3 a 13·5 cm, 10·5 cm *b* 50 mm, 46 mm *c* 42 m, 38 m

4 255 m, 245 m *5* 24 cm, 20 cm *6* 40·95 h, 40·25 h

7 97·5 mm, 82·5 mm *8 a* 75 cm *b* 54 cm *c* 22 cm

Page 183 Exercise 7

1 a 5 cm, 3 cm *b* 4 g, 2 g *c* 7 s, 5 s

d 5·3 cm, 5·1 cm *e* 1·4 kg, 1·2 kg *f* 0·53 m, 0·51 m

2 a 1 km *b* 1 cm *c* 0·1 g *3* 9·0 cm, 7·0 cm

4 39 kg, 37 kg *5* 40·0 cm, 38·0 cm

6 8·5 cm, 3·5 cm *7* 756 ml, 744 ml

Page 184 Exercise 8

1 24·75 m^2, 15·75 m^2 *2* 23·75 cm^2, 12·75 cm^2

3 38·44 mm^2, 33·64 mm^2 *4* 4·375 cm^2, 1·875 cm^2

5 7·875 km^2, 4·375 km^2 *6* 20·4 m, 19·6 m; 26·01 m^2, 24·01 m^2

7 1·1025 m^2, 0·9025 m^2; 0·05, 0·1 approx.

Arithmetic—Answers to Chapter 2

Page 192 Exercise 1

1 1, 10, 11, 100, 101, 110, 111, 1000, 1001

2 a 5 *b* 13 *c* 38 *d* 454

3 a 10111 *b* 100101 *c* 110000 *d* 1000001 *e* 1111111

4 1011 *5* 10100 *6* 110000 *7* 1 *8* 110 *9* 101

10 1111 *11* 1001110 *12* 10111101 *13* 11, rem. 10 *14* 101, rem. 10

15 11, rem. 110 *17* 1000 *18* 100 *19* 1111000 *20* 10

21a 10100 cm *b* 1101100 cm *c* 111100 cm

22a 11110 cm^2 *b* 1011011 cm^2

23a 1001011 *b* 100111 *24a* F *b* T *c* F *d* T *e* T

25a 1000110 *b* 10110 *c* 101, rem. 1100

26a 5 *b* 11_{two} *27a* ends in 0 *b* ends in 00

Page 196 Exercise 2

1 b 3 *c*

+	0	1	2
0	0	1	2
1	1	2	10
2	2	10	11

×	0	1	2
0	0	0	0
1	0	1	2
2	0	2	11

2 *a* $(2\times3^2)+(1\times3)+0=21$

b $(2\times3^3)+(1\times3^2)+(2\times3)+0=69$

c $(2\times3^3)+(2\times3^2)+(0\times3)+2=74$

3 *a* (*1*) 71 (*2*) 126 (*3*) 230

b (*1*) 2012 (*2*) 2020 (*3*) 100 000 *c* 10, 100, 1000

4 *a* 1021 *b* 2200 *c* 1011 *d* 21 *e* 20111 *f* 1, rem. 102

5 *a* 0, 1, 2, 3, 4

b (*1*) $(2\times5)+3=13$ (*2*) $(4\times5^2)+(1\times5)+2=107$

(*3*) $(2\times5^3)+(3\times5^2)+(1\times5)+0=330$

6 *a* (*1*) 69 (*2*) 40 (*3*) 225 (*4*) 369

b (*1*) 443 (*2*) 2040 (*3*) 101200 (*4*) 113000 *c* yes; no

7 *a* 1130 *b* 11040 *c* 41 *d* 102 *e* 21021 *f* 343

8 0, 1, 2, 3, 4, 5, 6, 7 9 *a* 8 *b* 35 *c* 86 *d* 448 *e* 537

10 *a* 12 *b* 33 *c* 301 *d* 652 *e* 1750 *f* 10000

11 *a* divisible by 8 *b* A multiple of 1000 is divisible by 8.

12 *a* 150 *b* 405 *c* 43 *d* 1

e 234 *f* 2176 *g* 42, rem. 2 *h* 111, rem. 1

13 1100100; 10201; 400; 144; 121; 91; 84; 50; 10

14 *a* 0, 1 *b* no. 15 *a* 4 *b* 7 *c* 5 *d* 4 *e* 5 *f* 7

16 35; 45; 104; 1002

Page 200 Exercise 3

1 *a* 63 *b* 108 *c* 106 *d* 1714 *e* 5015

2 *a* 23 *b* 84 *c* 130 *d* 6e4 *e* 20*te*

3 divisible by 12; divisible by 144

4 *a* 1219 *b* 1*t*653 *c* 3*t* *d* 2246 *e* 64*tt* *f* 54*e*2*e*

Arithmetic—Answers to Revision Exercises

Page 203 Revision Exercise 1

1 *a* 1 cm, 0·5 cm, 5·5 cm, 4·5 cm

b 0·1 m, 0·05 m, 4·85 m, 4·75 m

c 0·1 g, 0·05 g, 5·35 g, 5·25 g

d 0·1 cm, 0·05 cm, 28·25 cm, 28·15 cm

e 0·01 m², 0·005 m², 8·725 m², 8·715 m²

2 148·5 cm, 147·5 cm

3 a $\frac{1}{22}$, 4·5% *b* $\frac{1}{16}$, 6·2% *c* $\frac{1}{124}$, 0·8% *d* $\frac{1}{46}$, 2·2%

4 a 0·15 cm *b* 0·005 m *c* 0·0005 m² *d* 1 ml

5 a 1059 cm³, 1053 cm³ *b* 1·594 cm, 1·586 cm

6 a 10·6 g, 10·4 g; 4·8 g, 4·6 g *b* 7·9 m, 7·7 m; 2·7 m, 2·5 m

c 1568 km, 1566 km; 986 km, 984 km *d* 3·6 mm, 3·4 mm; 1·8 mm, 1·6 mm

7 42 cm to 46 cm *8* 4·1 h, 3·9 h *9* 10·2 cm, 7·8 cm

10 2 h, 1·53 h *11* 39, 44

12a 26·25 cm², 16·25 cm² *b* 1·3225 cm², 1·1025 cm²

Page 205 Revision Exercise 2

1 a 111 *b* 10110 *c* 100000 *d* 1000001 *e* 1010011

2 a 11 *b* 46 *c* 174 *d* 419 *3 a* 11120 *b* 173 *c* *t*3

4 a 10100 1000001 *c* 10, rem. 11

d 100 1111 *f* 111

5

twelve	*ten*	*eight*	*five*	*three*	*two*
23	27	33	102	1000	11011
2*e*	35	43	120	1022	100011
25	29	35	104	1002	11101

6 100000_{two} cm; 100111_{two} cm² *7* 2020_{three} mm; 22100_{three} mm²

8 a 13 *b* 70 *c* 255 *9 a* 1110 *b* 112 *c* 12012 *d* 10011 *e* 1002 *f* 101121

10a 0, 1, 2, 3, 4, 10, 11, 12, 13, 14, 20, 21, 22 *b* 163, 3001

11a 411 *b* 41 *c* 1341 *d* 13 *e* 4103 *f* 111, rem. 2

12a 1050 *b* 13 *c* 1227 *d* 6122 *e* 666 *f* 504

13a zero placed after original digits *b* divisible by eight

c if it ends in four or zero

14a *e*44 *b* 162 *c* 959*t* *15* six; 35

16 eight; 525 *17* $x = 7, y = 5, z = 4$ *18* $x = 8, y = 7$ *21* no

Arithmetic—Answers to Cumulative Revision Section

Page 209 Exercise 1

1	191	*2*	26·0	*3*	0·0357	*4*	32·1
5	0·311	*6*	0·153	*7*	33·1	*8*	0·0610
9	16·0	*10*	0·0771	*11*	0·492	*12*	0·006 40
13	0·617	*14*	0·282	*15*	1020	*16*	0·542
17	0·838	*18*	0·377	*19*	57·9	*20*	0·706

Page 211 Exercise 2

1 243 *2 a* $13\frac{1}{2}$ *b* $2\frac{2}{3}$ *c* 80

3 a £18·45 *b* £0·48 *c* £11·05½

4 a $11\frac{3}{8}$ *b* $\frac{7}{12}$ *c* $2\frac{5}{12}$

d $1\frac{1}{2}$ *e* 6 *f* $1\frac{2}{5}$

5 a 180; 1·14; 876 500; 0·01 *b* 1·47; 0·917; 15·4; 2·4

6 a 0·75, 1·375, 4·05, 1·4 *b* $\frac{4}{5}$, $\frac{2}{25}$, $3\frac{1}{4}$

7 a 478·3 *b* 480 *c* 478·26 *d* 478·265

8 a 1000 *b* 20 *c* 2000 *d* 30 *e* 20

9 a 1·537 *b* 17·5 *c* 0·0425 *d* 0·004 *e* 0·06

10 2900

11a $1{\cdot}23 \times 10^3$ *b* $2{\cdot}8 \times 10^5$ *c* $5{\cdot}86 \times 10^6$

d $5{\cdot}6 \times 10^{-3}$ *e* 8×10^{-5} *f* $8{\cdot}3 \times 10^{-8}$

Page 212 Exercise 3

1 a 0·25 m, 3 m, 1·48 m, 0·345 m

b 2500 m, 1760 m, 540 m, 7·5 m

c 2·5 kg, 0·45 kg, 0·031 kg, 0·007 kg

d 5 litres, 4·8 litres, 0·125 litre

e 4000 cm^3, 3700 cm^3, 875 cm^3, 7 cm^3

f 300 000 m^2, 27 000 m^2, 5000 m^2

2 58, 46, 2668 *3 a* 62·8 cm *b* 129·4 cm^2 *c* 77 cm^3

4 16·8 ha *5* 270 *6* (i) 234 mm^2 (ii) 1100 cm^2 (iii) 220 m^2

7 48 *8* 225 m^2; 182·5 cm^2 *9* 5·75 m

10 6·93 cm; 27·7 cm^2 *11* 0·6 m^3

Page 215 Exercise 4

1 *a* 18·49, 15 129, 0·2025, 9 000 000

b 14·36, 80·28, 671, 608 400, 0·317, 0·009 604

2 226 m^2

3 *a* 120, 1·5, 0·1, 1·4

b 5 and 6; 13 and 14; 7 and 8

c 0·8, 0·3, 9

4 2·65, 4·12, 6·08, 6·86, 8·19, 9·85

5 *a* 8·60 cm, 1·89 cm *b* 13·9 cm, 41·7 cm

6 *a* 4·84 *b* 15·3 *c* 48·4 *d* 153

e 0·753 *f* 0·186 *g* 0·0846 *h* 0·001 89

7 *a* 3·74 *b* 24·3, 0·283 *8* 1·8628 *9* *a* 9 *b* 5·08

Page 216 Exercise 5

1 *a* $\frac{1}{100000}$ *b* $\frac{1}{10}$ *c* $\frac{35}{16}$

2 *a* *i* *b* *n* *c* *d* *d* *i*

3 3 hours *4* 9·90 francs *5* 156

6 8·40 dollars, £1·46 *7* 60·4 pesetas, 876 pesetas

8 84 mm, 42 mm, 48 mm *9* 18·5 cm, 25·5 km

10a 400 m, 170 m *b* 1 ha

11a 2:3 *b* 4:9 *c* 8:27

Page 218 Exercise 6

1 *a* 1 h 30 min *b* 8 h 5 min *c* 18 h 20 min

d 6 h 15 min *e* 10 h 35 min

2 *a* 60 km/h *b* 78 km *c* 24 min *d* 3·54 mm/s

3 93 km/h *4* 09 40 hours *5* 2·7 km/s

6 17 00 hours *7* 616 km

8 *a* 11 36; 100 km *b* 1 h 48 min

c $62\frac{1}{2}$ km/h; $33\frac{1}{3}$ km/h; at 14 18 hours, 130 km from Vienna

Page 220 Exercise 7

1 a $\frac{3}{10}$ *b* $1\frac{1}{8}$ *c* $\frac{1}{200}$

2 a 87·5% *b* 133·3% *c* 4%

3 £2·17 *4* £11·25 *5* £41·25

6 a £1·40 *b* 82p *7* $12\frac{1}{2}$%

8 a $66\frac{2}{3}$% *b* 40% *9 a* £3·22 *b* £2·38

10 £18·05 *11* £118·30; £2957·50

12a £3·25 *b* £2·91

13a £315·25 *b* £57·36

Page 221 Exercise 8

1 £121·80 *2* £0·79 *3* £19·80 *4* £30·60

5 £271·50 *6* £212·50 *7* 550 *8* £180

9 £10·80, £7·56 *10* ***b***

Page 222 Exercise 9

1 a 54 cm; 150 cm^2 *b* 41 mm; 102 mm^2 *c* 8·8 km; 3·2 km^2

2 1·26 cm *3* 617 cm^3; 274 cm^2 *4* 100 cm

5 2560 cm^2 *6* 314 cm^3, 204 cm^2 *7* 1360 cm^2; 4710 cm^3

Page 223 Exercise 10

1 a (*1*) 14, 17, 20 (*2*) 25, 36, 49 (*3*) $\frac{1}{2}, \frac{1}{4}, \frac{1}{8}$

(*4*) 13, 21, 34 (*5*) 24, 35, 48 (*6*) 17, 19, 23

b (*1*) add 3 (*2*) square of next natural number

(*3*) divide by 2 (*4*) add previous 2 terms

(*5*) add next odd number difference (*6*) next prime number

2 1, 4, 9, 16, 25; 1, 3, 6, 10, 15; 19th and 20th

3 a 1, 4, 10, 20 *b* 1, 5, 14, 30

4 a 3, 5, 7 *b* 0, 3, 8 *c* 1, 3, 6 *d* $\frac{1}{2}, \frac{1}{3}, \frac{1}{4}$ *e* 1, 5, 14

c gives 'triangular' numbers; ***e*** gives 'square pyramid' numbers

5 a n *b* $n-1$ *c* $2n-1$ *d* $2(n-1)$ *e* $\dfrac{1}{n+2}$ *f* n^2+1

6 a 7, 8 *b* 27, 19 *7* (*4*), 0·618

Page 224 Exercise 11

1 a 0·3, 0·2, 0·18, 0·17, 0·17 *b* yes *c* 0·17

2 a $\frac{2}{9}$ *b* $\frac{5}{22}$ *3 a* $\frac{1}{11}$ *b* $\frac{2}{11}$ *c* $\frac{4}{11}$ *d* $\frac{9}{11}$ *e* $\frac{7}{11}$

4 (1, 1) (1, 2) (1, 3) (1, 4) (1, 5) (1, 6)
(2, 1) (2, 2) (2, 3) (2, 4) (2, 5) (2, 6)
(3, 1) (3, 2) (3, 3) (3, 4) (3, 5) (3, 6)
(4, 1) (4, 2) (4, 3) (4, 4) (4, 5) (4, 6)
(5, 1) (5, 2) (5, 3) (5, 4) (5, 5) (5, 6)
(6, 1) (6, 2) (6, 3) (6, 4) (6, 5) (6, 6)

a $\frac{1}{36}$ *b* $\frac{11}{36}$ *c* $\frac{5}{36}$ *d* $\frac{31}{36}$

e 0 *f* 1 *g* $\frac{1}{3}$ 7, 24

5 a 1 *b* 0 *c* $1-a$

6 a $\frac{1}{8}$ *b* $\frac{1}{8}$ *c* $\frac{3}{8}$

7 a $\frac{1}{6}$ *b* $\frac{1}{6}$

8 a $\frac{1}{4}$ *b* $\frac{1}{2}$

Page 226 Exercise 12

1 a 2; 2 *b* $\frac{2}{15}$ *c* 49 *2 a* 12·9 *b* 83% *3* 2·2

4 b 6 to 10; 20·8 *c* 0·08

5 a 52·0, 63·9, 56·6 *b* 15, 10, 22 *c* *B*, *C* *d* 42, 58, 45

Arithmetic—Answers to Cumulative Revision Exercises

Page 228 Cumulative Revision Exercise 1A

1 0·01 *2* $3{\cdot}9 \times 10^4$; $3{\cdot}6 \times 10^{-4}$ *3* 44 m *4* £19·68

5 62; 111110 *6* 3·2 cm *7* 1:50 000 *8* 942; £1554

9 £3 990 000; £147 000; £1·04

Page 229 Cumulative Revision Exercise 1B

1 2×10^{-5}; 4×10^2 *2* 24 500 cm^2

3 19·4 *4* 22 cm and 17 cm; 374 cm^2

5 £167·20 *6* See Exercise 11, question *4*. *a* $\frac{5}{36}$ *b* $\frac{5}{12}$

7 110111; 55 *8* 12%

9 a £48·55 *b* £35

Page 230 Cumulative Revision Exercise 2A

1 0·625 *2* $5\frac{1}{4}$%; £147 *3* 5·6214 *4* $\frac{2}{3}$

5 36% *6* 48·8% *7* 19 m/s *8 a* £42 *b* £1250

9 a Fibonacci *b* 10101, 100010

c 1, 2, 3, 5, 8, 13; 1, 2, 10, 12, 22, 111

Page 231 Cumulative Revision Exercise 2B

1 30%, 23·1%, 52·5, 34·4 *2 b* 6·4 *c* 6·6, 4·2, 8·4; 2·1

3 a 59·6 cm^3 *b* 176 g *4 a* 2200; 2 *b* 9, 8, 9, 8, 7

5 1·2 km *6.* 10 cm

Page 232 Cumulative Revision Exercise 3A

1 a 0·251 *b* 0·530 *2* 60 km/h, 102 km/h, 95 km/h

3 2 km, 6 km, 4 km; 16 km^2

4 a (*1*) 2, 5, 8, 11; 29 (*2*) 2, 5, 10, 17; 101 (*3*) 3, 5, 9, 17; 1025

b (*1*) 11, 13, 15; add 2 (*2*) 48, 96, 192; multiply by 2

(*3*) 35, 48, 63; add next odd number difference

5 a 40 200 km *b* 14 h 53 min *c* 33 000 km/h

6 £0·88, £6000; £171·60 *7* 7·82 cm, 3·58 cm

Page 233 Cumulative Revision Exercise 3B

1 a 1050 *b* 73·2 *c* $2{\cdot}87 \times 10^{-3}$, or 0·002 87 *d* 0·131

2 £173·30 *3* 16·1 cm *4* after 40 seconds

5 0·220 m^3; 246 kg *6* £3369·50 *7* 12 45 hours; 225 km

Trigonometry—Answers to Chapter 1

Page 240 Exercise 2

1 P(3, 3), Q(−1, 1), R(−2, −1), S(2, −3)

2 a T(4, 0, 2), S(0, 0, 2), V(0, 3, 2) *b* (*1*)(4, 1·5, 1) (*2*)(2, 1·5, 1)

3 a E(5, 5, 0), B(10, 0, 0), C(10, 10, 0), F(10, 5, 0) *b* 13

4 B(40, 0, 0), C(40, 26, 0), N(28, 13, 0), Q(28, 13, 10)

5 B(1, 1, 0), V(1, 1, 3)

Page 242 Exercise 3

1 a three at each corner *b* 6 cm, 7 cm, 8 cm *c* 4 cm, 6 cm, 3·5 cm

2 a 40 mm, 32 mm, 50 mm *b* 20 mm, 16 mm, 25 mm

3 a T *b* F *c* T *d* T *e* T *4 a* 10 m *b* 12 m

5 a 24 mm *b* 23·2 mm

Page 244 Exercise 4

1 a (*1*)CBR (*2*)QRB *b* (*1*)DAS (*2*)PSA

c (*1*)DBS (*2*)QSB *d* (*1*)RBS (*2*)ASB

2 a CBW *b* DBX *c* EBY *d* CEW *e* YWE *f* VEW

3 a 45° *b* 35·3° *c* 35·3° *4 a* 25 km *b* 10·6 km *c* 27·9°

5 a 5 cm *b* 6 cm *c* 67·4° *d* 45·2° *6* 71·6° *7 a* 63·4° *b* 6·71 cm

8 a 50·2° *b* 7·81 *9 a* 50 m *b* 23·6° *c* 43·8°

10a 45° *b* 12·2 cm *11a* 7 cm *b* 59°

Page 248 Exercise 5

1 a (*1*) GFY or HEZ (*2*) XYF or WZE (*3*) XFY or WEZ (*4*) GYF or HZE
b (*1*) YZX or GHF (*2*) WXZ or EFH (*3*) WZX or EHF (*4*) YXZ or GFH

2 a ∠BAB′ or ∠CDC′ *b* 23° or 23·2°

3 a 26·5° or 26·6° *b* 36·9° *c* 73·8° *4* 14° or 14·1°

5 33·7°, 112·6° *6 a* 56·3° *b* 73·8° *7 a* 60° *b* 90° *c* 49°

8 a 10 m, 17·3 m *b* lines perpendicular to HK *9* 70·5°

Page 250 Exercise 5B

1 b (*1*) 70·4 or 70·5° or 70·6° (*2*) 54·7° or 54·8°

2 a VM is drawn to the midpoint of the base of isosceles △VAB; SM is part of the diameter through the midpoint of chord AB.

b 65·7 or 65·8° *3 a* 25 cm *b* 53·1

4 29 cm; $\cos a° = \frac{21}{29}$, $\cos b° = \frac{16}{29}$, $\cos c° = \frac{12}{29}$, etc.

5 a $\sqrt{75}$ or 8·66 cm *b* 54·7° or 54·8° *c* 90° *d* 109·4° or 109·5°

6 a 30° *b* 60° *c* 60° or 120°

Page 254 Exercise 6

1 (i)–(iv)

2–3 *4*

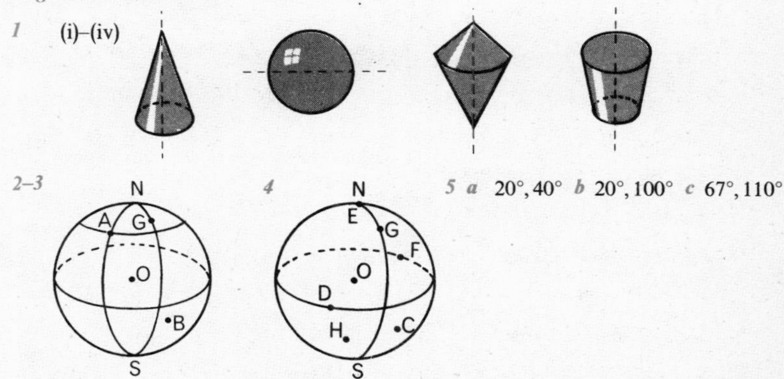

5 a 20°, 40° *b* 20°, 100° *c* 67°, 110°

Page 255 Exercise 7B

1 3190 km, 5220 km *2* 1110 km, 6970 km

3 60°N or S *4 a* 14 200 km *b* 90°; $\frac{1}{4}$; 10 000 km; 4200 km

5 6080 km

6 a 3190 km *b* 3190 km *c* 29 *d* 3230 km

7 6370 km

Trigonometry—Answers to Revision Exercise

Page 259 Revision Exercise 1

1 a O(0, 0, 0), A(8, 0, 0), B(8, 6, 0), C(0, 6, 0), D(0, 0, 4), E(8, 0, 4), F(8, 6, 4), G(0, 6, 4); (4, 3, 2) *b* 10 units, 10·8 units

2 a PAD, PAC, PAB, DAB *b* (*1*) PA (*2*) PQ *c* (*1*) PCA (*2*) BPC

3 a F *b* F *c* T *d* T *e* T *f* T *4 b* A, P; S *c* (*1*) 5 cm, 5·39 cm (*2*) 2 cm

5 a 8·08 m *b* 15° *c* 12·6° *6 a* 14·5° *b* 14·4°

7 a 15 cm *b* 10 cm *c* (*1*) 53·1° (*2*) 59°

8 a 10 cm, 14·1 cm *b* (*1*) 45° (*2*) 45° *9 a* 90° *b* 90° *c* 35·3° or 35·4°

10a 7·5 cm, 6·5 cm, 7·5 cm, 8·3 cm *b* 72·1° *c* 23·2 cm^2

11a AOU, COA, DQU *b* 0°, 106° *c* 3°, 106°, 109°, 90°, 31° *12* 13 h 42 min

Trigonometry—Answers to Cumulative Revision Section

Pages 263 Exercise 1

1 0·8, 0·6, 0·75; 36·9° *2* 0·772, 1·213 *3 a* 1 *b* $3\frac{1}{3}$ *c* $\sqrt{3}$

4 (i) 4·23 cm (ii) 52·5 cm *5 a* 3·29 cm *b* 21·4 cm^2

6 a ∠OCA, ∠OQP, ∠MSN *b* 33·6, 16, 12·9

7 a −0·342 *b* 0·500 *c* 0·839 *d* 0 *e* 1 *f* −2·605 *g* 0 *h* 0·707

8 a 70·4, 109·6 *b* 42·9, 317·1 *c* 73·7, 253·7 *d* 115, 245

e 30, 150 *f* 48·2, 311·8

Page 264 Exercise 2

1 29° *2 b* 6·5 cm^2 *3 a* 75°, 10·6 cm *b* 66·5 cm^2

4 51·7°, 128·3° *5 a* 1·81 cm *b* 1·51 cm *6* 10·6 cm

7 a 4·77 m *b* 8·34 m^2 *8 a* $\sqrt{3}/2$ *b* 30°, 4

9 79·8°, 52·6°, 47·6°, 135 cm^2 *10a* 38·2 cm^2 *b* 10·7 cm *c* 58·3, 48·8

11 B: 25·9 km, 96·6 km. C: 217·1 km, 38·2 km *12* 6·4 km

13a 13·8 cm *b* 23·1° *14a* 13 cm, 17 cm, 5 cm *b* 137° *c* 22·2 cm^2

15 121 km, 076° *16* 16 200 m, 350·5°

Trigonometry—Answers to Cumulative Revision Exercises

Page 266 Cumulative Revision Exercise 1A

1 (3, 4), (−3·4, 9·4) *2 a* 85·7, 274·3 *b* 95·7, 275·7 *c* 14·5, 165·5 *d* 0, 360 *e* 41·8, 138·2 *f* 20·6, 200·6

3 17·8 cm, 9·1 cm *4 a* T *b* T *c* T *d* T T F F T

6 a 41·0 km, 18·9 km *b* N 65·3° E *7 a* 5, 13, 12·4 cm *b* 36·9°, 13·3°

c 14·1° *8* 125·1°, 73·6 cm^2

Page 267 Cumulative Revision Exercise 1B

1 4 *2 a* {210, 330, 570, 690} *b* {30, 210, 390, 570} *c* {90, 270, 450°, 630°}

3 a 4·7 cm *b* 38 cm^2 *5 a* $v \cos b°$ *b* $v \cos (b-a)°$ *c* $v \sin (b-a)°$

6 a (5, 0, 3), (0, 0, 3), (0, 4, 3) *b* 36·9°, 25·1° *c* (2·5, 2, 1·5)

d 90° *e* 53·1° *8 a* 1·89 m *b* 67·1°

Page 268 Cumulative Revision Exercise 2B

1 (5, 53·1°), (17, 118·1°), (13, 247·4°) *2 a* 5·96 cm *b* 13·3 cm^2

4 a 50·2° 63·4° *7* 21·2, 10·5 km/h *9* 72·3°, 107·7°; 13 cm *10* 225 km

Computer Studies—Answers to Chapter 1

Page 272 Exercise 1

1

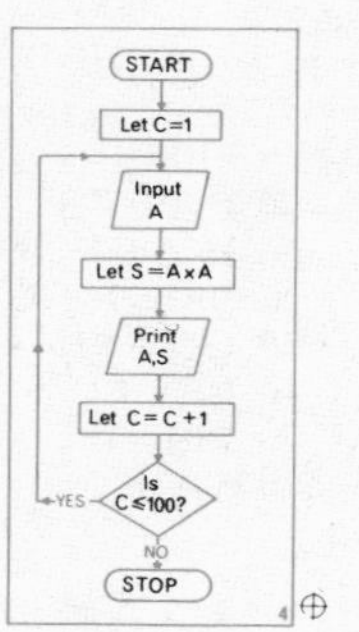


```
10 REM MEAN OF 3 NOS
20 INPUT A, B, C
30 LET M = (A+B+C)/3
40 PRINT M
50 STOP
```

(Flow charts for 2, 3, 4 are similar)

2

```
10 REM AREA OF CIRCLE
20 INPUT R
30 LET A = 3·14*R*R
40 PRINT A
50 STOP
```

3

```
10 REM VOL OF SPHERE
20 INPUT R
30 LET V = 4/3*3·14*R↑3
40 PRINT V
50 STOP
```

4

```
10 REM SKELETON CUBOID
20 INPUT L, B, H
30 LET T=(L+B+H)*4
40 PRINT T
50 STOP
```

5

```
10 REM INSURANCE
   PREMIUM
20 INPUT V
30 LET I = V/100*60/100
40 PRINT I
50 STOP
```

6

```
10 REM DISCOUNT
20 INPUT S
30 LET P=S-(15/100*S)
40 PRINT P
50 STOP
```

7

```
10 REM RATES
20 INPUT V
30 LET R = V*111/100
40 PRINT R
50 STOP
```

Page 274 Exercise 2

1 a

```
10 REM DIFFERENCE
20 INPUT X, Y
30 IF X>Y THEN 70
40 LET Z = Y-X
50 PRINT Z
60 GOTO 90
70 LET Z = X-Y
80 PRINT Z
90 STOP
```

b

```
 10 REM PASSMARKS
 20 INPUT P
 30 LET C = 0
 40 INPUT M
 50 IF M = -1 THEN 100
 60 IF M GE P THEN 80
 70 GOTO 40
 80 LET C = C+1
 90 GOTO 40
100 PRINT C
110 STOP
```

2

```
10 REM COMMISSION
20 INPUT S
30 IF S>5000 THEN 60
40 LET C = 10/100*S
50 GOTO 70
60 LET C=(10/100*5000)
   +(5/100*(S-5000))
70 PRINT C
80 STOP
```

3

```
10 REM INCOME TAX
20 INPUT S
30 IF S GE 1000 THEN 60
40 PRINT 'NO TAX'
50 GOTO 80
60 LET T = (S-1000)*35/100
70 PRINT T
80 STOP
```

4

```
10 REM AREA OF RECTANGLE
20 LET C = 1
30 INPUT L, B
40 LET A = L*B
50 PRINT A, L, B
60 LET C = C+1
70 IF C LE 100 THEN 30
80 STOP
```

5

```
 10 REM RALLY
 20 INPUT Q
 30 LET C = 0
 40 INPUT T
 50 IF T<0 THEN 100
 60 IF T LE Q THEN 80
 70 GOTO 40
 80 LET C = C+1
 90 GOTO 40
100 PRINT C
110 STOP
```

6
```
 10 REM ASSURANCE
 20 INPUT P
 30 IF P<0 THEN 110
 40 IF P GE 25 THEN 80
 50 LET A = 10
 60 PRINT P, 'PREMIUM', A,
    'ALLOWANCE'
 70 GOTO 20
 80 LET A = 2/5*P
 90 PRINT P, 'PREMIUM', A,
    'ALLOWANCE'
100 GOTO 20
110 STOP
```

7
```
 10 REM CAR ALLOWANCES
 20 LET C = 1
 30 INPUT N, E, M
 40 IF E>1500 THEN 100
 50 IF E>1100 THEN 80
 60 LET T = M*6
 70 GOTO 110
 80 LET T = M*7
 90 GOTO 110
100 LET T = M*8
110 PRINT N, T
120 LET C = C+1
130 IF C LE N THEN 30
140 STOP
```

8
```
 10 REM CAR ALLOWANCES
 20 LET C = 1
 30 LET S = 0
 40 INPUT N, E, M
 50 IF E>1500 THEN 110
 60 IF E>1100 THEN 90
 70 LET T = M*6
 80 GOTO 120
```
(*continued on right*)
```
 90 LET T = M*7
100 GOTO 120
110 LET T = M*8
120 PRINT N, T
130 LET S = S+T
140 LET C = C+1
150 IF C LE 11 THEN 40
160 PRINT S, 'TOTAL'
170 STOP
```

9
```
 10 REM EXAM GRADES
 20 LET C = 1
 30 INPUT N, M
 40 IF M>79 THEN 130
 50 IF M>59 THEN 110
 60 IF M>44 THEN 90
 70 PRINT N, M, 'D'
 80 GOTO 140
```
(*continued on right*)
```
 90 PRINT N, M, 'C'
100 GOTO 140
110 PRINT N, M, 'B'
120 GOTO 140
130 PRINT N, M, 'A'
140 LET C = C+1
150 IF C LE 30 THEN 30
160 STOP
```

10
```
 10 REM EXAM GRADES
 20 LET C = 1
 30 LET S1 = 0
 40 LET S2 = 0
 50 LET S3 = 0
 60 LET S4 = 0
 70 INPUT N, M
 80 IF M>79 THEN 200
 90 IF M>59 THEN 170
100 IF M>44 THEN 140
110 PRINT N, M, 'D'
120 LET S4 = S4+1
```
(*continued on right*)
```
130 GOTO 220
140 PRINT N, M, 'C'
150 LET S3 = S3+1
160 GOTO 220
170 PRINT N, M, 'B'
180 LET S2 = S2+1
190 GOTO 220
200 PRINT N, M, 'A'
210 LET S1 = S1+1
220 LET C = C+1
230 IF C LE 30 THEN 70
240 PRINT S1, S2, S3, S4
250 STOP
```

Page 278 Exercise 3

1
```
10 REM VALUES OF FUNCTION
20 FOR X = -4 TO 4
30 PRINT X, 10*X↑2
40 NEXT X
50 STOP
```

2
```
10 REM ATTENDANCE
20 LET S = θ
30 FOR C = 1 TO 200
40 INPUT A
50 LET S = S+A
60 NEXT C
70 LET M = S/200
80 PRINT M
90 STOP
```

3
```
10 REM SUM OF NUMBERS
20 LET S = 0
30 FOR C = 1 TO 100
40 LET S = S+C
50 NEXT C
60 PRINT S
70 STOP
```

4
```
10 REM CUBES
20 FOR C = 1 TO 50
30 INPUT A
40 PRINT A, A↑3
50 NEXT C
60 STOP
```

5
```
10 REM TEMPERATURES
20 FOR F = 0 TO 212
30 LET C = 5/9*(F-32)
40 PRINT C
50 NEXT F
60 STOP
```

6
```
10 REM VALUES OF FUNCTION
20 FOR X = 0 TO 10 STEP 2
30 LET A = X↑3-3*X↑2+5
40 PRINT A
50 NEXT X
60 STOP
```

7
```
10 REM EVEN NUMBERS
20 LET S = 0
30 FOR C = 502 TO 1000 STEP 2
40 LET S = S+C
50 NEXT C
60 PRINT S
70 STOP
```

8
```
10 REM EXCHANGE
20 FOR P = 5 TO 100 STEP 5
30 LET D = P*2·8
40 PRINT D
50 NEXT P
60 STOP
```

9
```
10 REM SQUARES
20 FOR L = 10 TO 1000 STEP 10
30 LET A = L*L
40 PRINT A
50 NEXT L
60 STOP
```

10 Replace line 20 by 20 FOR C=1 TO 100, and lines 60 and 70 by 60 NEXT C

11 Replace line 20 by 20 FOR C=1 TO 11, and lines 120 and 130 by 120 NEXT C

12 Replace line 20 by 20 FOR C=1 TO 30, and lines 140 and 150 by 140 NEXT C

Page 280 Exercise 4

1
```
10 REM SIDE OF SQUARE
20 INPUT A
30 LET L = SQR(A)
40 PRINT L
50 STOP
```

2
```
10 REM RADIUS
20 INPUT A
30 LET R = SQR(A/3·14)
40 PRINT R
50 STOP
```

3
```
10 REM PYTHAGORAS
20 INPUT C, A
30 LET B = SQR(C↑2 − A↑2)
40 PRINT B
50 STOP
```

4
```
10 REM QUADRATIC
20 INPUT P, Q, R
30 LET X =
   (−Q + SQR(Q↑2 − 4*P*R))/(2*P)
40 PRINT X
50 STOP
```

5
```
10 REM METRES TO KM AND M
20 INPUT M
30 LET K = INT (M/1000)
40 LET M = M − K*1000
50 PRINT K, 'KILOMETRES', H, 'METRES'
60 STOP
```

6
```
10 REM P TO POUNDS AND PENCE
20 INPUT N
30 LET P = INT (N/100)
40 LET N = N − P*100
50 PRINT P, 'POUNDS', N, 'PENCE'
60 STOP
```

7
```
10 REM TIME UNITS
20 INPUT M
30 LET H = INT (M/60)
40 LET M = M − H*60
50 LET D = INT (H/24)
60 LET H = H − D*24
70 PRINT D, 'DAYS', H, 'HOURS', M, 'MINS.'
80 STOP
```

8
```
10 REM TRIANGLE
20 INPUT B, C, C1
30 LET B2 = B*3·14/180
40 LET C2 = C*3·14/180
50 LET B1 =
   C1*SIN (B2)/SIN (C2)
60 PRINT B1
70 STOP
```

9
```
10 REM TRIANGLE
20 INPUT A, B1, C1
30 LET A2 = A*3·14/180
40 LET A1 =
   B1↑2 + C1↑2 − 2*B1*C1*COS (A2)
50 LET A3 = SQR (A1)
60 PRINT A3
70 STOP
```

10
```
10 REM AREA OF TRIANGLE
20 INPUT A, B1, C1
30 LET A2 = A*3·14/180
40 LET S = 0·5*B1*C1*SIN (A2)
50 PRINT S
60 STOP
```

11
```
10 REM TAN (A)
20 INPUT A
30 LET A2=A*3·14/180
40 LET T=SIN (A2)/COS (A2)
50 PRINT T
60 STOP
```

Page 283 . Exercise 5

1
```
 5 REM SUM OF SQUARES
10 DIM Y(50)
15 LET S = 0
20 FOR C = 1 TO 50
25 INPUT Y(C)
30 LET S = S + Y(C)*Y(C)
35 NEXT C
40 PRINT S
45 STOP
```

2
```
 5 REM MEAN
10 DIM B(500)
15 LET S = 0
20 FOR C = 1 TO 500
25 INPUT B(C)
30 LET S = S + B(C)
35 NEXT C
40 LET M = S/500
45 PRINT M
50 STOP
```

3
```
 5 REM COMPETITION
10 DIM A(83)
15 FOR C = 1 TO 83
20 INPUT A(C)
25 LET P = A(C)/70*100
30 PRINT P
35 NEXT C
40 STOP
```

4
```
 5 REM LARGEST NUMBER
10 DIM Z(20)
15 LET M = 0
20 FOR C = 1 TO 20
25 INPUT Z(C)
30 IF Z(C) < M THEN 40
35 LET M = Z(C)
40 NEXT C
45 PRINT M
50 STOP
```

5
```
 5 REM FUNCTION
10 DIM F(9)
15 LET X = 1
20 FOR C = -4 TO 4
25 LET F(X)=C↑3-C↑2-6
30 LET X=X+1
35 NEXT C
40 FOR X = 1 TO 9
45 PRINT F(X)
50 NEXT X
55 STOP
```

6
```
 5 REM FUNCTION SIGN
10 DIM F(9)
15 LET X = 1
20 FOR C = -4 TO 4
25 LET F(X)=C↑3-C↑2-6
30 LET X=X+1
35 NEXT C
40 FOR X=1 TO 8
45 IF (F(X+1)*F(X)) GE 0 THEN 55
50 PRINT X, X+1
55 NEXT X
60 STOP
```